Rethinking Feminist Theories for Social Work Practice

Christine Cocker • Trish Hafford-Letchfield
Editors

Rethinking Feminist Theories for Social Work Practice

palgrave
macmillan

Editors
Christine Cocker
School of Social Work
University of East Anglia
Norwich, UK

Trish Hafford-Letchfield
School of Social Work and Social Policy
University of Strathclyde
Glasgow, UK

ISBN 978-3-030-94240-3   ISBN 978-3-030-94241-0 (eBook)
https://doi.org/10.1007/978-3-030-94241-0

This Palgrave Macmillan imprint is published by the registered company Springer Nature Switzerland AG.
The registered company address is: Gewerbestrasse 11, 6330 Cham, Switzerland

Christine dedicates this book to Adi, Rivka, Frania and Lydia and feminists from all generations in her family and friendship networks. Trish dedicates this book to the beautiful Nora Grace Letchfield, born in 2021 during the COVID-19 lockdown; she has high hopes for her influence on this world.

Foreword

As a feminist social work scholar, I am always heartened when I come across scholarly texts engaging with feminist inquiry in social work, and *Rethinking Feminist Theories for Social Work Practice* does just that. This is a ground-breaking work that centres feminist theories, values and knowledge as they apply to social work practice, theory and education, and the timely book's key aim is to turn the feminist gaze on the core principles and values of social work, thus inviting readers to interrogate how gender manifests in social work practices. This key concept is especially important, as social work is a site where gendered ideologies are played out in practice, be it with older adults, children and families, youth justice or mental health. Additionally, although women constitute the bulk of the workforce in the social work profession, they are still disadvantaged in terms of representation in senior leadership roles. In the current era where feminism has fallen out of focus in social work, and indeed there is pushback to critical ideas, *Rethinking Feminist Theories for Social Work Practice* is to be welcomed because it disrupts the dominant discourses that foster a gender-neutral social work practice. Most notably, the standout feature of this book is that it engages with feminist thinking to re-emphasise the centrality of gender and its intersections with other axes of identities, such as social class, race, disability, sexuality and age, for understanding and critiquing social work practice. In doing so, we are introduced to a diverse group of feminist social work scholars through their powerful counter-narratives addressing issues such as gender and ageing, lesbian parenting, women in prison, feminist social work pedagogy, disability and gender, black feminisms and activism, anti-racist practice, gender identity and expression, transfeminist theory and feminist leadership in social work, to name but a few of the topics tackled. Each chapter offers a rich source of knowledge that weaves together multiple marginalised identities and experiences for social work educators, students and practitioners to draw on to develop feminist-informed social work practice.

In many ways, by foregrounding debates on gender and social work through a feminist lens, the scholars in this collection spotlight the dominant construction of gender in the profession. Indeed, an important feature of *Rethinking Feminist Theories for Social Work Practice* is that it encourages social workers to pay particular attention to the deeply embedded gender assumptions that still predominate much social work practice. Ultimately, their aim is to undermine the gendered norms, beliefs and assumptions that prevail in social work for developing anti-oppressive practice. That is, not only does it address how gender operates at different levels in social work practice, it also acts as a springboard to examine the engrained gendered assumptions that underlie the organisational culture and delivery of social work. Most importantly, this book challenges us to attend to the ways that much social work practice can reproduce, rather than challenge dominant discourses about gender. Thus, the overarching goal is twofold, namely, to open a discursive space to enable social work practitioners to challenge the entrenched gender normativity in social work and, more fundamentally, to facilitate deeper discussions of different ways of knowing for understanding the intersecting systems of inequalities that underlay much social work practice.

In recent years, intersectionality has received increased attention in social work as a feminist theoretical lens, and as a useful way to make sense of contexts of oppression, most specifically, the interlocking systems of oppression (Bernard 2021; Mattsson 2014; Mehrotra 2010; Murphy et al. 2009; Nayak & Robbins 2018). In other words, we need a good deal of more critical feminist inquiry in social work as much more remains to be said. Therefore, *Rethinking Feminist Theories for Social Work Practice* has greater importance for challenging the single-story analysis of gender and will undoubtedly play an essential role in expanding feminist scholarship in social work. Indeed, this book offers important new insights into how subjugated service users navigate the multiple intersecting oppressions that impact their lived experiences. The scholars in this collection have thus provided a compelling book that essentially furthers our thinking of critical feminist practice for more nuanced understandings of the enduring consequence of normative gender ideas in social work. More specifically, this book's distinctive and important contribution is that it powerfully reminds us of the relevance of feminist scholarship in social work; the authors reaffirm the pursuit of feminist inquiry in social work, thus making this collection a must read for anyone engaged in theory and practice that seeks to challenge the gender ideologies that underscore the profession.

Professor Claudia Bernard
London, UK

References

Bernard, C. (2021). *Intersectionality for social workers: Introduction to theory and practice*. Routledge.

Mattsson, T. (2014). Intersectionality as a useful tool: Anti-oppressive social work and critical reflection. *Affilia*, *29*(1), 8–17.

Mehrotra, G. (2010). Toward a continuum of intersectionality theorizing for feminist social work scholarship. *Affilia*, *25*(4), 417–430.

Murphy, Y., Hunt, V., Zajicek, A. M., Norris, A. N., & Hamilton, L. (2009) *Incorporating intersectionality in social work practice, research, policy and education*. NASW.

Nayak, S., & Robbins, R. (2018). *Intersectionality in social work: Activism and practice in context*. Routledge.

Acknowledgements

Thanks to the clever, co-operative and well-organised contributing authors who all did their stuff brilliantly!

Permissions

The authors of Chap. 11 JD Drummond and Shari Brotman wish to thank Springer Nature for permission to reproduce material from the publication.

'Intersecting and Embodied Identities: A Queer Woman's Experience of Disability and Sexuality' (2014) *Sexuality and Disability* 32(4): 533–549.

Licence no 4721340225208

Contents

1	**Rethinking Feminist Theories for Social Work Practice** Christine Cocker and Trish Hafford-Letchfield	1
Part I	**Feminist Theories for New Challenges in Social Work**	15
2	**Feminisms: Controversy, Contestation and Challenge** Barbara Fawcett	19
3	**Feminist and Empowerment Theory and Practice: A Powerful Alliance** Tina Maschi, Sandra G. Turner, Smita Ekka Dewan, Adriana Kaye, and Annette M. Hintenach	37
4	**Feminist Research and Practice: Reorienting a Politic for Social Work** Sam Harrell, Ben Anderson-Nathe, Stéphanie Wahab, and Christina Gringeri	59
5	**A Pedagogy of Our Own: Feminist Social Work in the Academy** Sarah Epstein, Norah Hosken, and Sevi Vassos	77

6 **Collaborative Autoethnography for Feminist Research** 97
Jennifer Dyer, Sarah Pickett, Jennifer Davis, Kathleen Hackett, Cindy Holmes, Julie James, Daze Jefferies, Kimberley Manning, Annie Pullen Sansfaçon, and Julie Temple Newhook

Part II Feminisms and Intersectionalities 117

7 **Afrocentric Feminism and Ubuntu-Led Social Work Practice in an African Context** 123
Kathomi Gatwiri and Sharlotte Tusasiirwe

8 **Tears of Shame: Sri Lankan Mothers Negotiating Experiences of Caregiving and Disability** 141
Fiona Kumari Campbell

9 **Voices of Syrian Refugee Women in Jordan Living with Exacerbated Gender-Based Violence During COVID-19: Conceptualizing a Feminist Perspective for Social Work** 159
Dina Pervez Sidhva

10 **The Transformative Potential of Transfeminist Social Work Practice** 175
Jama Shelton, Maggie Dunleavy, and Kel Kroehle

11 **Exploring the Intersection of Queer Disability as Life Story: A Feminist Narrative Approach to Social Work Research and Practice** 189
Jennifer JD Drummond and Shari Brotman

12 **Invisible Women: Critical Perspectives on Social Work and Gender in Later Life** 207
Trish Hafford-Letchfield

Part III Gender in Social Work Practice 229

13 Using Sex Worker Feminisms in Practice to Promote a Peer-Based Methodology; Exploring Personal and Professional Identities in a Research Alliance Centring Sex Worker Lived Experience 235
Roxana Diamond, Priscilla Dunk-West, and Sarah Wendt

14 Does Feminist Social Work Practice Need Time? Gender, Parenting and Changing Times for Social Work 255
Linda Bell

15 Lesbian Parenting: Rebellious or Conformist? 271
Christine Cocker

16 Child Sexual Exploitation, Victim Blaming or Rescuing: Negotiating a Feminist Perspective on the Way Forward 287
Jane Dodsworth

17 Social Work Men as a Feminist Issue 303
Jason Schaub

18 A Relational Approach to Work with Couples Where Men Have Been Violent Towards Women: Feminist Dilemmas and Contributions to Social Work Practice 321
Rebecca Infanti-Milne, Richard Mc Kenny, and Lee Walton

19 Feminist Perspectives on Social Work Leadership 339
Adi Cooper and Lyn Romeo

Index 357

Notes on Contributors

Ben Anderson-Nathe is a Professor in the Child, Youth, and Family Studies programme in the School of Social Work at Portland State University, USA. His research and scholarship focuses on youth and young adults, critical and qualitative epistemologies, gender and sexuality and professional development of youth workers. Anderson-Nathe has written three books, including *Youth Workers, Stuckness, and the Myth of Supercompetence*; *Compassionate Confinement: A Year in the Life of Unit C*; and *Feminisms in Social Work Research: Promise and Possibilities for Justice-based Knowledge* (an edited collection exploring critical feminist research).

Linda Bell was until recently an Associate Professor in the Department of Mental Health and Social Work at Middlesex University, UK, where she is now an honorary research fellow/visiting academic. She has researched and written on various aspects of social care, health, gender issues, parenting and research ethics. She has experience of teaching research methods to social workers and of supervising masters and doctoral dissertations. Bell is an anthropologist, whose doctoral work from the University of London focused on families in the UK. She is a longstanding member of the international 'Women's Workshop on Qualitative Family/Household Research'. Her recent publications include two books, one in 2017 *Research Methods for Social Workers* with Palgrave Macmillan and in 2020, *Exploring Social Work: An Anthropological Perspective.*

Shari Brotman is an Associate Professor at the McGill University School of Social Work, Canada. She has worked extensively as an educator, researcher and practitioner in the fields of gerontology and anti-oppression social work practice. Her scholarly activities centre on questions of access and equity in the design and delivery of health and social care services to people from marginalised communities (LGBTQ and immigrant) and their caregivers. Her work incorporates a feminist intersectional lens exploring multiple identity and interlocking oppressions as lived by people and communities. She undertakes qualitative, community and arts-based research.

Fiona Kumari Campbell is a Professor in the School of Education and Social Work at University of Dundee, Scotland. She was Deputy Head of School (Learning and Teaching Scholarship) at the Griffith Law School, Australia, until July 2014. Previous to this, she was Convenor of Disability Studies, School of Human Services and Social Work, Griffith University (2001–2010). Campbell is Adjunct Professor of Disability Studies at the Faculty of Medicine, University of Kelaniya, Sri Lanka; she is a person with disability and is associated with several minority groups. Campbell has written extensively on issues related to Global South theory, disability—a philosophy of ableism, disability in Sri Lanka, dis/technology, and is recognised as a world leader in scholarship around studies on ableism. After the successful publication of *Contours of Ableism: The Production of Disability and Abledness* (Palgrave Macmillan) in 2009, she is working on three book manuscripts: *#Ableism: An Interdisciplinary Introduction to Studies in Ableism*; *Textures of Ableism: Disability, Voice and Marginality*; and *Jewlanka: Footprints of Jewish Presence in Sri Lanka from Ancient Times to the Mid-1950's.*

Christine Cocker is Professor of Social Work and head of the School of Social Work at the University of East Anglia, Norwich, UK. She is also a qualified social worker. Prior to academia Cocker practiced in child and family social work. She continues to have strong links with practice as an independent member of a Local Authority Fostering Permanence Panel and as the Independent Chair for a Local Authority's Children's Academy. Her research and publications are in the area of social work with looked after children, LGBT+ issues in social work, and Transitional Safeguarding. Amongst her many books are *Rethinking Anti-Discriminatory and Anti-Oppressive Theories for Social Work Practice*, with Trish Hafford-Letchfield (Palgrave Macmillan, 2014) and the third edition of *Social Work with Looked After Children* (2019). Recent research projects Cocker has been involved with include a systematic literature review examining transparenting through the lifespan, supporting Local Authorities to develop Transitional Safeguarding approaches to working with young people aged 16–24, and lesbians working in senior social care leadership positions. Cocker is on the editorial board of the journal *Practice: Social Work in Action*.

Adi Cooper, OBE, is the Care and Health Improvement Advisor for London for the Local Government Association, and lead on adult safeguarding. She is the Independent Chair of the City and Hackney Safeguarding Adults Board and the Haringey Safeguarding Adults Board. She is also a Visiting Professor at the University of Central Bedfordshire. She holds a PhD in South Asian Studies. Cooper was the Strategic Director of Adult Social Services, Housing and Health in the London Borough of Sutton for nine years. She is a qualified social worker. Cooper has been a social work manager and professional leader in adult social care for over 30 years and has worked in several London Boroughs, in performance, health improvement, adult social services and commissioning. She has contributed to national policy development, service improvement and Care Act 2014 guidance in adult safeguarding, for example developing and promoting the Making Safeguarding Personal programme.

Jennifer Davis is Instructor in Psychology at Lethbridge College, Canada. She holds a PhD in Psychology from McMaster University and her research areas include parenting and the effects of parenting on child outcomes.

Smita Ekka Dewan is an Assistant Professor in the Human Services Department at New York City College of Technology, City University of New York (CUNY), USA. Her teaching and scholarly interests include social work practice with immigrants and refugees, violence against women, international social work, human rights and social justice and programme evaluation.

Roxana Diamond is a PhD candidate within the College of Education, Psychology and Social Work at Flinders University, Australia. She is conducting peer research looking at the everyday lives of South Australian sex workers. In 2016 Diamond graduated with First-Class Honours in Social Work from the University of South Australia and explored the criminalisation of sex workers in South Australia. Her work has been published in the *Women's Studies International Forum* and The Conversation. She sits on the Sex Industry Network (SIN) board and has served as president, vice president and now is acting as a general member. SIN is by and for sex workers and the only peer-based sex work organisation in South Australia. Diamond is passionate about sex worker rights and is heavily involved in the campaign to decriminalise sex work in South Australia. She has been a sex worker since 2012 and will finish the final stages of her PhD unfunded and living off sex work.

Jane Dodsworth is an Associate Professor at the University of East Anglia, UK, and is a qualified social worker with experience of working with children and young people at risk of and experiencing child sexual exploitation, children in the child protection system and children in care. She was previously involved in policy development on child sexual exploitation at Local Children's Safeguarding Partnership level and became interested in exploring issues and preventative strategies for this vulnerable group further. Dodsworth's doctoral research was a qualitative study exploring the narratives of young and adult women involved in sex work in the UK to determine whether it was possible to identify key risk and protective factors influencing involvement. Her key research interests are in sexual exploitation, sex work, sex workers as mothers, issues of victimhood and agency and the perspectives of those directly involved.

Jennifer JD Drummond is a registered social worker and the manager of the Sexual Assault Resource Centre at Concordia University, Canada. Prior to founding the Centre in 2013, she worked as a research coordinator on various projects and with an advocacy organisation in Montreal serving women who work in the sex industry. Drummond received her Master of Social Work (MSW) from McGill University and her research interests include women's sexuality, identity, gender-based violence, feminist and narrative methodologies.

Priscilla Dunk-West is an Associate Professor at Charles Darwin University, Australia. She has held senior academic positions in universities in England and Australia. Committed to social justice principles, critical thinking and analysis, she is experienced in transforming and leading innovative pedagogical design in university settings. Her sole-authored seminal book *How to be a Social Worker: A Critical Guide for Students* develops a new theoretical and pedagogical approach designed to equip students to consolidate and grow their professional selves. Dunk-West researches in the areas of intimacy, relationships and parenting and is regularly called upon as an expert in these areas in print, radio, podcast and television media. She has undertaken national and international research and continues to publish, conduct empirical studies and translate the findings to lay audiences. She is passionate about 'real world' research and translational processes and recently won an Australian award for media engagement.

Maggie Dunleavy (they/them/theirs) is a Master of Social Work (MSW) candidate at Silberman School of Social Work, USA. They are interested in research and clinical work that supports and honours the courage, creativity and power of trans people and communities.

Jennifer Dyer is Department Head of Gender Studies at Memorial University of Newfoundland, Canada. Her research includes interdisciplinary studies, queer theories of aesthetics, the social value of art and semiotics.

Sarah Epstein is Lecturer in Social Work at Deakin University, Australia. Her research and teaching interests are in the area of feminist critical social work education, the intersection between feminism and masculinity and human rights-based social work practice.

Barbara Fawcett is Professor of Social Work and co-head of the School of Social Work and Social Policy at the University of Strathclyde, UK. Previously she was Professor of Social Work at the University of Birmingham, UK, and Professor of Social Work and Policy Studies at the University of Sydney, Australia Prior to joining academia, Fawcett spent a number of years in the field as a senior social work practitioner, social work hospital manager, contract researcher and head of mental health services for a large Local Authority. She has written widely and is the author and co-author of 11 books. Her research focuses on gender studies, disability, older age and mental health, as well as participatory research methodology.

Kathomi Gatwiri is an award-winning trauma and race researcher, senior social work academic and psychotherapist. In her current role as Senior Lecturer in Social Work and Community Welfare at Southern Cross University, Australia, Gatwiri teaches across a range of social work and social policy subjects and conducts interdisciplinary research on trauma, race and gender. She investigates how the politics of race in Australia and in particular how racial micro-aggressions and racial trauma manifest in the workplace. Gatwiri works closely with organisations to develop practice guidelines, train and foster conversations on cultural safety. Gatwiri is also a regular writer for SBS and The Conversation, and the founder of Healing Together: a service that provides accessible, culturally safe therapy for Black people and people of colour in Australia. Gatwiri is also a senior research fellow in the Centre for Excellence in Therapeutic Care (CETC) and a researcher in the Centre for Children and Young People (CCYP) at Southern Cross University.

Christina Gringeri is Professor and Director of Doctoral Studies at the College of Social Work, University of Utah, USA. Her research has largely focused on women and work, especially in lower-income households. In recent years she has contributed to research examining feminisms in social work research, rigour and quality in qualitative research, trauma in the life narratives of low-income women and building bridges between classroom and field education in the areas of policy and diversity. Gringeri is working on a major feminist project examining the narratives of ordained Catholic women seeking changes in patriarchal traditions and institutions.

Kathleen Hackett is a parent of two elementary school-aged children and a graduate student in Geography at Memorial University of Newfoundland, Canada. The focus of their studies is critical queer health geographies, specifically understanding the influence of geographical landscapes of colonial, heteronormative, cisnormative governance and policy in shaping LGBTQIA2S+ health discourse.

Trish Hafford-Letchfield is Professor of Social Work and co-head of the School of Social Work and Social Policy at the University of Strathclyde, Glasgow, Scotland, prior to which she was Professor of Social Care at Middlesex University, London, UK, from 2016 to 2019. Hafford-Letchfield is a qualified nurse, social worker and educator and has 18 years of practice experience mostly with adult social care. Her main research interests are in ageing and care in marginalised communities particularly in co-produced and applied research. She has had more than 100 publications since 2004. The latest are two edited volumes of a series on 'Sex and Ageing' with Paul Simpson and Paul Reynolds: *Sex and Diversity in Later Life* and *Desexualisation in Later Life: The Limits of Sex and Intimacy* published in 2021. Hafford-Letchfield's recent research is focused on ageing and suicide prevention, LGBT+ ageing and problematic substance use in later life. Hafford-Letchfield is a research associate of the Stellenbosch University and a board member of the UK Chapter of the International

Society of Substance Use Professionals and Research Committee of British Association for Counselling and Psychotherapy.

Sam Harrell is a doctoral student in social work and social research at Portland State University's (PSU) School of Social Work, USA. They have worked in anti-violence, child welfare, homelessness, queer youth and prison and jail re-entry services. They are exploring histories and practices of carceral social work, with a focus on mandatory reporting. They are an associate instructor at PSU and Indiana University, where they teach undergraduate and graduate courses on research, organisational leadership and macro practice.

Annette M. Hintenach is a fourth-year doctoral student at the Fordham University Graduate School of Social Service, USA, and is a gerontological social worker, emerging health services researcher and scholar. Her research interests include compulsive hoarding in older adults, high costs of prescription drugs and policy implications for the elderly and ageing people in institutional and long-term care facilities including prisons.

Cindy Holmes is an Assistant Professor in the School of Social Work at the University of Victoria, British Columbia, Canada. Holmes' interdisciplinary scholarship grows out of over 25 years of community-based work and is informed by decolonial and anti-racist feminist, queer and trans theories and social justice movements. Her research interests include violence, health and place; wellbeing of lesbian, gay, bisexual, queer, trans and Two-Spirit (LGBTQ2S) people; white settler colonialism and decolonisation; and community-based participatory action research methodologies.

Norah Hosken is Senior Lecturer in Social Work and Course Director for the Bachelor of Social Work programme at Deakin University, Australia. Her research and teaching interests are in the areas of feminism, social work education, poverty, class, addiction and shame.

Rebecca Infanti-Milne started working with children and young people 26 years ago as a play worker. She is a registered social worker and family therapist and has worked in a range of 'frontline' child protection and safeguarding roles, as an advocate for children in secure accommodation, in residential care and in specialist domestic violence services. Infanti-Milne has had a number of leadership roles in social work and works as Systemic Lead in the London Borough of Hackney. Infanti-Milne has taught systemic ideas and has taught at Middlesex University, Frontline and on numerous systemic social work courses. From 2017 to 2020, with Lee Walton, Infanti-Milne was co-lead psychotherapist responsible for developing and leading a local authority domestic abuse service, which included responsibility for assessment, risk assessment and the delivery and supervision of therapeutic interventions for couples who wished to remain together after violence and abuse. Lee, Rebecca and Richard continue to work together offering assessment, therapy, consultation and advice where domestic violence is a concern.

Julie James is an Assistant Professor in the School of Child and Youth Care at Ryerson University, Toronto, Canada. Her work focuses on supporting young trans and gender non-conforming people, youth-led activism and youth-led Indigenous resurgence.

Daze Jefferies is a graduate student at Memorial University of Newfoundland, Canada. She is an artist and writer from the Bay of Exploits. Her research-creation touching embodiments, geographies and histories of trans women (and) sex workers in Ktaqmkuk (Newfoundland, Canada) have been exhibited and performed nationally. Co-author of *Autoethnography and Feminist Theory at the Water's Edge: Unsettled Islands* (2018), she has recent or forthcoming publications in *Atlantis: Critical Studies in Gender, Culture, and Social Justice* (2020), *The Dalhousie Review* (2020), *Arc* (2021), *Feral Feminisms* (2021) and *Journal of Folklore Research* (2021). Her forthcoming poetry collection 'We Hold a Body of Water Together' explores hidden histories and fishy futures of trans women and sex workers in Atlantic Canada.

Adriana Kaye is a social worker and research assistant at the Fordham University Graduate School of Social Service, USA, and a doctoral student at the Tulane University School of Social Work, USA. She engages in community, volunteer and contemplative activism.

Kel Kroehle (they/them) is a doctoral candidate at the University of Pennsylvania's School of Social Policy and Practice, USA. Kroehle is a practitioner-researcher, integrating their work within the youth organising field with a programme of research covering youth sociopolitical development, trans equity and the treatment of youth power by schools, nonprofits and philanthropy. Kroehle holds their Bachelor's Degree in Gender Studies and LGBTQ Studies from the University of Wisconsin-Madison before receiving their Master of Social Work degree from the University of Pennsylvania, USA.

Kimberley Manning is Associate Professor of Political Science at Concordia University, USA. A former principal of the Simone de Beauvoir Institute (2016–2021), Manning analyses political life through the lens of feminist theory, with particular attention to the affective enactment of family ties in social movements and state formation. Manning's research includes studies of China's revolutionary past as well as the more recent Canadian mobilisation of transgender children and youth. She is the author of the monograph *The Party Family: Revolutionary Attachments and the Gendered Origin of Chinese State Power*, to be published with Cornell University Press.

Tina Maschi is a Professor in the Graduate School of Social Service (GSS) at Fordham University, New York, USA. Her teaching and scholarly interests include diversity and inclusion, human rights, equality and justice. Maschi is a mental health practitioner, artist, writer, researcher, scholar and advocate with over 100 peer-reviewed journal publications and book chapters that address contemporary issues. Her book publications include *Aging Behind Prison Walls: Studies in Trauma and Resilience*, *Practitioner as Researcher*, *Forensic Social Work* and *A Human Rights Approach to Social Work Research and Evaluation: A Rights Research Manifesto.*

Richard Mc Kenny is a registered social worker, family therapist and a clinical supervisor. He worked for many years in specialist family assessment services and in national Child and Adolescent Mental Health Services (CAMHS), and for the last ten years has taught systemic theory and practice. From 2017 to 2020, with Rebecca Infanti-Milne and Lee Walton, he helped establish a local authority domestic abuse service and provided clinical supervision to the project. Mc Kenny works as a family therapist, as a consultant and as a clinical supervisor for a number of organisations, and together with Rebecca and Lee he offers assessment, therapy, consultation and advice where domestic violence is a concern.

Julie Temple Newhook is Instructor in Gender Studies and a research associate in the Janeway Pediatric Research Unit at Memorial University of Newfoundland, Canada.

Sarah Pickett is an Assistant Professor in the Faculty of Education at Memorial University of Newfoundland, Canada. She is a registered psychologist and the faculty founder/advisor to the Gender and Sexuality Alliance (GSA) in Education at Memorial. Her research has focused on affirmative sexuality and gender practice and pedagogy in education and healthcare.

Lyn Romeo, CBE, is the Chief Social Worker for Adults in England, a post she took up in September 2013. Previously, Romeo worked as the Assistant Director for Adult Social Care in the London Borough of Camden. She has also worked as an inspector with the Social Services Inspectorate, as well as working in Yorkshire for over 20 years as both a field social worker and in a variety of management roles across children and adults.

Annie Pullen Sansfaçon is a Professor at the School of Social Work, University of Montreal, Canada, and the holder of Canada Research Chair on transgender children and their family: transforming knowledge into action. Through her research, she draws from her background in ethics and anti-oppression in order to understand the experiences of those two groups and to propose ways for working with them in social work and other care profession.

Jason Schaub is Lecturer in Social Work at University of Birmingham, UK. He has practised as a counselling social worker in the USA and as a children's social worker in Ireland and later in the UK for several Children's Social Care departments. His research explores sexuality and gender and social work practice with children, with a particular focus on LGBT young people.

Jama Shelton (they/them) is an Assistant Professor at the Silberman School of Social Work, Hunter College, USA, and the Chief Strategy Officer for the True Colors Fund. Shelton's research examines the needs and experiences of LGBTQ youth experiencing homelessness and the service providers with whom they work. In particular, Shelton is interested in identifying and addressing systemic barriers rooted in hetero/cisgenderism that frequently constrain the successful transition out of homelessness for LGBTQ youth and young adults. Recently, Shelton co-edited *Sexual Orientation, Gender Identity/Expression and Child Welfare*, a special issue of Child Welfare and the peer-reviewed text *Where Am I Going to Go? Intersectional Approaches to Ending LGBTQ2S Youth Homelessness in Canada and the US.*

Dina Pervez Sidhva is Lecturer in Social Work at the University of the West of Scotland, Scotland. Her research focuses on the lives of forced migrants in Scotland and in Jordan, with a special focus on issues of gender-based violence. She sits on the Board of Directors of the Village Storytelling Centre in Glasgow. Sidhva's most recent research funded by the Chief Scientist Office (Scotland) looked at the impact of COVID-19 on the family carers of care home residents in Scotland. Further afield (in Jordan) she has just completed a pilot project on empowering Syrian refugee girls and women living with gender-based violence through enterprise education.

Sandra G. Turner is an Associate Professor at the Fordham University Graduate School of Social Service, USA, and Director of the Women and Girls Institute. Her teaching and scholarly interests include feminist and empowerment theory and practice, trauma, mental health and addictions. Turner has extensive practice experience in working with female victims of sexual and other forms of violence and abuse. She also has over 30 peer-reviewed journal publications and book chapters on these issues.

Sharlotte Tusasiirwe is a Ugandan-born, internationally educated social worker. After completing a Bachelor's Degree in Social Work in Uganda and a Master's Degree in Social Work and Human Rights from University of Gothenburg, Sweden, Tusasiirwe pursued her PhD from Western Sydney University from 2016 to 2020. Tusasiirwe is very interested in researching social work education and her PhD has been focused on how to decolonise social work education and practice to create a culturally appropriate and contextually relevant profession. She has researched how African knowledges and *Obuntu/Ubuntu* philosophies can inform social work. She is interested in theorising about how diverse epistemologies from our diverse cultures can be at the centre of social work education and practice. She loves teaching and researching indigenous knowledges, ageing and age-old wisdom, community-led initiatives, community development and advocacy and gender, among others.

Sevi Vassos is Lecturer in Social Work and Course Leader for the Master of Social Work programme at Deakin University, Australia. Her research interests are in the areas of critical feminist approaches in social work education and more specifically practice-based learning. Her teaching interests are in direct practice skill development, critical social work policy and organisational practice.

Stéphanie Wahab is a Professor at Portland State University's School of Social Work, USA, and Honorary Research Associate Professor at the University of Otago, Social and Community Work, New Zealand. Her body of work, rooted in critical, post-structural and feminist studies, centres on structural violence related to social inequality, sex work and intimate partner violence. She teaches courses focused on social justice, philosophies of science, qualitative inquiry, intimate partner violence and motivational interviewing. She is the co-editor of *Feminisms in Social Work Research: Promise and Possibilities for Justice-based Knowledge*, and the co-editor-in-chief for *Affilia: Women and Social Work*.

Lee Walton is a registered social worker, family therapist and a clinical supervisor. He has worked in 'frontline' child protection and safeguarding roles, in child and adolescent mental health services and in specialist domestic violence services. From 2017 to 2020, with Rebecca Infanti-Milne, Walton was a co-lead psychotherapist responsible for developing and leading a local authority domestic abuse service. This included responsibility for assessment, risk assessment and the delivery and supervision of therapeutic interventions for couples who wished to remain together after violence and abuse. Walton works as a family therapist, as a consultant and as a clinical supervisor for a number of organisations, and together with Rebecca and Richard he offers assessment, therapy, consultation and advice where domestic violence is a concern.

Sarah Wendt is Professor of Social Work at Flinders University, Australia. Prior to academia Wendt practiced in the field of domestic violence. She has taught in social work for over a decade. She has written on violence against women and social work practice. Her particular research projects explore the impact of domestic violence on women's citizenship, service provision in the field of domestic violence, young women's experiences of violence and abuse and engaging men to address domestic violence. More recently she has been researching collaboration across child protection and domestic violence sectors. Wendt is Director of the Social Work Innovation Research Living Space (SWIRLS) at Flinders University. https://www.flinders.edu.au/swirls. SWIRLS engages with complex problems that need nuanced, evidence-based responses. In doing so, members of SWIRLS embrace innovation, creativity, risk-taking, collaboration and fresh perspectives.

List of Figures

Fig. 3.1	A human rights-based approach: Empowerment theory and feminist theory in practice	39
Fig. 3.2	Empowerment theory and practice	43
Fig. 3.3	Incarcerated older women—Assessment and intervention using a life course systems power analysis model	47
Fig. 6.1	Developing a collaborative autoethnography process	107
Fig. 11.1	Concepts and connections shaping Josie's experiences	197

List of Tables

Table 3.1	Human rights framework	38
Table 11.1	Step-wise analytic concepts connected	196

CHAPTER 1

Rethinking Feminist Theories for Social Work Practice

Christine Cocker and Trish Hafford-Letchfield

Introduction

The global position of women remains far from equal when compared to men. Less than 50% of working-age women are in paid employment. This figure has not changed over the last 25 years. Every day, women globally spend approximately three times as many hours on unpaid domestic and care work compared to men (4.2 hours compared to 1.7) and in some areas of the world this figure is higher. Women remain underrepresented in managerial positions across the globe, occupying only 28% of manager roles. This has remained static since 2005. Among Fortune 500 corporations, only 7.4% of women are chief executive officers, indicating a 'glass ceiling' for women gaining middle and senior management roles. In

C. Cocker (✉)
School of Social Work, University of East Anglia, Norwich, UK
e-mail: Christine.cocker@uea.ac.uk

T. Hafford-Letchfield
School of Social Work and Social Policy, University of Strathclyde, Glasgow, UK
e-mail: Trish.hafford-letchfield@strath.ac.uk

C. Cocker, T. Hafford-Letchfield (eds.), *Rethinking Feminist Theories for Social Work Practice*,
https://doi.org/10.1007/978-3-030-94241-0_1

political life, while women's representation in parliaments has increased, this has still not crossed the barrier of 25% of parliamentary seats in 2020 (United Nations, 2020).

The one area of global success in gender equality is in education. Girls and boys equally participate in primary education in most countries across the world and perform better than boys in terms of their academic achievement. In tertiary education, female students now outnumber male students across the board, but women remain underrepresented in science, technology, engineering and maths (STEM) subjects (United Nations, 2020).

As a social and political movement, feminism has made its mark on the world in many different ways over a number of generations. Feminism provides a critical frame for examining gender disparities in everyday life. These gender disparities are relevant to almost any given social issue. Additionally, when thinking about the lived experiences of women across the globe, other factors affect a woman's ability to participate in society. These include a history of colonialism, geography, ethnicity, religion, class, disability and sexual orientation. Structural inequalities are multiple and interconnected, and using an intersectional lens is vital to unpacking this complexity, but it should not be used solely as an additive frame. Rather women's experiences should be understood as multidimensional, as gender inequalities interface with other inequalities, which will have an impact on each woman in different ways (Bernard, 2022; Crenshaw, 1991).

Mary Wollstonecraft's publication *A Vindication of the Rights of Women* (Wollstonecraft, 1792) is often cited as the first feminist text. However, it was not until the twentieth century that many social changes began taking place, including women's suffrage (the nineteenth century in Aotearoa/New Zealand), changes to employment law, property rights and health, the opportunities for education and better representation at all levels of the workforce, including in areas of public life, such as politics. The 1960s and 1970s saw further social changes occur in women's reproductive and sexual rights and the feminist movement of that day exerted sizeable influence through protest. The seven demands of the Women's Liberation Movement in the UK in the 1970s comprised:

- equal pay;
- equal education and jobs;
- free contraception and abortion on demand;
- free 24-hour nurseries; financial and legal independence;

- the right to a self-defined sexuality and an end to discrimination against lesbians;
- freedom for all women from intimidation by the threat or use of violence or sexual coercion regardless of marital status; and
- an end to the laws, assumptions and institutions which perpetuate male dominance and aggression to women (Sisterhood and After Research Team, 2013).

More than half a century later, although progress has been made, women still do not enjoy full equality with men in the UK, despite earlier changes to legislation setting out equal pay and equal rights. The gender pay gap remains and the #MeToo movement helped to reignite awareness of sexual harassment in the workplace. Gender-based violence figures also remain disturbingly high. Violence against women and girls remains a global issue (United Nations, 2020). For the fifth year in a row, on International Women's day, the British Labour MP Jess Phillips read out in the UK House of Commons the names of UK women and girls who were killed by men over the previous year (n = 118). She said, "Violence against women and girls is an epidemic ... if as many people died every week at a sporting event ... there would be national outcry ... all of these women mattered" (Jess Phillips, MP 8 March 2021).

In terms of feminism's relationship with social work, there have been a number of crucial studies on social issues, where gender as a biological fixed entity has been analysed for its social, historical, economic and political construction (Dominelli, 2002; Orme, 1998; White, 2006). Most of this discourse has focused on service user perspectives, their experiences and the nature of their relationships with social workers and their organisations. It is reasonable to say, however, that mainstream research, education and practice in social work have not been explicit about gender and any specific analysis is far from being well established within mainstream curricula (Cree & Dean, 2015; Lauve-Moon et al., 2020). There are a number of other publications which have addressed feminism and social work over the years (Butler-Mokoro & Grant, 2018; Dominelli, 2002; Dominelli & McLeod, 1989; Fawcett, 2000; Wendt & Moulding, 2016; White, 2006). This publication builds on the work of these earlier volumes.

Gender-powered relations within social care organisations themselves remain an issue of concern. Gender and sexuality issues remain defining features in most care organisations, for example, in gendered patterns of hierarchy, occupational segregation, the predominance of heterosexuality, harassment and discrimination, and in the questioning of work-life balance, particularly in relation to family responsibilities (Hafford-Letchfield, 2011). These are in turn defined by, and instrumental in reproducing, social relations of age, class, disability, culture and ethnicity.

Similarly, major changes in legislative frameworks to promote the rights of women and other areas of gender equality have challenged and continue to challenge long-standing heteronormative and heterosexist frames of reference in both social work practice and professional education and the way these are organised. Developments in legislation and promotion of rights within particular areas, such as employment, crime, partnerships and family law, have gone some way to address this. However, there is still an absence of any systematic approach to address gender (and sexuality issues) in social work. This includes acknowledgement of the complexities of identities and lives within the current dynamic and changing social environment in which care organisations operate.

Theoretically, a move towards more pluralist approaches within the post-structuralist and post-modernist turns has also given rise to the intersectionality of women within the feminist movement (Crenshaw, 1991) and how rights interact with other multiple social divisions and differences. These third wave feminists have critically questioned the notion of coherent identities and view freedom as resistance to categorisations of identity (Mann & Huffman, 2005). Evolving discourses frame and determine social knowledge and our subsequent understanding of power as well as our understanding about the concentration of power in relation to the nature of personal identity, organisational life and the way our societies are structured. These 'institutions' continue to carry gendered meanings and reinforce gender inequalities (Foucault & Sheridan, 1991). Globalisation and individualisation together with increasingly individualised consumer cultures within the modernisation of care services have demanded continuous innovation and performance improvement in all aspects of social care. Within this context, we face a particularly complex and contradictory picture of what progress is actually being made towards gender and sexual equality.

This book provides a range of topics and positions relating to these themes. Authors reinforce messages that whilst a number of gains have been achieved, if left critically unexamined, sexism, heterosexism, genderism and gender biases will continue as more subtle practices embedded within social work's core processes and activities. Addressing this requires increased reflexivity in order to negotiate boundaries within an institutionalised cultural script (Hafford-Letchfield, 2011). This book intends to elaborate on the range of ways in which gender identities are performed and enacted, how women are positioned and subjugated and how these inequalities are sustained and generated across the structures and practices of care services. It reviews some of the broader key discourses, when thinking about the relationships within social work, and addresses new debates in feminist theories, for example, in relation to diverse gender identities, sexual identities, racism and colonialism. Feminist theory is both an explanatory and critical theory for social work (Allan et al., 2009) and sits alongside other critical frameworks central to its research, practice and teaching activities.

We have structured the book into three themes that cut across key feminist perspectives and the relevance to social work. These themes are:

1. Feminisms and theorising for social work
2. Intersectionality
3. Gender in social work practice

We are delighted to include contributions from a number of different international perspectives. One of the key themes in the book is intersectionality (Crenshaw, 1991); this comprises one section in the book, but all authors have explored intersectional issues across the topics they cover in their respective chapters. These pluralist perspectives reflect the social and political environment in which people are working, but there is still so much for social work to learn about its own practices around diversity and equality (Cocker & Hafford-Letchfield, 2014).

Chapter Synopses

The first five chapters on feminisms and theorising for social work develop our knowledge of how feminist theories and approaches can be applied to social work education, research and practices. In Chap. 2, Barbara Fawcett provides an overview of the development of feminisms, including exploring the tensions brought about by attempts to incorporate the very different experiences of Black women, lesbians and working-class women into 'second wave' feminism. The chapter also explores the contribution of 'third' and 'fourth wave' and 'choice' feminisms. It also considers the contribution made by postmodern feminism and the effect that deconstruction of language, culture and power/knowledge dynamics has had on how we understand 'truth'. It also appraises the contribution of 'Islamic feminism' to consider the connections and disconnections between faith, culture and patriarchy. The chapter then reflects on the contribution of these various feminist influences on social work in terms of current practice and future directions.

In Chap. 3, Tina Maschi, Sandra G. Turner, Smita Ekka Dewan, Adriana Kaye and Annette Hintenach explore the relationship between feminist and empowerment theory and practice with social work. Writing from the USA, they explore how social work professional identities have aligned with the core values and perspectives of feminist and empowerment practice theories. The psychological, social, and structural rights and needs of historically and emerging underrepresented and underserved populations, such as women and children living in poverty, racial ethnic minorities, immigrants, and older and lesbian, gay, bisexual, trans and queer (LGBTQ) people, are discussed using feminist and empowerment ideals. Maschi and colleagues draw on a case example of a 50-year-old incarcerated woman to illustrate how feminist and empowerment approaches can address the individual and collective needs of incarcerated women.

In Chap. 4, Sam Harrell, Ben Anderson-Nathe, Stéphanie Wahab and Christina Gringeri explore what makes research 'feminist', noting the alignment between feminist research and social work values and principles. It uses a metaphor of a braid to analyse how the complex and mutually reinforcing threads of neoliberalism, professionalisation and criminalisation have constrained the profession of social work. They explore the profession's ongoing complicity in durable and mutable institutions, systems and discourses of marginalisation and oppression. The authors suggest that feminisms can help researchers turn a critical gaze on structures of gender, including how genders are developed and maintained, and the ways they shape peoples' lived experiences. Critical feminisms in particular expand this critical gaze to other power relations as well, bridging analyses of gender to those of race, class, sexuality, disability and other social categories. Harrell and colleagues believe that feminism's contribution to social work is that it nudges us to question the status quo, moves beyond limiting binaries and creates space for collaborative actions.

Sarah Epstein, Norah Hosken and Sevi Vassos in Chap. 5 examine the way in which western social work is practised and taught. They suggest that it is shaped within an institutional context of a higher education sector and neoliberal societal context that valorises the individual. Feminist critical social work educators face constraints and challenges when trying to imagine, co-construct, enact and improve ways to engage in the communal relationality of feminist critical pedagogy. In this chapter, Epstein and colleagues, based in Australia, articulate their response to these challenges—through the development of a feminist collaborative, research, teaching and advocacy centre called Critical Edge Women (CrEW). CrEW draws upon feminist social work and pedagogical literature, and its members own experiences, to share their development of a feminist social work pedagogy as a distinct (explanatory) theory and practice model. The chapter concludes with a discussion of how a feminist critical social work framework in social work pedagogy and practice is foundational in the enactment of social work values and ethics.

In Chap. 6, Jennifer Dyer, Sarah Pickett, Jennifer Davis, Kathleen Hackett, Cindy Holmes, Julie James, Daze Jefferies, Kimberley Manning, Annie Pullen Sansfaçon and Julie Temple Newhook write from the perspective of parents of trans and gender diverse children or their allies. They have all engaged in advocacy work and have developed the Collaborative Autoethnography (CAE) project to document their experience of advocacy. CAE is a feminist, qualitative methodology that promotes community building between participants and can, by its process, become transformative for participants. It simultaneously produces knowledge and action by combining ethnography, biography and self-analysis and by producing data about interpretations of self and social phenomena involving the self. This chapter presents how CAE was deployed and provides a reflection about their experiences in participating in such processes.

The next six chapters explore intersectionality. In Chap. 7, Kathomi Gatwiri and Sharlotte Tusasiirwe explore the roles of both feminism and social work in Africa for Africans, as they remain highly contested. Feminism is mostly sidelined and positioned as a western ideology imposed upon Africans, while social work is criticised for the underlying whitewashed, colonised knowledge it inculcates in students and practitioners. Without proper contextualisation and theorisation, both feminism and social work can function as colonising ideologies, which do more harm than good in the African space. Gatwiri and Tusasiirwe argue that defining the needs of Africans in Africa through feminist lenses requires us to comprehend the pervasiveness of both colonialism and colonisation and their unrelenting impacts on Black African bodies. Feminist social work must also critique global power constructs, contextualising both feminism and social work within their historical and political context. This chapter positions African feminism as an Afrocentric and decolonising method and framework that prioritises African knowledge, theories, practices and ways of "being, doing and knowing". It is within this context that the authors articulate its role in social work practice in Africa, while also problematising the epistemological contradiction it presents. They argue that African feminism is a construction of knowledge that is about returning to forgotten knowledge, forgotten ways of "being, doing and knowing" and reclaiming a 'somehow lost or disfigured identity' which was and is continually altered by colonialisation and its ever lingering presence.

Fiona Kumari Campbell in Chap. 8 presents original research about configurations of shame (lajja-baya) as it is understood within a Sri Lankan Buddhist framework and its impact on mothers and partners where a family member experiences Disability. Understandings of care ethics in Western societies are formulated through a prism of possessive individualism and a demarcation of the private and public spheres. In order to engage with anti-racist practice, Kumari Campbell argues that it is critical that practitioners and researchers embark on the journey of comparative travel, that is to "shift our frames altogether so we see things differently from another angle" (Butnor & McWeeny, 2014, 11). The author critiques the work of American anthropologist Bambi Chapin (2014) on Sri Lankan child-rearing practices, before presenting her own research around Buddhist cosmologies around fear-shame and the operation of an ethic of lajja-baya in women and girls' lives and how this might impact on caregiving roles. These insights are then brought together through a consideration of what cultural humility rather than cultural competence might mean for epistemological and methodological approaches to research and practice.

In Chap. 9, Dina Pervez Sidhva also presents original research about the gender-based violence experienced by Syrian refugee women living in Jordan. This chapter explores how a feminist framework that takes into account interpersonal, social and structural factors can assist in understanding the complexities of multiple inequalities experienced by these women. Sidhva uses intersectionality (Crenshaw, 1991) as an underpinning factor in her analysis; the aim is to provide social workers with a deeper understanding and empathetic grasp of the intersectional susceptibilities experienced by refugee women.

In Chap. 10, Jama Shelton, Maggie Dunleavy and Kel Kroehle provide an overview of transfeminism and document the inherent potential in a partnership between transfeminism and the social work profession. There is an absence of literature about this topic, which is a concern given the high rates of discrimination faced by trans people and trans women of colour in particular, and the implication for how social workers are committed to social justice and equitable service delivery. The chapter explores where transfeminism 'fits' within feminist theories, before going on to consider transfeminism's relevance to social work.

In Chap. 11, the authors tell the story of Josie, a gender non-conforming, queer woman living with chronic illness and disability. Jennifer (JD) Drummond and Shari Brotman use a feminist-narrative methodology to elicit Josie's description of her experiences of school, health care and the ways in which she negotiates sexual and gender identity and expression in the context of the many challenges she experiences in living with pain and disability. Through attention to critical moments and turning points in her lived experience of disability, sexuality and gender, Drummond and Brotman examine the ways in which intersecting identities both shape and are shaped by context and experiences of the body as both complex and fluid. The ways in which bodily performance and negotiation are used to resist multiple forms of discrimination are also highlighted. Josie's narrative provides a rich account of her experiences embodying multiple marginalised and intersecting social locations and offers important insights on how to engage in a social work practice informed by a feminist-narrative approach, particularly when addressing the intersecting social locations of disability and queer identities. This chapter situates the usefulness of an intersectional lifeline and feminist-narrative interviewing as tools for the development of critical, transformative research and practice in social work.

Trish Hafford-Letchfield in Chap. 12 examines the literature on gender and ageing through the lens of critical feminist gerontology. She draws on multidimensional concepts comprising power, economic, social and symbolic relations operating simultaneously at intrapersonal, interpersonal, institutional and societal levels. Hafford-Letchfield applies an intersectional lens and feminist perspective to social work practice with women in later life to help us move beyond gender as one way of explaining differences towards ensuring advocacy and empowerment to address inequalities and shape supportive social work interactions. She also refers to the role of relational theory in social work practice to provide insights into the complexity of gender relations in later life, with a clear focus on the uniqueness of a person's circumstances within the context of ageism. The wider social work profession and its policy makers have not yet grasped the challenge and importance of gerontological social work and its evidence base and this chapter reviews some key areas of structural gender inequalities in later life and draws on some of the empirical evidence. Hafford-Letchfield aims to facilitate a better more nuanced understanding of the specific challenges faced by women in later life to help social workers improve support through purposive gender-responsive and age-sensitive practice.

In this final section, contributors focus on social work practice through the lens of performativity and gender. In Chap. 13, Roxana Diamond, Priscilla Dunk-West and Sarah Wendt explore the everyday experiences of sex workers in South Australia. In taking a structural approach, they use radical feminist theories relating to sex work to argue against women's agency by positioning choice as antithetical to structural oppression. Conversely, radical social work theories take into account both structure and agency, seeing each as nuanced and non-dichotomous. The authors argue that feminist theories, in which the everyday highlights both structure and agency, points to the need to employ new forms of theory to better understand sex work. Honouring lived experience is central to these understandings, and this chapter explores one of their authors' experiences of this, alongside the other authors' reflections on what this means for the academy and for reflexive feminist scholarship.

Linda Bell in Chap. 14 explores what is meant by 'feminism(s)' in current social work contexts in England. Bell explores the relevance of gender and particularly 'gender specificity' to feminism now, and whether being 'gender-specific' is sometimes seen as discriminatory by social workers today. She uses her own collaborative research with social workers and their colleagues to explore the meaning of 'gender-specific', using a project which provides support for women whose children have been taken into public care. In the context of other examples, Bell examines how changed ways of working could contribute towards altering social workers' sensitivities towards using feminist approaches in practice.

Chapter 15 by Christine Cocker explores the feminist theoretical discourses around lesbian parenting to consider whether the 'gayby boom' and lesbian parenting in particular remains 'progressive' as an alternative family form, or whether it is now so commonplace that it occupies another different space altogether. Cocker provides a brief overview of the history of lesbian parenting in the UK over 50 years before considering how various discourses over this time attached meaning to lesbians having and raising children. Reflections are offered about how social work practice with lesbian parents has changed over time, and what learning from this can be applied to social work practices with queer families.

Chapter 16 explores issues of victimhood and agency, consent or constraint for those who have experience of sexual exploitation. Jane Dodsworth explores discourses on this issue, which are largely derived from a gendered, patriarchal narrative that seeks to victimise or blame. Whilst there have been some shifts in thinking in terms of how child sexual exploitation is perceived, that is, as abuse/exploitation rather than child prostitution/criminality, the wider narratives informing perceptions still result in victim blaming by professionals and the wider public and often self-blaming by those exploited. Using feminist theory and values, this chapter examines the impact of the dominant narratives on the lives of several women who experienced sexual exploitation as children and sex work as adults. It explores the impact of gendered narratives on their lives, their environments and their families. Dodsworth argues that it is crucial, from a feminist perspective, to listen to the voices of those directly involved; they are the experts in their own lives.

In Chap. 17, Jason Schaub explores the position and effect of men in social work practice and education as a feminist issue. Men are a significant numerical minority in social work, but hold disproportionately more (and greater) positions of power. The complication of a profession that has numerically more women but fewer women in positions of power causes issues for the workplace and raises questions for the profession. Schaub considers the 'glass escalator' concept, which suggests that men in women majority occupations are disproportionately represented in management and other positions of power. Given the challenges with retention rates in the social work profession currently, including the difficulty in recruiting men into the profession, a more detailed and nuanced examination of this issue is missing from the knowledge base, and this chapter, drawing on Schaub's own research, addresses this oversight.

Chapter 18 written by Rebecca Infanti-Milne, Richard Mc Kenny and Lee Walton describes the relationship between feminist theory and family therapy practice. Infanti-Milne, Mc Kenny and Walton provide an overview of feminism and family therapy before describing aspects of their own systemic social work and therapeutic practice, working together with parents where at least one parent has used violence and abuse. The work described was undertaken as part of a larger project in an inner city area in east London from 2017 to 2020, which is one of the most diverse in the UK. Families live with the consequences of poverty, inadequate housing, ill-health, the influence of criminal gangs, racism, sexism and Islamophobia. Infanti-Milne and colleagues explore their own positionality in relation to the skilled work they undertake and consider the gaps in knowledge and skills that social workers have to assess and work with men where violence and abuse is a factor in family relationships.

Finally, in Chap. 19, Adi Cooper and Lyn Romeo discuss the contributions of feminist ideas to 'leadership' in social work, with particular emphasis on the role of women as leaders and managers. For a profession heavily comprised of women, men remain over-represented in senior leadership and management positions in the sector. The number of Black men and women in senior leadership positions is also limited. Cooper and Romeo explore feminist theories around intersectionality, diversity and equality, as they apply to leadership and management, to consider the reasons for this imbalance and discuss the benefits to the profession of changing this dynamic, so that managers and leaders are more representative of their profession and the communities served. The chapter also explores developments in practice leadership and strategic leadership from a feminist perspective.

Conclusion

This book offers a rich variety of perspectives about what feminism and feminist theories mean for social work practice in a global context. It provides a useful addition to our existing knowledge about feminist ideas and offers challenges and examples to social workers about how these theories apply across many different areas of practice. Women make up half the world's population, and for too long women's views and experiences have not been heard and valued—they have been seen as secondary to men's. All the contributors to this edited book have worked or are working as social workers and are involved in working alongside social workers. The contributions across research, policy, practices and pedagogies shared give considerable insight into the many different roles social workers have. One of the overarching commonalities is that feminist theories can assist in understanding the lived experiences of women, but it is crucial to consider how intersectionality affects people's experiences. There is not one feminism but many feminisms and in order for feminism to remain relevant for social work, it cannot continue to be confined to a Eurocentric perspective. Embracing the diversities of feminisms in the world is crucial for developing social work practices that respond to women's individual stories and support communities where women can flourish as leaders. We hope that this book will inspire you to think differently, visualise and envisage a future where women occupy a space in the various societies where they live, that is not only equal to men, but reflect the potential of their lives.

References

Allan, J., Briskman, L., and Pease, B. (eds) (2009) *Critical Social Work: Theories and Practices For A Socially Just World (2nd Edition)*. Allen & Unwin, Crows Nest, NSW.

Bernard, C. (2022). *Intersectionality for social workers: A practical introduction to theory and practice*. Routledge.

Butler-Mokoro, S., & Grant, L. (2018). *Feminist perspectives on social work practice: The intersecting lives of women in the twenty-first century*. Oxford University Press.

Butnor, A., & McWeeny, J. (2014). Feminist Comparative Methodology Performing Philosophy Differently. In Ashby Butnor and Jennifer McWeeny (Eds.), *Asian and Feminist Philosophies in Dialogue: Liberating Traditions*. Columbia University Press.

Chapin, B. (2014). *Childhood in a Sri Lankan Village: Shaping Hierarchy and Desire*. New Brunswick: Rutgers University Press.

Cocker, C., & Hafford-Letchfield, T. (2014). *Rethinking anti-discriminatory and anti-oppressive theories for social work practice*. Palgrave.

Cree, V. E., & Dean, J. S. (2015). Exploring social work students' attitudes towards feminism: Opening up conversations. *Social Work Education, 34*(8), 903–920. https://doi.org/10.1080/02615479.2015.1081884

Crenshaw, K. W. (1991). Mapping the margins: Intersectionality, identity politics, and violence against women of color. *Stanford Law Review, 43*(6), 1241. Retrieved from https://search.ebscohost.com/login.aspx?direct=true&db=edsglt&AN=edsgcl.15019782&authtype=sso&custid=s8993828&site=eds-live&scope=site

Dominelli, L. (2002). *Feminist social work theory and practice*. Palgrave.

Dominelli, L., & McLeod, E. (1989). *Feminist social work*. Macmillan.

Fawcett, B. (2000). *Practice and research in social work: Postmodern feminist perspectives*. Routledge.

Foucault, M., & Sheridan, A. (1991). *Discipline and punish: The birth of the prison*. Penguin Books.

Hafford-Letchfield, T. (2011). Sexuality and women in care organizations: Negotiating boundaries within a gendered cultural script. In P. Dunk-West & T. Hafford-Letchfield (Eds.), *Sexual identities and sexuality in social work: Research and reflections from women in the field* (pp. 11–30). Routledge.

Lauve-Moon, K. R., Enman, S., & Hentz, V. (2020). Mainstreaming gender: An examination of feminist methodology in social work research. *Social Work, 65*(4), 317–324. https://doi.org/10.1093/sw/swaa039

Mann, S., & Huffman, D. (2005). The decentering of second wave feminism and the rise of the third wave. *Science and Society, 69*(1), 56–91.

Orme, J. (1998). Feminist social work. In R. Adams, L. Dominelli & M. Payne (Eds.), *Social Work: Themes, Issues and Critical Debates* (pp. 218–228). Basingstoke, London.

Sisterhood and After Research Team. (2013). *Women's liberation: A national movement.* Retrieved from https://www.bl.uk/sisterhood/articles/womens-liberation-a-national-movement#

United Nations. (2020, Tuesday, October 20). *The world's women 2020: Trends and statistics.* Retrieved from https://worlds-women-2020-data-undesa.hub.arcgis.com/

Wendt, S., & Moulding, N. (2016). *Contemporary feminisms in social work practice.* Routledge.

White, V. (2006). *The state of feminist social work.* Routledge.

Wollstonecraft, M. (1792). *A vindication of the rights of women: With strictures on political and moral subjects.* Thomas and Andrews.

PART I

Feminist Theories for New Challenges in Social Work

Trish Hafford-Letchfield and Christine Cocker

When referring to the wide range of divergent and diverse opinion of feminism, Lena Dominelli (2002) used the term 'feminisms' (p. 2) to highlight that any particular interpretation and construction of feminism may be just one of many. Dominelli's own perspective comprised an amalgamation of elements borrowed from a broad range of feminist views. Dominelli however asserted that despite any fragmentation, feminists will always have shared goals and principles to which they will return.

Feminist social work has very clear goals; to expose and critically analyse power as a dynamic, historic, and structural concept embedded in our world and to mobilise and take social action to challenge that power which is integral to a commitment to the core values of the social work profession: human rights, social justice, and professional integrity.

Chapters 2, 3, 4, 5, and 6 of this book provide a robust and firm foundation for articulating and grappling with 'feminisms', their theoretical origins, meanings, differentiations, and intersections with social work and how these theories inform what social work actually does. The contributors to these five diverse, yet coherent chapters address the origins and progressions of feminisms through their temporal social movements. The contributors provide us with numerous exemplars of feminist research, pedagogy, and practice, central to understanding social and economic inequality and assert the significance of standpoint for feminist enquiry. Across the five chapters, the authors have scoped out the ideological foundations for how social work might engage and utilise feminist theories and how these theories in turn inform ethical practice and professional interventions. Contributors provide authentic examples of what it means to engage with feminist research and make feminist social work pedagogical practice visible as both a theory and practice model. In short, more robust engagement with feminisms and theorising for social work is posited for sound knowledge-building communities that drive social activism and capacity for transformational change.

In Chap. 2, Barbara Fawcett challenges the idea that feminism is one coherent area and she takes us through the different phases of feminism that reflect contemporary society and encapsulate women's movements over time. As she points out, these phases are not linear and such progression involves ongoing circularities and interpretations. Fawcett describes the different 'waves' of feminism aligned to civil rights, other societal struggles, and the politicisation of public and private spheres. Fawcett explores how these movements are challenging social work such as through resistance to capitalism, socialism, and the role of identity politics. The deconstruction of power and knowledge in postmodern feminism, for example, has provided a more central focus on social relations and how power operates to oppress particular groups or communities at that moment in time.

The role that feminist research plays in helping us to see the centrality of gender equality is made accessible through the conceptual model of social work practice presented by Tina Maschi, Sandra G. Turner, Smita Ekka Dewan, Adriana Kaye, and Annette M. Hintenach in Chap. 3. Their whole systems model incorporates human rights, social, economic, and environmental justice, and person-in-environment perspective integrated with feminist and empowerment theories. This intersectional framework demonstrates how to work at each level (macro, meso, and micro). Through the analysis of Mary's story, an older woman who has experienced incarceration, the framework is applied and its combination of theoretical approaches is used to address accumulated life course disadvantages as well as developing interventions for Mary. Maschi et al. explore the empowering nature of feminist and empowerment ideals to address the psychological, social, and structural rights and needs of historically and emerging underrepresented and underserved populations within social work.

These themes are interrogated further in Chap. 4, where Sam Harrell, Ben Anderson-Nathe, Stéphanie Wahab, and Christina Gringeri use the concept of a 'braid' as a metaphor to articulate how complex and mutually reinforcing threads of neoliberalism, professionalisation, and criminalisation have constrained social work practitioners' and researchers' ability to meaningfully interrupt and address domestic and sexual violence. Using concrete practice examples, Harrell et al. propose that tenets of feminist research open a new politic for social work in which social work scholars, practitioners, and educators might interrupt dominant paradigms in the field and open space for collaborative action and a new social work future. This is necessary, they argue, to challenge the profession's ongoing complicity in discourse of marginalisation and oppression. This argument echoes what happens at the macro, meso, and micro institutions and systems described in Maschi et al.'s earlier framework. They talk about the need to actively interrupt, subvert, and resist the impacts of the 'braid' on social work research and practice.

From within a different but equally significant institution, that of the contemporary neoliberal academy, Sarah Epstein, Norah Hosken, and Sevi Vassos call for us to 'speak out against standardised, gender neutral or individualised education' in Chap. 5. They advocate and demonstrate their own activism in developing feminist pedagogies, which allow for both operationalising the category of woman and in keeping women's voices central. This, they propose must be the starting analytic entry point for learning about social work, at the same time as accounting for the diverse and heterogeneous social locations that women (service users, practitioners, and academics) occupy including class, age, race, sexuality, ability, and culture. The practice of consciously bringing women's experiences into the teaching, research, and administrative domains of the university is essential to challenge traditional and privileged knowledge, which reflects the accepted power relations of men. Epstein et al. provide brief explanations of pedagogic principles interspersed with illustrative examples from their practice as feminist social work pedagogues. Their mission is to make feminist social work pedagogical practice visible as a theory and practice model that is distinct from other pedagogies.

In Chap. 6, Jennifer Dyer, Sarah Pickett, Jennifer Davis, Kathleen Hackett, Cindy Holmes, Julie James, Daze Jeffries, Kimberley Manning, Annie Pullen Sansfaçon, and Julie Temple Newhook show us how feminist research can be participative and transformative. Collaborative auto-ethnography (CAE) is a feminist, qualitative methodology that simultaneously produces knowledge and action by combining ethnography, biography, and self-analysis. These authors use their experiences as parents of trans or gender diverse children or their allies to show how CAE works as a feminist research method in knowledge creation.

Social work has always retained a strong tradition in its theory-practice alliance, and throughout this book, contributors will examine how feminist social workers can respond to theories in developing their practice as well as critique and theorise in response to the challenges of practice. These five chapters that follow this introduction pay testament to this commitment.

Reference

Dominelli, L. (2002). *Feminist social work theory and practice*. Palgrave Macmillan.

CHAPTER 2

Feminisms: Controversy, Contestation and Challenge

Barbara Fawcett

Introduction

Social work and feminism have always had a somewhat mixed relationship. There are clear influences, but also elements of contestation, with those working in the fields of social work and social care being as susceptible to interpretations and trends in popular culture as the rest of the population. In this chapter, I examine the development, impact and tensions of different feminisms. We explore the place of experience and the schisms brought about by attempts to incorporate the very different experiences of Black women as well as those from disadvantaged backgrounds into 'second-wave' feminism. We consider the contribution made by postmodern feminism and investigate the influence of what has been termed 'choice' feminism on so-called third- and fourth-wave movements. We also appraise what is sometimes and controversially called Islamic feminism to consider the connections and disconnections between Islam, culture and patriarchy.

B. Fawcett (✉)
University of Strathclyde, Glasgow, UK
e-mail: Barbara.fawcett@strath.ac.uk

C. Cocker, T. Hafford-Letchfield (eds.), *Rethinking Feminist Theories for Social Work Practice*,
https://doi.org/10.1007/978-3-030-94241-0_2

Finally, we look at the effects of these various influences on social work in terms of current practice and future directions.

At the outset, it is important to highlight that there has never been one single form of feminism, there have always been many and accordingly, it is the term 'feminisms' that can be seen to best encapsulate the wide-ranging nature of women's movements. Similarly, although feminisms have been described in terms of 'waves', with the numbering implying a linear historical progression, ideas and understandings cannot be viewed in such a clear-cut manner and there are ongoing circularities, identifications and interpretations. Throughout, we find that in relation to popular representation, particular aspects have been highlighted and other facets either downplayed or ignored. An example can be taken from what has been referred to as 'first-wave' feminism(s). This is largely regarded as a fight of predominantly middle-class women for the vote. Walby (1997), Lorber (2011) and Matos (2015) contest this and maintain that it was far broader and more nuanced. Although the emphasis was on White women, Matos (2015) argues, that like second-wave feminism(s), there were both liberal and radical versions with a multifaceted range of participants and demands. These included campaigns for access to basic and higher education and training, for employment opportunities, for equal property rights for married women and for measures to control violence against women. At this time, women were also active in labour and union movements and campaigned for public provision for health care as well as for basic human rights across the board.

'Second-Wave' Feminism(s)

The political challenge mounted by what has become known as 'second-wave' feminism(s) is associated primarily with the political movements taking place in the 1960s and 1970s. Many women, involved in political activity such as the Civil Rights Movement and Anti-Vietnam War campaigns, became increasingly frustrated with the male-dominated nature of these groups. As Rogan and Budgeon (2018) highlight, when women tried to name their oppression, they tended to be dismissed and called to account for introducing personal matters into the public arena. Women at this time were also challenging centuries of being denied rights and equal status on the basis of arguments emanating from assertions of the 'natural order' and patriarchal religious ideology. A primary justification related to the separation of political power in the public sphere, which was the domain of men, from the family sphere, which was private and women orientated, but also subject to male patriarchic hegemony. Free will was the preserve of men, whilst, as a result of assertions about 'female nature', dating back to Greek and Roman times, women were regarded as requiring control, management and supervision. Second-wave feminism(s) politicised the private sphere in a range of groundbreaking and integrally compelling ways and demanded equal rights and recognition. As a result, second-wave feminism(s) was based around the pithy and powerful slogan 'the personal is political' (Rogan and Budgeon, 2018; Mendes, 2011; Crook & Gutnick Allen, 2014).

A key aspect of second-wave feminism(s), drawing from the influential work of Simone de Beauvoir's *The Second Sex* (1949), was the recognition and assertion that masculinity and femininity were not natural or predetermined in any way but were socially constructed, fashioned by religion, culture, socialisation processes and by education or the lack of it. This drew attention to masculinity and femininity not being fixed but rather being fluid and open to re-interpretation. This understanding enabled enduring notions of women as emotional, irrational, non-competitive and dependent and those of men as rational, emotionally controlled, competitive, autonomous and decisive, to be challenged. It also opened the door for established roles for women, such as that of 'housewife', to be resisted (West & Zimmerman, 1987; Risman, 2004).

Second-wave feminism(s) has often been depicted as a unitary movement, with critics drawing attention to the lack of attention paid to difference and diversity. However, within second-wave feminism(s) there were Marxist, socialist, liberal, standpoint and radical strands, all adhering to the central political slogan, but diverging in terms of the accompanying messages. Ramazanoglu (1989), for example, highlights that socialist and Marxist feminisms were particularly influential in Europe with radical feminism(s) being more prominent in the USA. She asserts that the former identified class and economic distinctions between women as key areas and challenged capitalism as well as patriarchy. Radical feminism(s), in contrast, although recognising differential power relations, focused on viewing all women as being subject to a common form of oppression, namely male patriarchy. Within this, there was a tendency to concentrate on the oppression of all women by all men and on the linking of private and public spheres. Doude and Tapp (2014) assert that radical feminism(s) generally challenged the patriarchal ideal of the private sphere for women as being about childrearing, marriage and the maintenance of the household. They also took issue with its political patriarchal equivalent which they associated with the harming of women through rape, domestic violence and prostitution. Standpoint feminism(s), shaped by the work of Smith and Hartsock (1983) amongst many others, has clear links with both Marxist and radical feminism(s). It is based on women's experiences of the workings and oppressions of patriarchy producing a different world view and knowledge framework. It is this experiential standpoint which becomes the basis for challenge and change. Liberal feminism(s) had a strong rights basis and was primarily concerned with rights for women and equality in relation to voting, economic independence, education and the entitlements of citizenship. Ascribed roles were challenged on the basis that they were imbued with inequality, but the emphasis was very much on rights in the public sphere rather than on politicising issues from the private realm (Saulnier, 1996; Lay & Daley, 2007).

Although second-wave feminism(s) is associated with the era of the 1960s and 1970s, identifications, ideas and indeed actions continue to permeate and inform. An example relates to the ongoing campaigns against rape which draw from all feminisms. It is not so long ago, for example, that rape was viewed as a biological predisposition amongst men. The various feminisms have ensured that the objectification of women has been foregrounded and subjected to continual challenges, although overt sexist practices have always to be continually contested and named as abusive manipulation, exploitation and rape. The #MeToo movement which gained prominence in 2019 illustrated that engagement is ongoing and battles have to be continually re-fought. More covert forms of discrimination, relating to gender roles and workplace disparities, associated with pay, promotion and the constructive tackling of career breaks and reproductive rights, also remain on the agenda for many feminisms. However, second-wave feminism(s) provided a platform for action by women and for women and, despite political circularity and recidivism, has made a historical difference and substantially altered understandings about 'women's place', 'women's position' and women's agency. Although there have been many different strands (which also continue to be further developed), the strong and clear message encapsulated in the slogan 'the personal is political' has been very successfully utilised. It has served to effectively foreground women's oppression, change attitudes and promote action.

All political movements have to have a central unifying message in order to be effective. However, a consequence of this is an accompanying assumption of unity and shared experiences. Second-wave feminism(s) has been criticised for these universalising tendencies and for assuming a similarity of experiences of oppression. In the USA particularly, second-wave feminism focused attention on White middle-class women and whilst in Europe, socialist and Marxist feminisms took account of the influence of class and economic deprivation within capitalism, it was these universalising tendencies which resulted in the increasing alienation of Black feminists, those from economically impoverished backgrounds and lesbian feminists who took issue with heteronormative assumptions.

Lesbian Feminism(s) and Standpoint

Lesbian activists played a major role in nuancing and developing the messages of second-wave feminism(s). Feminist lesbian politics added a deeper dimension in terms of critical perspectives and cultural orientation. They focused, not just on politicising power imbalances between men and women, but on exploring women's sexuality outside of both heterosexual and male-influenced gay liberation orientations. Feminist standpoint was generally used to demonstrate how injustice and oppression could be better analysed and understood from the position of those who were marginalised (Hartsock, 1983; Harding, 2004). However, lesbian activists used it to explore how sexuality is also historically, culturally and socially situated. They turned sexual marginalisation into a standpoint of epistemic privilege, a springboard for both critique and collective action within and outside second-wave feminism(s).

Black Feminism(s)

The Feminine Mystique, by Betty Friedan published in 1963, is viewed as groundbreaking in relation to the development of second-wave feminism(s). However, as Smith (2014) points out, Friedan's suggestion that women go out to work and hire domestic workers to perform their daily household chores failed to resonate with those Black women and women from economically pressurised backgrounds who would be undertaking the relinquished domestic tasks. Ironically, Black women at this time were also being criticised for going out to work and the 'Black matriarchy' as it was called were blamed for a range of social problems precisely because of their apparent economic independence (e.g. in the Moynihan Report of 1965). Accordingly, Black women were charged with:

emasculating Black men, whilst at the same time releasing them to be sexual predators; for promoting juvenile delinquency amongst young Black people by not being full-time homemakers; and for holding back the economic success of the Black population generally as a result of their deviation from prevailing American values (Smith, 2014).

Black women's history of slavery, exploitation, enforced sterilisation and entrenched racism clearly differed markedly from that of the women Freidan (1963) was referring to. However, Rogan and Budgeon (2018) argue that whilst second-wave feminism(s)' assumed homogeneity appeared to ignore the individual and collective experiences of Black women, its central slogan 'the personal is political' paradoxically proved to be central to the genesis of Black feminist politics. Rogan and Budgeon (2018) maintain that second-wave feminism(s) was the first radical movement to both base and create its politics on personal experience and, as a result, to extend critical understandings of the ways in which power operates. If, for example, politics is seen to be about power being exercised, not only at macro levels but also at micro levels in everyday life, then the range of activities contributing to a broader political movement expands exponentially. Rogan and Budgeon (2018) contend that, as part of this process, activities such as consciousness raising not only facilitated the collectivising of social experiences but also made connections to prevailing political forces. Accordingly, Black women could theorise from their own experiences and develop forms of Black feminism which not only confronted their experiences of sexism within their own communities but also challenged the racism they experienced within White second-wave feminism(s) (Phipps, 2016; Rogan and Budgeon 2018).

Rogan and Budgeon (2018) argument supports the formation and politicisation of Black feminism(s), but Black women in the USA highlight that their struggle has a long history. Barbara Smith (1983), Alice Walker (1984) and Patricia Hill Collins (1990), amongst many others, stress that Black women's knowledge relates to enduring multifactorial oppressions. Collins (1990) contends that this collective wisdom has resulted in the development of Black feminist thought as critical social theory. Smith (2014) points out that the influential Combahee River Collective, which originated in 1974, was made up of women who had been involved with the Black Panther Party and other antiracist organisations. They established a tradition that rejected the prioritisation of women's oppression over racism and racism over women's oppression. They rebuffed middle-class feminisms that failed to take account of the centrality of class and economic deprivation in poor and working-class women's lives. In 1989 Kimberlé Crenshaw first used the term 'intersectionality' to refer to the ways in which multiple oppressions are simultaneously experienced by Black women (Crenshaw, 1989; Smith, 2014). She made it clear that discrimination and oppression are not a series of add ons in terms of sexism, plus racism, plus ageism, plus disablism, but multiply experienced subjugations and repressions. Patricia Hill Collins (1990) went on to adapt Crenshaw's theory on intersectionality to the social sciences. Similarly with regard to sexual identities, Black first or lesbian first became linked by 'both and' rather than 'either or', with context playing a part in the sexual identity expressed. It is notable that, over the past thirty years, the concept of intersectionality has become both a means of understanding power dynamics and also a focus for action that applies to all women. As such it has become central to understanding and naming oppressive forces. Reni Eddo-Lodge (2020) notably asserted that to be an intersectional feminist, one had to speak up when confronted with an injustice and with regard to Black Lives Matter and injustice globally, this has become a prime intersectional activist statement.

Postmodern Feminism(s)

Concepts which incorporate the intersectionality of gendered power dynamics, as well as the importance of context, can be seen to be central to postmodern feminism(s). Postmodern perspectives have generally been viewed as elitist and pretentious, containing a form of relativism which makes it possible to support almost anything and to criticise nothing (Brodrib, 1992; Jackson, 1992). However, the work of Barrett (1987, 1992), Weedon (1987), Sawicki (1991), Flax (1992), Fraser and Nicholson (1993), Fraser (1993) and Fawcett (2000), amongst others, highlight how postmodern feminism(s) can embrace difference and diversity whilst also recognising and addressing intersectional power imbalances and the effects of these in different contexts. Rather than examining differences between groupings, such as between men and women, there is an emphasis on how differences have been constructed and how categories are created through difference.

Postmodern analysis is concerned with deconstruction, and there is a concomitant recognition that there is no such thing as 'the truth' or 'real' knowledge. There is an embracing of antifoundationalism and an acceptance of pragmatism, variety, contingency and most of all, uncertainty.

Accordingly, nothing is 'fixed' or certain or is seen to have any foundation in truth, logic or facts. The grand, progressive narratives of the modern era such as Marxism, liberalism and second-wave feminism(s) are rejected as foundationalist stories. There is an understanding that power and knowledge are intricately interwoven and play out in discursive contexts, with discourse defined as a critical analysis of the everyday or the 'taken for granted'. Within discourse analysis there is an exploration of apparent accepted 'truths' and 'facts' with an examination of how these came to be viewed in this way together with an exploration of the resulting implications.

Feminist postmodern perspectives are wide ranging and have variously focused on deconstructions of language, culture and power/knowledge dynamics. Judith Butler, for example, early on deconstructed gender, sexuality and identity, maintaining that all are culturally and social constructed and fluid. As Featherstone and Green (2013) point out, her emphasis has been more on the 'undoing' rather than the 'doing' of gender (Featherstone and Green, 2013; Butler, 1990; Butler, 1994). However, despite the many variations, feminist postmodern orientations can be seen to have subtly changed postmodern conceptualisations. Postmodern feminism(s) promotes deconstructive analyses of prevailing power/knowledge frameworks, as well as explorations of the ways in which power circulates in different social relations. However, it also emphasises the importance of context and draws attention to how positions can be fixed, weighted and responded to in different contexts. As a result, although essentialist, foundational 'truths' or 'facts' are rejected, the emphasis on the importance of context and of 'fixing' positions in context means that pertinent criteria can be evaluated, decisions made and action taken. So, for example, in a particular situation it becomes possible to cite exploitation and to exercise agency when covert pressure is placed on a woman to do something that is clearly in that context, not in her interest (Fawcett, 2016).

'Third-Wave', 'Fourth-Wave' and 'Choice' Feminisms

In many respects, postmodern feminism(s) stands outside what has controversially been referred to as 'third-wave' and 'fourth-wave' feminisms. These remain contested labels with both emphasising individualism and celebrating difference. They are generally seen as spanning the period from the mid-1990s until the present day. 'Third-wave' feminism(s) has been characterised by an emphasis on multiplicity and difference and on the fluidity of sexuality with the foregrounding of queer theory and bisexual and trans identities. There has also been a focus on cultural production, on micro politics and on individual emancipation (Gillis et al., 2007). Contradiction and paradox feature significantly with regard to both areas of exploration as well as to the ways in sexual identities that are either constructed or deconstructed (Matos, 2015). 'Fourth-wave' feminism(s) is very much associated with social media and the internet. Munro (2013) calls it the 'call out' culture, where misogynist or sexist views can be called to account. It is characterised by 'privilege checking' or looking at where an individual is coming from, ideologically and epistemologically. However, it also features 'doxing' where personal files are accessed and distributed and cyberbullying. It has fostered internet inclusion and action as well as exclusion and fragmentation (Munro, 2013; Kearney, 2012). 'Third-wave' and 'fourth-wave' feminisms have been associated with what have been called 'postfeminism(s)', postfeminist sensibility and 'choice' feminism(s) (Braithwaite, 2002; Gill et al., 2017), although some, such as Evans (2015, 2018), particularly in relation to 'third-wave' feminism(s), strongly contest this. Shared key features include a belief in individual personal autonomy, the achievement of personal goals through unilateral action, self-expression through bodily image and an acceptance that it is the responsibility of the individual to make things happen (Piepmeier, 2006; Budgeon, 2015; Gill et al., 2017). Prominence is given to individual experiences, but unlike second-wave and postmodern feminism(s) the purpose is not to collectivise and theorise, but to make personal statements. Full attention is paid to difference and diversity, but without the emphasis on contributing towards a coherent and cohesive narrative as with the Combahee River Collective (Smith, 2014). All stress the importance of personal choice and it is this overarching aspect which can be seen as the determining feature of these forms of feminisms.

According to Snyder-Hall (2010), 'choice' feminism(s) enables women to map out their own courses of action through the many contradictory discourses they encounter. It addresses false universalism, embraces difference and takes issue with prescriptions of what 'feminism' should be about. Budgeon (2015) counters these arguments by drawing from Hirshman (2006) and asserting that 'choice' feminism(s), rather than forming a new kind of feminist politics, actually demonstrates a 'fear of politics' by its refusal to make critical judgements. This can result in further limiting the participation of women in changing the conditions which define the choices they have available to them.

Gill et al. (2017) use the term 'postfeminist sensibility' to refer to the practices and the consequences of 'choice' feminism(s). They maintain that a contemporary emphasis on individualism and personal choice goes hand in hand with a concomitant disregard or lack of understanding of structural inequality and a general downplaying of the operation of prevailing and pervasive sexist forces. Budgeon (2015) contends that 'choice' feminism(s) can be used to justify a range of behaviours, including women's participation in sexualised cultures, in pornography, in the adoption of a gendered division of labour and in the exploitative celebration of beauty and the body, on the basis of the belief that a personal and unfettered choice is being made.

'Choice' feminism(s) can be seen to represent an 'anything goes provided it is my choice' stance and whilst this promotes freedom of expression (within prevailing laws) it also can be seen to ignore history and negate the persistent and insidious themes which have particularly affected women. As a result, there is no analysis about how 'free' choice may be constrained and restrained by the operation of overt and covert power relations, by entrenched and unacknowledged patriarchal values and by enduring gendered limitations and restrictions. In terms of the media, it opens the door to standard and stereotypical female images being reshaped and redefined and presented as what women want. The setting is contemporary, but the underlying gendered messages present old style exploitation repackaged as women's free choice. There can be seen to be something akin here to Black feminism(s)' charges against what they perceived as White middle-class second-wave feminism, namely a complete disregard of unequal power relations and intersectional subjugations. It is also the paradox of critically analytical forms of feminism(s) being directed towards manifestations of feminism(s) which embrace old-style constraints on the basis of new style, personalised justifications. We now turn to look at forms of feminism(s) which have additional issues to address.

Islamic Feminism(s)

Islamic feminism(s), although influenced by the various feminisms, has to take on board matters of faith as well as women's relationship to Islam. At the centre of ongoing discussion and debate is the view that the range of feminisms generally are secularly orientated and have their basis in Western political movements. This has led to feminist analyses being regarded as representing Western values and, for some, being seen as a betrayal of

Islam. Accordingly, there are those who see feminism and Islam as incompatible in terms of the different views of equality posed. However, there are others who make associations between the two or use feminism as a point of critique and there are many who want to create space for feminist-orientated conversations and challenges within Islam itself. McGinty (2007), for example, argues that Islamic feminism is becoming a global movement and that a religious identity as a Muslim can also serve to create a space where resistance to patriarchal ideas can take form

Moghadam (2001) distinguishes between Islam as a political movement, between Islam as a set of religious beliefs varyingly practiced in different counties and between fundamentalist Islam, which again varies in precepts but not in the force with which these are applied She maintains that Islam cannot be isolated as an example of extreme patriarchy and that as a set of religious beliefs it is no more or less patriarchal than Judaism or Christianity. She highlights the tensions in Islamic feminism(s) and maintains that these contribute to the contemporary heady mix.

Seedat (2013) points out that a widely held normative view is that Islam, as a religion, corrected pre-Islamic gender bias and, as a result, affirmed women's spiritual equality with men. Equality is therefore enshrined in complementary difference, with men and women being regarded as ideally suited to their divinely ordained, but very different roles. Adherence to these roles is seen as both empowering and also inherently necessary to being a Muslim woman.

Since the late 1980s and early 1990s, this normative understanding has been contested by a questioning of the historical legacy of the patriarchy of early Islamic society as well as by contemporary Islamic practices. Ahmed (1992), for example, distinguishes between two distinct trends within Islam. On the one hand, embedded pragmatic social regulations serve to preserve male hierarchies and patriarchy, whilst equally entrenched social morality has a more egalitarian emphasis, carrying with it the potential for transformational change.

Within current discussion and debate, the notion of 'Islamic feminism(s)' remains controversial. Cooke (2001) used this term to call for greater Islamic feminist analysis of the Qur'an without the interjection of male religious scholarship and authority, emphasising the Qur'an's potential for the achievement of social justice. Cook (2001) does not view Islamic feminism(s) as a coherent entity, rather she refers to "a strategy of multiple-critique … where parallel commitments to multiple ideological frameworks are possible, even when they appear to be contradictory" (Cooke (2001) quoted in Seedat, 2013: 409–410).

However, the term Islamic feminism has been rejected by others as being too secular and confrontational (e.g. Wadud, 1992; Barlas, 2002). Najmabadi (2000) draws attention to the utility of adopting a progressive and non-confrontational approach so that pragmatic social relations and pre-ordained roles are questioned within a space within Islam (Najmabadi, 2000). Seedat (2013) also points to a widely held school of thought which stresses that, despite being subverted in a variety of ways, in a range of contexts by patriarchy, the 'divine truth' of Islam is anti-patriarchal and is therefore open to different interpretations and practices.

Overall, with regard to the various feminisms, the contemporary picture is mixed, with different forms of feminism coming into the frame both simultaneously as well as at different points. Many men also identify as pro-feminist, but as with many women, when asked what they mean by this, responses are varied and wide ranging. There are also male movements which cite sexist discrimination as rallying calls. The #MeToo movement highlights that collective action is still a powerful tool, although the key focus can be seen to be largely personalised with emphasis being placed on rooting out individual 'perpetrators' rather than upon consistent critical and deconstructive analysis. Nonetheless, the wide-ranging feminisms have proved influential for social work and we will now turn to consider their current and ongoing legacy.

Feminism(s) and Social Work

The many feminisms continue to inform social work in a variety of ways. As highlighted, in terms of feminist ideas, there is not a steady march of progress, but rather circularity, change and adaptation. Liberal feminism(s), which informed the suffragette movement or 'first-wave' feminism(s), with the emphasis on rights and changing the law, still carries considerable weight, as does making public and politicising what takes place in private domains. The considerable work that has taken place in addressing violence against women is testimony to this. Other areas where there is a distinct resonance relate to valuing experience. The understanding that women and men can be experts through experience is a valuable part of social work practice as is the importance of collectivising experiences and theorising through experience. Drawing particularly from postmodern feminism(s), there is also the imperative of a commitment to ongoing analysis and critical reflection. This relates to questioning 'taken for granted' assumptions, and exploring how a situation has developed and how, working with those central to it, it can be most usefully responded to and addressed. However, three conceptual areas which deserve particular attention at this point are 'choice', intersectionality and uncertainty.

'Choice' as we have seen is a tricky concept. It can be foregrounded as an empowering force, but can also serve to mask enduring structural restrictions and oppressions. It can be used to sanction inequality in relation to position and in the use and the deployment of resources. It can also be employed to justify exploitation and the continued operation of unequal and pervasive power dynamics. Within social work, justifying cuts to direct payments or personal budgets on the basis of personal choice or citing the exercise of personal choice when there are no alternatives available serve as pertinent examples. 'Choice' is important, but as a free rolling, non-rooted, unanalysed driver for action, it carries with it caustic and contrary elements. The ongoing critical appraisal of terms such as 'choice' are a key aspect of what social work can take from the various feminisms and usefully utilise on an everyday basis.

Intersectionality is another aspect which has permeated anti-oppressive practice within social work. The 'Black Lives Matter' movement, which draws global attention to systemic abuses, highlights the importance of coalescing challenge and resistance around a centralising message, whilst also appreciating that experiences differ, power imbalances intersect and context varies. Within social work, an appreciation that categorises, for example, 'older people', comprising a range of very different individuals with varied experiences, interconnections and intersections, is an important anti-oppressive element. Similarly, recognising individual diversity whilst also being cognisant of enduring forms of discrimination and oppression which require collective action is crucial in the commitment made by social work to achieving social justice. Social work's values and principles relating to anti-oppressive and anti-discriminatory practice, taking on board the importance of intersectionality, can all be seen to have been influenced by the various feminisms. However, as with the various feminisms, these also can play out in diverse ways with processes and consequences sometimes diverging from initial principles and underpinning ethics.

Another area which comes into the frame draws from postmodern feminism(s) in particular and relates to an appreciation and understanding of uncertainty. Social work has evolved as a modernist project with a belief that risk and uncertainty can be broken down into component parts and managed effectively. This belief has been challenged by authors such as Featherstone et al. (2009) and White (2009), but remains embedded in many operating procedures and practices within social work agencies. COVID-19 has served to further erode confidence in a 'certain' world. Care homes in the UK, for example, were supposed to be protected and to protect vulnerable people, yet up to half of the deaths attributed to COVID-19 in England by June 12, 2020, took place in care homes (Savage, 2020; Laing, 2020). This was largely because care homes were placed under pressure to accept frail older people discharged from hospital without individuals being tested for COVID-19. Social workers were largely caught between government policy to 'support the NHS' which meant in some cases that 'guidelines' were interpreted as regulations with regard to transferring older people from hospital and supporting older people and their families through restrictions caused by social distancing measures. Everything became very uncertain and those social workers who were able to make a difference had to work with uncertainty and use it proactively to achieve results.

The various feminisms have influenced theory, values and practice in social work, but ironically social work's relationship with gender inequality and the theoretical perspectives drawn from feminism(s) have often been insufficiently acknowledged. Given social work's emphasis on social justice, the lack of greater overt recognition and the naming of key influencing factors drawn from specific feminisms is perhaps surprising. The reasons for this can be seen to be many and varied and include 'professionalisation' and 'managerialism' as well as the emphasis placed on political viability. Nevertheless, social work has managed to retain a critical edge and although at times this does appear to be a little blunted, as highlighted in this chapter, its retention is due in no small measure to feminism(s) influence.

Concluding Remarks

The many forms of feminism have served to challenge, to bring about change, to identify and name oppression and discrimination and to fight for social justice. They have challenged prevailing attitudes, ways of operating and entrenched behaviours and have brought about significant transformations. However, gains can be reversed and re-interpretation can provide space for enduring inequalities, inequities and discrimination to continue under the guise of a different name. The celebration of individual choice, without recognition of how choices can be constrained and reframed to serve very different agendas, serves as a prime example here. Social work can be seen to carry similar tensions, contradictions and challenges. There is a strong commitment to social justice and tackling oppression and discrimination which sit side by side with increasingly rigid and constraining procedures and forms of micro management. The taking of risks and protection from risks are enduring aspects of the human condition, however, a preponderant emphasis on risk management within social work can stifle flexibility, innovation and productive working partnerships. Nonetheless, there remain many connections with the strongest being the emphasis which social work places on deconstructive analysis and critical reflection. These crucial aspects can be side-lined, but are embedded in the fabric of social work and are indispensable in continuing to do what social workers do.

References

Ahmed, L. (1992). Women and Gender in Islam: Historical Roots of a Modern Debate. New Haven: Yale University Press.

Barlas, A. (2002). "Believing Wormen" in Islam: Unreading Patriachal Interprettions of the Qur'an. Texas: Texas University Press.

Barrett, M. (1987). The Concept of Difference. *Feminist Review, 26*: 29–41.

Barrett, M. (1992). Words and Things: Materialism and method in Contemporary Feminist Analysis. In M. Barrett & A Phillips (Eds), *Destabilising Theory: Contemporary Feminist Debates*, 201–219. Cambridge: Policy Press.

Braithwaite, A. (2002). The personal, the political, third wave and post feminism. *Feminist Theory*, (33), 335–344.

Brodrib, S. (1992). *Nothing mat(t)ers: A feminist critique of postmodernism*. Spinifex.

Budgeon, S. (2015). Individualised femininity and feminist politics of choice. *European Journal of Women's Studies, 22*(3), 303–318.

Butler, J. (1990). *Gender trouble: Feminism and the subversion of identity*. Routledge.

Butler, J. (1994). *Undoing gender*. Routledge.

Cooke, M. (2001). *Women Claim Islam: Creating Islamic Feminism Through Literature*. London: Routledge.

Crenshaw, K. (1989). Demarginalizing the intersection of race and sex: A black feminist critique of antidiscrimination doctrine, feminist theory and antiracist politics. *University of Chicago Legal Forum, 1989*, Article 8.

Crook, S., & Gutnick Allen, S. (2014). In Conversation with the Women's Liberation Movement. *History Workshop Journal, 77*(1): 339–342.

Doude, S. B., & Tapp, K. E (2014). Radical Feminism. https://doi.org/10.1002/9781118517390.wbetc090

Eddo-Lodge, R. (2020). *On speaking up: Intersectionality in 'The Feminist Question'*, medium.com. Accessed 10 Oct 2020.

Evans, E. (2015). *The politics of third wave feminism, neoliberalism, intersectionality and the state in Britain and the US.* Palgrave Macmillan.

Fawcett, B. (2000). *Feminist perspectives on disability.* Prentice Hall.

Fawcett, B. (2016). Feminism and disability. In S. Wendt & N. Moulding (Eds.), *Contemporary feminisms in social work practice.* Routledge.

Featherstone, B., & Green, L. (2013). Judith Butler. In M. Grey & S. A. Webb (Eds.), *Social work theories and methods* (2nd ed.). Sage.

Featherstone, B., Morrris, K., & White, S. (2009). A marriage made in hell: Early intervention meets child protection. *British Journal of Social Work, 44*(7), 1735–1749.

Flax, J. (1992). Beyond equality, gender, justice and difference. In I. Boch & S. James (Eds.), *Beyond equality and difference*, 193–210. London: Routledge.

Fraser, N. (1993). Unruly Practices: Power, Discourse and Gender. *Contemporary Social Theory.* Cambridge: Policy Press.

Fraser, N., & Nicholson, L. (1993). Social criticism without philosophy: An encounter between feminism and postmodernism. In M. Docherty (Ed.), *Postmodernism: A Reader*, 415–432. Hemel Hempstead: Harvester Wheatsheaf.

Friedan, B. (1963). *The Feminine Mystique.* New York: Norton.

Gill, R., Kelan, E. K., & Scharff, C. M. (2017). A postfeminist sensibility at work. *Gender, Work and Organization, 24*(3). https://doi.org/10.1111/gwao.12132

Gillis, S., Howie, G., & Munford, R. (Eds.). (2007). *Third wave feminism: A critical exploration* (Expanded 2nd ed.). Palgrave.

Harding, S. (Ed.). (2004). *The feminist standpoint theory reader.* Routledge.

Hartsock, N. (1983). *The feminist standpoint: Developing the ground for a specifically feminist historical materialism.* Springer.

Hill Collins, P. (1990). *Black Feminist Thought: Knowledge, Consciousness and the Politics of Empowerment.* Boston: Unwin Hyman.

Hirshman, L. R. (2006). *Get to work: A manifesto for women of the world.* Viking.

Jackson, S. (1992). The amazing deconstructing woman. *Trouble and Strife, 25*, 25–23.

Kearney, M. C. (2012). *The gender and media reader.* Routledge.

Laing, W. (2020). Care Homes for Older People: Market Analysis and Projections, Healthcare Intelligence, LaingBuisson, May 2020 https://www.laingbuissonevents.com/wp-content/uploads/2017/05/William-COP.pdf. (Accessed: 6.6.2020).

Lay, K., & Daley, J. (2007). A critique of feminist theory. *Advances in Social Work, 8*(1).

Lorber, J. (2011). *Gender inequality: Feminist theories and politics.* Roxbury Publishing Company.

Matos, C. (2015). Postmodernism, equality and feminism: Current contemporary issues. *Research.* https://doi.org/10.13140/RG.2.1.2853.5761

McGinty, A. M. (2007). Formation of alternative feminities through Islam: Feminist approaches among Muslim converts in Sweden. *Women's Studies International Forum, 30*(6): 474–485.

Mendes, K. (2011). Reporting the women's movement: News coverage of second wave feminism in the UK and US newspapers, 1968–1982. *Feminist Media Studies, 11*(4), 483–498.

Moghadam, V. M. (2001). Organizing Women: The new women's movement in Algeria. https://doi.org/10.1177/092137400101300201

Munro, E. (2013). Feminism: A fourth wave, political insight. *Political Studies Association.* https://doi.org/10.1111/2041-9066.12021

Najmabadi, A. (2000). (Un) Veiling Feminism. *Social Text*, Vol 18, No 3 (64) Duke University Press. https://doi.org/10.1215/01642472-18-3_64-29

Phipps, A. (2016). Whose personal is more political? Experience in contemporary feminist politics. *Feminist Theory.* https://doi.org/10.1177/1464700116663831

Piepmeier, A. (2006, March 17). Postfeminism vs the third wave. *Electronic Book Review.* https://electronicbookreview.com/essay/postfeminism-vs-the-third-wave/

Ramazanoglu, C. (1989). *Feminism and the contradictions of oppression.* Routledge.

Risman, B. J. (2004). Gender as a social structure: Theory wrestling with activism. *Gender and Society, 18*(4), 429–450.

Rogan, F., & Budgeon, S. (2018). The personal is political: Assessing feminist fundamentals in the digital age. *Social Sciences, 7*(8).

Saulnier, C. (1996). *Feminist theories and social work: Approaches and applications.* Howarth.

Savage, M. (2020). More than half of England's coronavirus related deaths will be people from care homes. *The Guardian.* www.guardian.co.uk (Accessed: 7.6.2020).

Sawicki, J. (1991). *Disciplining Foucault: Feminism, power and the body.* Routledge.

Seedat, F. (2013). When Islam and feminism converge. *Muslim World, 103*, 404–420.

Smith, B. (1983). Black Feminism Divorced From Black Feminist Organizing. *The Black scholar*, 14(1): 38–45.

Smith, S. (2014). Black feminism and intersectionality. *International Socialist Review*, (91) www.isreview.org

Snyder-Hall, R. C. (2010). Third-wave feminism and the defense of 'Choice'. *Perspectives on Politics, 8*(01).
Wadud, A. (1992). *Qur'an and Woman.* Kuala Lumpur: Fajar Bakhti Press.
Walby, S. (1997). *Gender transformations.* Routledge.
Walker, A. (1984). *In search of our mother's garden.* The Women's Press.
Weedon, C. (1987). *Feminist practice and poststructuralist theory.* Blackwell.
West, C., & Zimmerman, D. (1987). Doing gender. *Gender and Society, 1*(2), 125–151.
White, S. (2009). Fabled uncertainty in social work: A coda to Spafford et al. *Journal of Social Work.* https://doi.org/10.1177/1468017308101824

CHAPTER 3

Feminist and Empowerment Theory and Practice: A Powerful Alliance

Tina Maschi, Sandra G. Turner, Smita Ekka Dewan, Adriana Kaye, and Annette M. Hintenach

Introduction

In contemporary social work discourse, scholars and practitioners are categorizing their social change efforts under the *Grand Challenges* (Uehara et al., 2013) initiative. Its goal is to promote dialogue and policy reform about different social justice issues, such as ending family and community violence, reducing health inequities, mass incarceration, unemployment and education, and promoting the conditions for health productive aging, racial and ethnic justice, and leveraging the power of technology for the greater good (Poe-Yamagata & Jones, 2000). Achieving this initiative relies on social innovation to increase health and well-being and

T. Maschi (✉) • S. G. Turner • A. Kaye • A. M. Hintenach
Fordham University, Bronx, NY, USA
e-mail: tmaschi@fordham.edu; sturner@fordham.edu; ahintenach@fordham.edu

S. E. Dewan
New York City College of Technology, Brooklyn, NY, USA
e-mail: sdewan@citytech.cuny.edu

C. Cocker, T. Hafford-Letchfield (eds.), *Rethinking Feminist Theories for Social Work Practice*,
https://doi.org/10.1007/978-3-030-94241-0_3

empowerment among vulnerable populations and improvement in environmental conditions for people of all ages across the globe. A critical function of leveraging the "power" of social work is to elevate the front-line workers who are women and come from places of varying intersectional differences based on race, gender and sexual identity, disability, or immigration status. Emerging social issues that are driven by various social movements such as the "Black Lives Matter" movement, "Me Too" movement, and the environment justice movement are also increasingly challenging the conventional roles of a social worker.

To this end, this chapter explores the role of feminist and empowerment theories in empowering professionals, such as the female-dominated social justice profession of social work and the vulnerable populations and organizational and community environments that they serve, such as older women with serious offence histories in prison. The central guiding conceptual models of social work practice, a human rights, social, economic, and environmental justice and person-in-environment perspective, are integrated with feminist and empowerment theories.

Human Rights Framework

A human rights framework is an overarching perspective that can inform social work interventions about the pathways to empowerment for individuals, families, and communities at the local or global level. It has underlying values and principles (UN, 2015) such as respect for all persons, the intrinsic value, dignity and worth of each person, and the duty of governments to their citizens (UN, 1948). Table 3.1 lists the basic principles of the human rights framework.

Figure 3.1 shows the elements of human rights, empowerment theory, and feminist theory and how they relate to social work practice.

Table 3.1 Human rights framework

Human rights framework
Focuses on:
Protection of civil, political, economic, social, and cultural rights
Promotion of personal and collective empowerment
Basic principles:
Universality
Nondiscrimination
The indivisibility and interdependence of rights (political, civil, social, economic, and cultural)
Participation
Accountability
Transparency

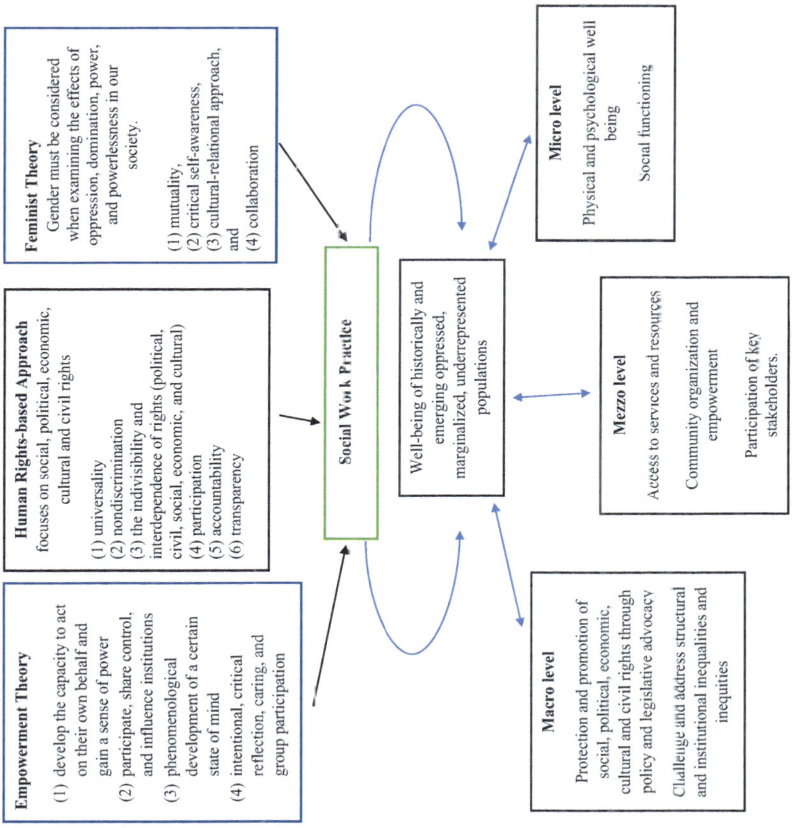

Fig. 3.1 A human rights-based approach: Empowerment theory and feminist theory in practice

Applying an Intersectional Rights Framework

There is a current trend in feminist and progressive social work to build the capacity of the profession to promote well-being, empowerment, rights, justice, diversity, equity, and inclusion. This is consistent with the International Federation of Social Workers' global definition of the social work profession—one that promotes "principles of social justice, human rights, collective responsibility and respect for diversities" (IFSW, 2020). Empowerment and feminist informed theories are commonly geared toward elevating the human rights and needs of individuals, families, groups, communities, and the collective that have experienced oppression, including historical and intergenerational trauma and oppression. The history, central concepts, and themes of empowerment and feminist informed theories are multidimensional and often related to multiple dimensions that include the intrapersonal, interpersonal, community, collective, and political domains. The close alliance of empowerment and feminist practice with human rights, oppression, and ecological systems theories has been a powerful force in assisting frontline practitioners to facilitate personal and collective liberation and empowerment.

Feminist and Empowerment Theories: The Power of Two Theories

Feminist Theory

Feminism emphasizes the importance of the social, political, and economic structures that shape human societies and stresses that gender must be considered when examining the effects of oppression, domination, power, and powerlessness in our society. In feminist theory, there are four core concepts: (1) mutuality, (2) critical self-awareness, (3) cultural-relational approach, and (4) collaboration. When working directly with clients, feminist empowerment theories and practice are useful in guiding social workers to help clients claim their power, build self-confidence, and self-esteem as they engage in mutual, non-hierarchical relationships (Crenshaw, 1991; Jordan, 2010). Those who espouse feminist theory and principles believe that the inferior status of women is derived from unequal political, economic, and social power relations. Although oppression and empowerment theories are now seen as more mainstream in social work, feminist theory and gender issues have still not become well established in social work curricula or research (Hafford-Letchfield et al., 2014; Lauve-Moon et al., 2020). Lauve-Moon et al. (2020) studied the extent to which gender and gender inequality are included in social work research and found very few studies of this in mainstream social work journals.

Concept of Mutuality in Feminist Practice

Scholars such as Jordan et al. (1991) were the originators of the relational model of practice, in which the practitioner develops a mutual relationship with her client as opposed to being an outside observer attempting to understand what the other person is dealing with or has experienced.

Feminist practitioners let themselves be seen by their clients, and the relationship itself is often the vehicle to engender strength and self-esteem (Turner, 2001).

A study of feminist interventions in Canada found that feminist models of working with clients are more effective in supporting mutual client-worker strategies to change larger system targets and thus are more effective than those based on other, even innovative, models of practice (Gorey et al., 2003). The authors of this review of feminist social work interventions found them highly effective and also underrepresented in social work literature (Gorey et al., 2003).

History and Development of Feminist Theory

Originally, feminism began as a primarily white middle-class movement focused on women's right to vote. Black women were also actively involved in this movement, but often not recognized (Dill, 1983; DuBois, 1999; King, 1988). After winning the right to vote, second-wave feminists focused on the domination and oppression built into the structure of the patriarchal society. Fighting for affordable childcare, equal pay for equal work, and freedom from oppression were some of the goals of feminist activists after World War II. Black and/or lesbian scholars such as Lorde (1984) and Collins (1991) were more interested in social justice for all LGBTQIA+ (esp. lesbian) and women's collective/solidarity rights as opposed to individual advancement. An example of feminist theorizing can also be seen more recently in the articulation of the struggles for social justice and unique oppressive experiences of BIWoC, that is, black, indigenous, and women of color. This highlights the oppression experienced especially by women of color as opposed to people of color which is reflected in the term BIPoC that refers to black, indigenous, and people of color.

When working directly with clients, feminist and empowerment theories and practice are useful in guiding social workers to help clients claim their power, build self-confidence, and self-esteem as they engage in mutual, non-hierarchical relationships (Crenshaw, 1991; Jordan, 2010). Mutual relationships based on a cultural-relational approach and collaboration help to build critical self-awareness and self-confidence and self-esteem (Jordan, 2010).

Traditional social work interventions focus primarily on the individual and the amelioration of individual deficits. In most of these social work interventions such as Cognitive Behavioral or even the Strengths Perspective, gender and unequal distribution of power between the genders are rarely explicitly addressed. Empowerment theory is also sometimes critiqued to focus on the empowerment of the individual woman through individual change.

Feminist interventions are more transformative in nature while traditional ones can be considered transactional. They may benefit the individual but not a larger system. Feminist theory indeed has much in common with empowerment theory as both focus on domination and subordination; however, feminist theory focuses on gender while empowerment theory looks at the influence of race and class. Feminist social workers believe that societal inequality is at the root of the oppression of women and, like empowerment, workers urge both women and men to engage in efforts to bring about broader social change (Turner, 2018; Turner & Maschi, 2015).

Central Concepts of Empowerment Theory and Research (Fig. 3.2)

Empowerment Theory

Next, we analyze how feminist and empowerment ideals address the psychological, social, and structural rights and needs of historically and emerging underrepresented and underserved populations, such as women and children living in poverty, racial ethnic minorities, immigrants, and older and LGBTQ people.

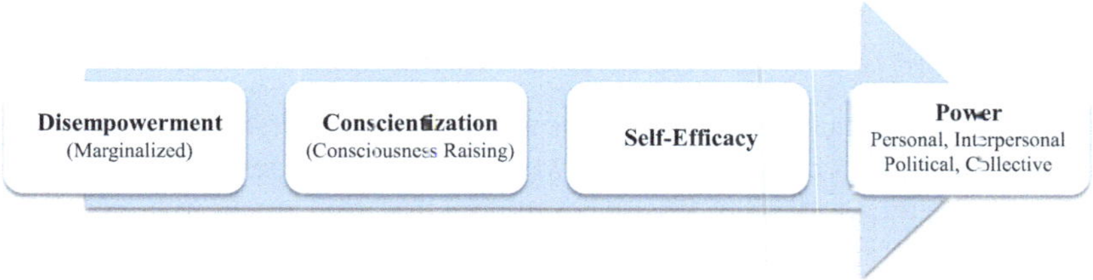

Fig. 3.2 Empowerment theory and practice

While empowerment practice is almost synonymous with social work practice, its meaning and application remain somewhat elusive and a matter of debate. Interpretations differ on whether empowerment is a philosophy, theory, or practice model (Blair-Medeiros & Nelson-Alford, 2021; Cornell Empowerment Group, 1989; Gutiérrez & Lewis, 1999; Swift & Levin, 1987; Presser & Sen, 2000). Empowerment as a theory for direct social work practice at individual, group, community, and political levels is a staple in many US-based social work texts that promote a generalist model of social work (Allen-Meares & Garvin, 2000; Miley et al., 2011; Robbins et al., 2012; Torre, 1985). However, in its application, one common critique of empowerment practice is that it tends to focus on individual change in process of empowerment as opposed to strategies to address structural changes.

Therefore, it is important to address the many "faces" of empowerment since as a concept it varies in meaning and is defined and operationalized in varied ways (Cattaneo & Chapman, 2010). While there are multiple definitions of empowerment, the one most relevant to social work practice is that it is a process by which

> individuals, groups, and communities develop the capacity and gain the skills to act on their own behalf and gain a sense of power in their personal, interpersonal, and environmental interactions. (Gutierrez et al., 1998; Reichert, 2006)

Empowerment social work is much like the feminist approach, which operates within a human rights and social justice framework with the goal of building a more compassionate world. The work is both clinical and community oriented (Lee, 2001).

Life Course History Power Analysis: An Integrative Theoretical Analysis

In order to conduct the life history power analysis, we first present the case study of an incarcerated older adult woman who describes her experiences before, during, and after prison. Approval for this qualitative research study was obtained from the Institutional Review Board at Fordham University. Directly following, we apply the life course history power analysis that integrates feminist and empowerment theories to the case of Mary. In her own words, she describes her life before, during, and her anticipation of release. We then analyze her case through an intersectional feminist, empowerment, and human rights lens.

Case Study: Mary: Excerpts from a life history narrative of an incarcerated woman from the IRB-approved Hartford Prison Study (see Maschi & Morgen, 2020)

1. **My life experiences**

I am a 54-year-old woman who has spent my life sheltered. When I was 12 years old my parents separated and eventually divorced. During this time, I was a mother, wife, sister, and housekeeper. My dad was a very strict Catholic who was mentally, emotionally, and physically abusive and an alcoholic. He not only beat my mother but also us children. My dad was very deranged. He drew a gun on my mom which warranted her to leave us children. Eventually we all, through torture and drama, left or were physically removed and reunited with my mom. To this day, my mom is very depressed, shamed, guilty, and remorseful for leaving us children. I believe she will suffer the rest of her life since she has already suffered 74 years of it. With me in here, I feel I have caused her more guilt and shame.

Since I have been in prison, I realized I chose men exactly like my father. I have been through several years of counseling prior to being locked up but was afraid to open up. At times, I am still the same and don't know why. I've lived my life, and still do, being scared, vulnerable, too trusting, and just a loving, caring, compassionate, empathetic, hide-it-all person. To avoid this, I keep to myself and give little advice even though I have a lot to say inside, but feel like I'm on display if I do. I don't like being criticized like I have always been.

I was married for almost 10 years. I loved him so much and no matter what he's put me through, I always will-without reason. I got pregnant and have a beautiful 15-year-old daughter who resides with my ex in Florida. Not a day goes by that my heart doesn't break because I haven't seen her really since March 2006. I also have a 29-year-old son. I raised him as a single parent. He served 8 years in the military and was a challenge during his school years, and in fact, still is today. I love and adore both my children and feel I have really put them through shame.

2. **How I see these experiences as influencing my current situation**

Well I'm not quite sure what you are asking me, but I will give you my interpretation of the question. Since I've been here in XXXX prison, I feel less worth a crap. Well, putting it bluntly, it's horrible. Some programs are helpful—Cage Your Rage/Anger Management. This is the program that really opened my eyes to see my life stems back from when I was a child. My choices in men are based on what I saw when I was raised, so-called, by my dad. I am a very quiet, frightful, self-introverted adult who gets sick in the stomach. I don't know how to explain but I don't want to be involved with people or this situation. I have faith that God has me here for a reason. I wanted revenge but not anymore. I was set up and am having a very difficult time getting someone to see it. I often wonder if I'm stupid or just plain naive. It's very difficult to try to find myself and explain myself where it's understandable. Shame, I deal with every day in every part of my life. I have good days and bad ones. I just want to go home and live my life in a bubble with the ones I love.

3. **My experience of prison**

I wouldn't recommend it to anyone of any age, color, culture, or human! In some respects, it has helped me in my faith and the strength it has given me. It has helped me realize I am important, but how do I change that? I have been exposed to several circumstances that my sheltered life prevented. I have learned to forgive but will never forget. It's just a shame I haven't had contact with my daughter. I worry about what others think of me and ask the question WHY? All the time. Prison is a hard place. Pure Hell! I found I am more reserved, introverted, scared, and don't want to communicate at all. I am blessed to be treated with respect because I give respect. Not always is this the case, authority plays a powerful part. As long as you are in khaki, you are considered non-human. I take every day at a time and in stride. I have learned not to take things personally, don't make assumptions, let go, be impeccable with my words, and do the best I can. But I am called scary Mary.

The Pathways of Cumulative Health and Justice Disparities

The pathway that results in prison for incarcerated people may vary in one or more cumulative disparities related to race, education, socioeconomic status, gender, disability sexual orientation, legal or immigration status, which may influence their access to health, behavioral health and social services, economic resources, and justice (Maschi, 2016; Maschi, Baer, Morrissey & Moreno, 2013; Maschi et al., 2012).

The Trauma of Incarceration: Prison as an Abusive System

As illustrated in Fig. 3.3, older adults and women in prison, such as Mary, are often an overlooked group who are at increased risk of many types of elder abuse and sickness in prison. The recent movement to release older prisoners who have committed non-violent crimes because of the risk of exposure to COVID-19 is positive but long overdue. Due to their increasing age-related frailty, older adults in prison are at increased risk of victimization and injury, medical and social care neglect, and exploitation of their resources (Maschi et al., 2014). Despite their vulnerabilities, there is also a growing body of evidence that documents their resilience, such as their use of cognitive, physical, emotional, spiritual, and social coping resources. These multidimensional domains of coping suggest there are avenues for prevention and intervention that promote health and positive human development and rehabilitation (Aday, 2003; Maschi et al., 2014; Maschi, Viola & Koskinen, 2015; Maschi, Viola & Morgen, 2013).

Assessment and Intervention Using a Life Course Systems Power Analysis Model

When considering the important task of assessment and intervention with incarcerated people, such as Mary, in the criminal justice system, conducting a holistic assessment of each person in a social and environmental context is essential. As shown in Fig. 3.3, a life course systems power analysis assists in identifying the assessment areas of Mary which more than likely will require an integrated clinical, case management, and/or advocacy response (Maschi et al., 2014).

The Life Course (The Length of the Model)

Consistent with integrated empowerment and feminist theories, the life course perspective gives central importance to the whole person or individual, their dynamically changing inner or subjective experiences and

Fig. 3.3 Incarcerated older women—Assessment and intervention using a life course systems power analysis model

meaning-making of life course events (e.g., objective event—victim of sexual assault; subjective response—adaptive or maladaptive response), and subjective well-being and is the central focus of the model. In the case of Mary, earlier life traumas and prison experience present a challenge for her pursuit of life course human agency.

The World Health Organization (1948) definition of health is a state of multidimensional well-being and not just the absence of disease. Specifically, well-being is defined by seven core domains: root (basic needs), physical, cognitive, emotional, social/cultural, spiritual, and participatory (political/legal) well-being. When cumulative determinants or social and environmental conditions are optimal during the life course, individuals express human agency through concern for self and others and sustain high subjective levels of well-being. These multidimensional domains of coping suggest there are avenues for prevention and intervention that promote health and positive human development and rehabilitation (Aday, 2003; Maschi et al., 2014; Maschi, Viola, Morgen & Koskinen, 2015).

However, when conditions are suboptimal, such as the experience of personal beliefs or attitudes, or when confronted with social and environmental barriers (poverty, low educational attainment, adverse neighborhood conditions, long prison sentences), a person's healthy expression of human agency may diminish his or her subjective well-being and negatively manifest as illness or offending behavior (Maschi et al., 2011, 2012; Maschi & Baer, 2013). As the case example shows, Mary has had adverse life experiences, including a prison placement that she describes as challenging her health and well-being.

Systems (The Width of the Model)

As in the case of Mary, her individual life course, her intersections with systems, such as family, service, and legal systems, changed over time. There is a lack of access to evidence-based services and justice during the course of her criminal justice system experience. For example, a central value and ethical principle of human rights philosophy is honoring the dignity and worth of all persons and respect for all persons, including people in prison (UN, 1948, 1977, 2012). In many cases, this principle is not honored for older adults in prison when there is a lack of programming that takes into consideration the intersectional experiences of inmates.

Power (The Height of the Model)

The history of access to power and privilege throughout the life course is important to assess, especially for older adults who have experiences of being victimized and have committed crimes. Power dynamics across the life course may be balanced (equitable) or imbalanced (oppressive).

According to this model and as illustrated in the case of Mary, individuals or groups can be oppressed at the personal (i.e., everyday interactions), structural (e.g., institutional), or cultural levels (e.g., societal attitudes, media). An individual's internalization of negative self-messages may influence behaviors toward others (Mullaly, 2010; Mullaly & West, 2017).

Interdisciplinary perspectives are commonly fragmented when addressing aging people in prison and are another social/environmental factor to consider. Given that social work perspectives vary from medical perspectives, assessments and interventions should address varying perspectives including social and environmental issues. Using a life course systems power analysis, a holistic and integrated theoretical base is essential to adequately address the process and outcomes of the crisis (Greenfield, 2012).

Embedding Restorative Justice and Risk, Need, and Responsivity (RNR)

Given that feminist and empowerment theories often do not unpack how issues related to crime and rehabilitation unduly affect oppressed and disempowered individuals and groups, we also integrate principles and practice of restorative justice (Maschi & Morgen, 2020). We particularly address risk, need, and responsivity (RNR).

RNR is based on three therapeutic principles, which focus on matching services to an individual's risk level (risk principle); addressing criminogenic (e.g., sexual drive, the sequelae of traumatic stress, and social rejection) and non-criminogenic needs (need principle); and tailoring interventions based on an individual's motivation, learning style, agency, identity, and systems context (responsivity principle) (Maschi & Morgen, 2020). The promotion of protective factors in restorative justice approaches (e.g., Circles of Accountability and Support or COSAs) (Wilson et al., 2007) is central in models that focus on offender reintegration that include fostering interrelationships, community engagement and support, as well as mutual accountability in addressing the needs of older adults in prison. Additionally, principles of risk, need, and responsivity (Andrews & Bonta,

2010) coincide with the shift from a punishment/criminalization approach to a rehabilitation model that is more attentive to social and psychological risk and protective factors in designing interventions. RNR underscores respectful and collaborative working relationships between clients and correctional agencies that promote the use of effective assessments and interventions, resulting in lower recidivism rates (Andrews & Bonta, 2010).

Application of the Life Course History Analysis Theory to Practice with Mary

What perhaps is most challenging for practitioners who work with clients like Mary (older adults with histories of victimization prior to and while in prison) is that there is no cut and dry distinction between being a "victim" and "offender". As illustrated by Mary, older adults with histories of victimization and criminal conviction histories are diverse. Mary's story reveals an accumulation or aggravation of life course disadvantages.

For Mary, enhancing social relations compromised by a history of marginalization and victimization and addressing the principles of risks, need, and responsivity are important to help instill empowerment (Andrews & Bonta, 2010). Additionally, infusing community-based supports will help reestablish connectedness that is sensitive to cultural and gender differences in each of these cases and may have a stabilizing effect, particularly post-release, and decrease the likelihood relapse for sexual and/or violent re-offense. Although practitioners (e.g., social workers and counselors) may work with individuals in the criminal justice system, it is imperative that they acknowledge their own personal empowerment as a means of building a mutual relationship and actively engage in leadership that involves facilitating inter-agency cooperation and mutual support. Feminist and empowerment-based practice takes into consideration justice-involved individuals of the criminal justice system as a practice setting.

Human Rights, Feminist, and Empowerment Assessment and Intervention Plan

In designing an intervention plan for Mary, a life course systems analysis was conducted. It consisted of a holistic assessment of biopsychosocial, spiritual, and legal aspects. Core assessment tools were used; instruments that assess traumatic stress (e.g., PTSD Checklist), as well as instruments that specifically estimate the risk for sexual offense (e.g., Static 99, designed only for males), and instruments that can assess a broader range of risk factors for both male and female adults (e.g., Level of Service Inventory, LSI-R) should be administered in order to help determine appropriate interventions.

In the case of Mary, there are considerations regarding history of victimization (domestic violence), coping resource issues, health care concerns, and disengagement with family, with the addition of gender-specific considerations and the fact that her only criminal history was related to the attempted murder of her abusive husband. Using a risk and needs assessment appropriate for females (e.g., the LSI-R) and gender-specific strategies (e.g. feminist) for establishing therapeutic engagement is crucial in this case.

Intervention Plan

Mary was referred to the specialized unit for seniors, which included specific programming for seniors. The programming highlighted in this chapter is based on the Nevada Department of Corrections, Senior Structured Living Program, or True Grit Program (Harrison et al., 2012). The program is staffed by an interdisciplinary team of doctors, nurses, social workers, chaplains, lawyers, advocates, and volunteers. The program infuses principles of human rights and social justice, such as dignity and worth of the person, and incorporates comprehensive structured services that foster biopsychosocial well-being among older adults in prison. Preliminary analysis of the qualitative data from participants suggests that they view the program as an invaluable part of their lives, helping them cope with daily prison stress while allowing them to offer restitution for their crimes and plan for community reintegration. Preliminary quantitative analysis suggests that participants who are released from prison have a less than 5% recidivism rate (Maschi & Morgen, 2020).

Although not specifically stated in the intervention plan, a feminist approach that incorporated the principles of mutuality, critical self-awareness (the ability not only to be self-aware but also to hold oneself accountable for social injustices), a cultural-relational approach, and collaboration for the work with Mary would have been beneficial for her. For example, an intervention plan using the feminist approach for Mary would provide continued individual and community support to her even after she leaves prison. Application of a feminist approach in the intervention would also require development and incorporation of processes that would allow Mary to engage in a process of critical self-awareness and use a cultural-relational approach to examine her own experiences within institutions that have traditionally been male dominated such as the traditional family structures, labor markets and economies, law enforcement and prison. The processes of critical self-awareness and use of cultural-relational approaches in programming are a crucial step toward empowerment—both during and after incarceration. Empowerment is also facilitated through continued programming and intervention planning process that are based on the two other essential core processes of feminist social work, that is mutuality and collaboration.

Empowerment theory, human rights approach, and feminist theory (as presented in Fig. 3.1) are an intrinsic ideological foundation of the social work profession and each framework contributes immensely to any intervention planning for individuals and communities who have complex social needs. In social work practice with such individuals and communities all three frameworks contribute immensely to enhancing the diverse domains of well-being at the micro, mezzo, and macro levels.

In the case of Mary, the health and justice disparities are a direct consequence of the violation of human rights principles as laid out in the human rights framework (Table 3.1). As described in the case vignette above, the pathway that results in prison may vary in one or more cumulative disparities related to race, education, socioeconomic status, gender, disability sexual orientation, legal or immigration status, which can influence their access to health, behavioral health and social services, economic resources, and justice. As the international human rights movement is gaining momentum in its efforts to advocate for the rights of incarcerated people of all ages, the forensic practice community is challenged to think creatively and out of the concrete "prison" box on how to respond effectively, especially as it relates to special populations who experience a lifetime of cumulative disadvantages, such as women and children, persons with physical and behavioral health issues, and older people (UN High Commissioner, 2012).

The human rights framework, empowerment theory, and the feminist approach not only help us understand why and how these disparities emerge, but also inform development of interventions to address these disparities. Social workers are best qualified to recognize, assess, and address these accumulative disadvantages while developing interventions. The accumulated life course disadvantages also warrant the application of combination of theoretical approaches. For Mary, the critical self-realization and clarity of her own marginalization and victimization might be empowering as she prepares for life outside prison. The core principles of the human rights framework and feminist social work provide that ideological foundation for any intervention plan developed by social workers to address the needs of older adults like Mary while being incarcerated and during the post-incarceration phase.

Conclusion

This chapter explored how the professional identity of social work has aligned with core values and perspectives of human rights, feminist, and empowerment practice theories. The empowering nature of feminist and empowerment ideals to address the psychological, social, and structural rights and needs of historically and emerging underrepresented and underserved populations, such as women and children living in poverty, racial ethnic minorities, people in prison, immigrants, and older LGBTQ people were delineated. A life course history power analysis was used to illustrate the interaction of life course development, power dynamics, and system interactions. We recommend the use of an integrated and intersectional practice model that infuses human rights, as well as feminist and empowerment approaches to address the needs of the individual and collective needs of vulnerable populations, such as incarcerated older women.

As illustrated in the case study of Mary, the biggest challenges for interdisciplinary professionals and programs to foster health and well-being and empowerment among incarcerated older adults are developing competencies in working with women with multiple issues related to power. Although the extent to which some skills are used depends upon where a professional is "positioned" in the system (e.g., clinical social worker in prison, reentry program administrator), it involves having competencies in aging (gerontological practice), physical and mental health assessment and intervention, case management, interdisciplinary collaboration, discharge planning, and legal and policy advocacy.

Comprehensive senior-specific programming, such as True Grit prototypes, bridges prison to community services, a leap forward toward addressing human rights-based best practices for older adults in prison, especially with histories of being victims and committing offenses. Promising practices with older adults in prison often honor the dignity and worth of the person, foster human agency and autonomy, and promote holistic well-being. Program components often include geriatric case management services for medical, mental health, substance abuse, family, social services, housing, education or vocational training, restorative justice (e.g., victim or victim-offender mediation services), spiritual counseling, exercise and creative arts programs, employment, and/or retirement counseling. Program-specific aspects include one or more of the following: "age" and "cognitive capacity," sensitive environmental modifications, interdisciplinary staff and volunteers trained in geriatric-specific correctional care, complementary medicine, specialized case coordination, the use of family and inmate peer supports and volunteers, mentoring, and self-help advocacy group efforts (Davidson & Rowe, 2010).

References

Aday, R. (2003). *Aging prisoners.* Praeger.

Allen-Meares, P., & Garvin, C. (2000). *The handbook of social work direct practice.* Sage Publications.

Andrews, D., & Bonta, J. (2010). Rehabilitating criminal justice policy and practice. *Psychology, Public Policy, and Law, 16*(1), 39–55.

Blair-Medeiros, S. L., & Nelson-Alford, C. (2021). Black faces, white spaces: Navigating a women's center as Queer Black Women leaders. *Vermont Connection, 42*(1), 13–24.

Cattaneo, L., & Chapman, A. (2010). The process of empowerment: A model for use in research and practice. *American Psychologist, 65*(7), 646–659.

Collins, P. (1991). *Black feminist thought.* Routledge.

Cornell Empowerment Group. (1989). Empowerment and family support. *Networking Bulletin, 1*, 1–23.

Crenshaw, K. (1991). Mapping the margins: Intersectionality, identity politics, and violence against women of color. *Stanford Law Review, 43*, 1241–1299.

Davidson, L., & Rowe, M. (2010). *Peer support within criminal justice settings: The role of forensic peer specialists.* Available at: https://www.samhsa.gov/gains-center.

Dill, B. (1983). Race, class, and gender: Prospects for an all-inclusive sisterhood. *Feminist Studies, 9*(1), 131–150.

DuBois, E. (1999). *Feminism and suffrage: The emergence of an independent women's movement in America, 1848–1869.* Cornell University Press.

Gorey, K., Daly, C., Richter, N., Gleason, D., & McCallum, M. (2003). The effectiveness of feminist social work methods. *Journal of Social Service Research, 29*(1), 37–55.

Greenfield, E. (2012). Using ecological frameworks to advance a field of research, practice, and policy on aging-in-place initiatives. *The Gerontologist, 52*(1), 1–12.

Gutiérrez, L., & Lewis, E. (1999). *Empowering women of color*. Columbia University Press.

Gutierrez, L., Parsons, R., & Cox, E. (1998). *Empowerment in social work practice*. Brooks/Cole Publishing Co.

Hafford-Letchfield, T., Lambley, S., Spolander, G., & Cocker, C. (2014). *Inclusive leadership in social work and social care*. Policy Press.

Harrison, M., Kopera-Frye, K., & Harrison, W. (2012). A promising practice – True Grit: A structured living program for older adults in prison. In T. Maschi et al. (Eds.), *Aging prisoners: A crisis in need of intervention* (pp. 57–69). Be the Evidence Project.

International Federation of Social Workers [IFSW]. (2020). *Global definition of social work*. Available at https://www.ifsw.org/what-is-social-work/global-definition-of-social-work/

Jordan, J. (2010). *Relational-cultural therapy*. American Psychological Association.

Jordan, J., Kaplan, A., Miller, J., Stiver, I., & Surrey, J. (1991). *Women's growth in connection: Writings from the stone center*. Guilford Press.

King, D. (1988). Multiple jeopardy, multiple consciousness: The context of a Black feminist ideology. *Signs: Journal of Women in Culture and Society, 14*(1), 42–72.

Lauve-Moon, K. R., Enman, S., & Hentz, V. (2020). Mainstreaming gender: An examination of feminist methodology in social work research. *Social Work, 65*(4), 317–324.

Lee, J. (2001). *The empowerment approach to social work practice*. Columbia University Press.

Lorde, A. (1984). Sister outsider: Essays and speeches. *The Crossing Press, 44*, 111.

Maschi, T. (2016). *Applying a human rights approach to social work research and evaluation: A rights research manifesto*. Springer Publishing.

Maschi, T., & Baer, J. (2013). The heterogeneity of the world assumptions of older adults in prison: Do differing worldviews have a mental health effect? *Traumatology, 19*(1), 65–72.

Maschi, T., Baer, J., Morrissey, M., & Moreno, C. (2013). The aftermath of childhood trauma on late life mental and physical health: A review of the literature. *Traumatology, 19*(1), 49–64.

Maschi, T., Kwak, J., Ko, E., & Morrissey, M. (2012). Forget me not: Dementia in prison. *The Gerontologist, 52*(4), 441–451.

Maschi, T., & Morgen, K. (2020). *Aging Behind Prison Walls: Studies in Trauma and Resilience*. New York, NY: Columbia University Press.

Maschi, T., Morgen, K., Zgoba, K., Courtney, D., & Ristow, J. (2011). Trauma, stressful life events, and post traumatic stress symptoms: Do subjective experiences matter? *The Gerontologist, 51*(5), 675–686.

Maschi, T., Morgen, K., Leibowitz, G., & Rees, J. (2018). Exploring the relationship between cumulative trauma and recidivism among older adults: Does race and offense history matter? Traumatology: An International Journal.

Maschi, T., Viola, D., & Koskinen, L. (2015). Trauma, stress, and coping among older adults in prison: Towards a human rights and intergenerational family justice action agenda. *Traumatology, 21*(3), 188–200.

Maschi, T., Viola, D., & Morgen, K. (2014). Unraveling trauma and stress, coping resources, and mental well-being among older adults in prison: Empirical evidence linking theory and practice. *The Gerontologist, 54*(5), 857–867.

Maschi, T., Viola, D., Morgen, K., & Koskinen, L. (2015). Trauma, stress, grief, loss, and separation among older adults in prison: The protective role of coping resources on physical and mental well-being. *Journal of Crime and Justice, 38*(1), 113–136.

Maschi, T., Viola, D., & Sun, F. (2013). The high cost of the international aging prisoner crisis: Well-being as the common denominator for action. *The Gerontologist, 53*(4), 543–554.

Miley, K., O'Melia, M., & DuBois, B. (2011). Assessing resource capabilities. In *Generalist social work practice: An empowering approach* (pp. 235–271).

Mullaly, B., & West, J. (2017). *Challenging oppression and confronting privilege* (3rd ed.). Oxford University Press.

Mullaly, R. (2010). *Challenging oppression and confronting privilege: A critical social work approach.* Oxford University Press.

Poe-Yamagata, E., & Jones, M. (2000). *And justice for some: Differential treatment of minority youth in the justice system.* Building Blocks for Youth.

Presser, H., & Sen, G. (2000). *Women's empowerment and demographic processes: Moving beyond Cairo.* Oxford University Press.

Reichert, S. (2006). *Understanding human rights.* Sage Publications.

Robbins, S., Chatterjee, P., Canda, E., Richardson, F., Franklin, C., Robbins, S., & Canda, E. (2012). Phenomenology, social constructionism, and hermeneutics. In S. P. Robbins, P. Chatterjee, & E. R. Canda (Eds.), *Contemporary human behaviour theory: A critical perspective for social work* (Vol. 3, pp. 322–344).

Swift, C., & Levin, G. (1987). Empowerment: An emerging mental health technology. *Journal of Primary Prevention, 8*(1–2), 71–94.

Torre, D. (1985). Empowerment structured conceptualization and instrument development. *Journal of Aging Studies, 17*, 427–446.

Turner, S. (2001). Resilience and social work practice: Three case studies. *Families in Society, 82*(5), 441–448.

Turner, S. (2018). Empowerment and feminist practice with forensic populations. In T. Maschi & G. S. Leibowitz (Eds.), *Forensic social work: Psychosocial and legal issues across diverse populations and settings* (2nd ed., pp. 377–388). Springer Publishing.

Turner, S., & Maschi, T. (2015). Feminist and empowerment theory and social work practice. *Journal of Social Work Practice, 29*(2), 151–162.

Uehara, E., Flynn, M., Fong, R., Brekke, J., Barth, R., Coulton, C., … Manderscheid, R. (2013). Grand challenges for social work. *Journal of the Society for Social Work and Research, 4*(3), 165–170.

United Nations. (1948). *The universal declaration of human rights*. Available at: http://www.un.org/en/documents/udhr/.

United Nations. (1977). *Standard minimum rules for the treatment of prisoners*. Available at: https://www.unodc.org/pdf/criminal_justice/UN_Standard_Minimum_Rules_for_the_Treatment_of_Prisoners.pdf.

United Nations. (2012, July 23–27). *Report of the United Nations High Commissioner for Human Rights*. Substantive session, Geneva

United Nations Human Rights [UN]. (2015). *Human rights framework*. Available at: https://www.ohchr.org/EN/Issues/Migration/Pages/HumanRightsFramework.aspx.

Wilson, R., Picheca, J., & Prinzo, M. (2007). Evaluating the effectiveness of professionally-facilitated volunteerism in the community-based management of high-risk sexual offenders: Part two – A comparison of recidivism rates. *The Howard Journal of Criminal Justice, 46*(4), 327–337.

World Health Organization. (1948). *Constitution*. World Health Organization.

CHAPTER 4

Feminist Research and Practice: Reorienting a Politic for Social Work

Sam Harrell, Ben Anderson-Nathe, Stéphanie Wahab, and Christina Gringeri

Introduction

In recent decades, social work research, practice, and education have become increasingly constrained by the braided influences of neoliberalism, professionalization, and criminalization. Using concrete practice examples, this chapter proposes that principles of feminist research rooted in understandings of power as everywhere, dynamic, historic, and structural offer a potential interruption to these macro influences on social work practice. Through their contributions to knowledge production, education, and practice, we hold that tenets of feminist research open a new politic for social work, in which social work scholars, practitioners,

S. Harrell (✉) • B. Anderson-Nathe • S. Wahab
Portland State University, Portland, OR, USA
e-mail: saha2@pdx.edu; banders@pdx.edu; wahabs@pdx.edu

C. Gringeri
University of Utah, Salt Lake City, UT, USA
e-mail: Christina.Gringeri@socwk.utah.edu

C. Cocker, T. Hafford-Letchfield (eds.), *Rethinking Feminist Theories for Social Work Practice*,
https://doi.org/10.1007/978-3-030-94241-0_4

and educators might interrupt dominant paradigms in the field and open space for collaborative action and a new social work future.

We write this chapter in light of the unexpected global upheaval starting in 2020 characterized by social unrest and uncertainty. The COVID-19 pandemic, which has hardest hit those communities already most targeted by the effects of systemic racism and classism, continues to threaten both lives and livelihoods, with the extent of infections that hit the US. This is coupled with profound social mobilization and activism surrounding the ongoing and public police killings of Black and brown people across the country (most recently George Floyd, Breonna Taylor, and Tony McDade), with calls for police divestment increasing amid a climate that has begun to take seriously the long-standing feminist calls for police abolition, decarceration, and creative revisioning of what constitutes accountability and "community safety." We offer this backdrop as context for the arguments we make that feminist principles offer social work research opportunities for building knowledge in the pursuit of a politic of accountability.

The Braid

We hold that social work practice and research occupies a precarious and complicated position related to the set of current circumstances. We draw on the metaphor of "the braid that binds us" from Mehrotra et al. (2016), which articulates how the complex and mutually reinforcing threads of neoliberalism, professionalization, and criminalization have constrained social work practitioners' and researchers' ability to meaningfully interrupt and address domestic and sexual violence. Originally conceptualized in the context of domestic and sexual violence, the braid constituted by these three strands constrains the entire profession of social work and articulates the profession's ongoing complicity in durable and mutable institutions, systems, and discourses of marginalization and oppression. Just as "the antiviolence movement's reliance on the state (for funding, policy engagement, and criminal legal response) has foreclosed a radical vision of domestic violence," we contend that the profession's historic and contemporary relationship with these same systems has limited social workers' potential to imagine and enact liberatory practices across the field (Mehrotra et al., 2016, p. 155). We suggest that feminist research principles, here understood to both inform and be informed by analyses of power rooted in critical, post-structuralist, queer, and trans theorizing, offer the potential to interrupt, subvert, and resist the impacts of the braid on social work research and practice (Lauve-Moon et al., 2020). Mehrotra et al. (2016) argued that "one of the biggest impacts of the braid is that it has depoliticized DV work" (p. 156); we propose that bolstering critical and post-structural feminist research in social work may help re-politicize the profession in a historic moment when such a politic is essential.

Specifically, feminist research rooted in a complex, historicized, and dynamic power analysis offers resistance to the constraining implications of the braid on social work research, practice, and education. Feminist research can interrupt (1) neoliberalism's embrace of individual change at the expense of social change, (2) an overreliance on objectivist epistemologies, and (3) perpetuation of carceral and policing logics that uphold white supremacy and reinscribe social workers' roles as agents of state control.

Neoliberalism and Social Work

The dominance of neoliberal ideology challenges the core values of social work practice and education, often reshaping the knowledge-building processes that underlie the academic programs preparing social work professionals (Reisch, 2013). Neoliberalism applies market values as the standards for assessing the worth of labor, goods, and services, and it rejects the notion of the collective public good (Gray et al., 2015; Pollack & Rossiter, 2010). Political power is exercised through and regulated by the market, rather than the state regulating the market; neoliberalism fundamentally shifts the relationship between the state and the economy. In practice, neoliberalism "deploys and manipulates the market and market-like rationalities to assess, measure, and decide state activity" (Gray et al., 2015, p. 380). Thus, neoliberalism grants greater autonomy to the market while subsuming state programs into the marketplace to ensure they operate by the market calculus (Schram, 2012).

The application of neoliberal ideology translates into managerialism in organizational practices, in which power accrues through the use of uniform procedures and requirements. Managerialism promotes an organizational culture of audit in which workers' competence is not assumed and where workers are surveilled to achieve benchmarks and performance indicators. Accountability becomes a central watchword of managerialism and workers must comply with a detailed accounting of time spent and outcomes achieved (Pollack & Rossiter, 2010).

Neoliberal success in defunding public education and promoting quasi-privatization of higher education has forced universities in general and social work programs in particular to become "accountable" for revenue-generating through competition for external grants and contracts, as well as public funding, for survival. Social service agencies and public education, including universities, assessed by market standards, experience rising pressure for accountability and efficiency; responsibility for revenue-generation; and an emphasis on personal—to the exclusion of social and moral responsibility. It is critical to note that universities and agencies compete for funds with ideological strings attached; contracts for services and grants for research determine interventions, topics, and methods, and, hence, shape the production of knowledge for practice. Universities comply and only value large-scale research projects that garner seven-figure grants; the "gold standard" in research is often considered to be the randomized control trial method, which is appropriate only for individual clinical intervention studies. Federal, state, and local contracts determine which services are valued, who is eligible, and which outcomes matter. Autonomy to serve clients in ways that support their best interests or to deliver training that prioritizes professional values is rarely funded; research that centers social justice or issues of equity is likewise rarely considered by funders (Acker & Wagner, 2017). The positivist notion that what cannot be measured does not exist has come home to roost, powered by the market economy. In social agencies and higher education, if the work is not funded, it has no value and is left either undone or unacknowledged. Thus, professional autonomy in social work practice and academic freedom in higher education become severely curtailed, and social work education becomes complicit in preparing professionals who see "social justice in individual rather than structural terms" and understand personal responsibility and choice as the pathway to justice, rather than building solidarity and targeting social systems for change (Reisch, 2013, p. 716).

Social work, in its ongoing quest for professionalization, has become complicit with neoliberalism in many ways. It has pushed for third-party reimbursements, particularly emphasizing individual clinical interventions at the expense of social change; it has developed and institutionalized competency appraisals and evidence-based practice without questioning the funding sources that control the production of knowledge; it has acquiesced to the surveillance of clients and workers within the culture of audit. The profession is also complicit with this ideology in its support of the carceral state. Neoliberalism, as institutionalized through the market, penalizes those who do not comply. Education funding is retrenched, social support programs disappear, and poor and minoritized communities are surveilled and incarcerated while suffering disproportionately from high unemployment, poor health, violence, and environmental degradation.

Critical feminist theories open the door for researchers to analyze these social problems from their roots in structures and systems that reproduce inequities and injustices, problematizing their definition as individual ills that stem from personal deficiencies. Feminist frameworks highlight the constraints placed on social work practices when the profession ties itself to public funding for research and practice: the profession is then regulated according to market values which require only those forms of practice and knowledge-building that produce value and/or cost savings for the state. Forms of practice—including research—as well as who receives services are limited by accountability, managerialism, and uniformity. Feminist principles in social work research challenge us to build knowledge that renders these processes and their effects visible and pushes practitioners and academics toward a politic of resistance and disruption within oppressive systems, thus creating space to reimagine systems and practices that uphold the dignity and worth of human beings and our fundamental right to self-determination.

Professionalization and Social Work

As national, regional, and local calls to defund, reform, and abolish the police lead the way to calls for replacing police with social workers (among other service providers), some may argue that this attention and spotlight represents the pinnacle of professionalization of the field. While we unequivocally support Abrams and Dettlaff's (2020) open letter to *NASW and Allied Organizations on Social Work's Relationship with Law Enforcement*, urging social workers to resist the calls to join law enforcement, we use this moment of national recognition and spotlight on social work to highlight the ways in which Social Work has made strides in the "struggles for supremacy of the welfare state" (Dominelli, 1996, p. 160).

While professionalization is not universally defined, Weis (2008) suggests that, fundamentally, professionalization refers to a "pattern of behaviors and attitudes that transforms individuals with respect to their vocation" (p. 222). As the complex history of social work's professionalization goes beyond the scope of this chapter, the move toward professionalization has included a shift away from moral and religious volunteer-driven work conducted by loosely organized to more organized networks of paid professionals with criteria for training, specific knowledge (typically in institutions of higher education), codes of ethics, and formal organizations for practice (Heite, 2012; Kanuha, 1998; Kunzel, 1993; Weiss-Gal and Welbourne 2008). Some have argued that moves toward professionalization in social work practice contribute to the micro-macro binary, where the critical traditions of social work have taken somewhat of a backseat to more micro-oriented clinical work (Reisch, 2013; Reisch & Andrews, 2002; Wenocur & Reisch, 1989).

Understanding the professionalization of social work demands a robust gender analysis given that social work labor, predominantly performed by cis-women (Kunzel, 1993), has been consistently undervalued and lacked social, political, and economic capital (McPhail, 2004; Snyder & Green, 2008). Also necessary is a strong colonial and racialized analysis to understand how social work, from its infancy, fed settler colonialism and white supremacy agendas, including through the Western legacy of professionalization of Social Work in developing countries (Nadkarni & Joseph, 2014; Mupedziswa & Sinkamba, 2014). Finally, also necessary is a solid class and political analysis (Wenocur and Reisch, 1989) for grasping the impacts of neoliberalism on social work discourses around the world (Spolander et al., 2014).

The professionalization of social work is complicated. It has facilitated advancements such as a living wage, legitimation, social capital, recognition, better working conditions for some, and a growing knowledge base, while simultaneously further aligning the profession with the interests of the State, and its governing and disciplinary technologies (Soss et al., 2011) as evidenced by the recent calls to replace police with social workers. Professionalization has certainly privileged and marginalized certain bodies, voices, identities, and discourses from knowledge production through a range of processes including but not limited to the application of expert knowledge to control worker/client relationships (Reisch, 2011), credentialing and registration (Baines and Clark, 2020), privileging of certain ways of knowing (Reisch, 2013; Staller, 2013; Wahab et al., 2014), and embracing the evidence-based practice movements. Ultimately, we propose as have others (Bacchi, 2012; Baines & Clark, 2020) that professionalization represents a battle for how we wish to be governed, in addition to how power will be practiced and shared, including through knowledge production. Consequently, we argue that feminist research can facilitate a form of knowledge production that creates counter-discourses to the hegemony of professionalism in social work. Built on and through feminist discourse as innovative, autonomous, collaborative, reflexive, wise/experienced, and collective (Osgood, 2006), feminist research can create knowledge in service of practices and policies that resist the types of managerial professionalism that values coemption, production, and governance at the expense of an ethic of care, social justice, and collective well-being.

Criminalization and Social Work

As mentioned above, social work appears to be preparing itself, once again, to ally with a law-and-order agenda. In a June 2020 brief from the Brown School of Social Work at Washington University, Michael Sherraden asserted that police are "necessary and highly valued" and that calls for abolition are "fanciful" and "misguided," while also naming policing as a "core institution of White supremacy" (p. 1). Sherraden echoes Trump's calls for co-responder programs (Trump, 2020, June 16) and NASW CEO Angelo McClain's (2020, June 15) suggestions that with enough social work collaboration, policing can be done "fairly."

As the racial uprising in the U.S. unfolds in 2020, the field is considering the president's Executive Order on Safe Policing for Safe Communities (Exec. Order No. 13929, 37235-37328, 2020). This order instructs the Attorney General to increase the capacity of social workers to work directly with the police, develop and implement co-responder programs, and direct more funding to nonprofits that help improve community-police relations. Just as we saw the union of anti-domestic violence movements and law enforcement in the 1990s, accompanied by a dramatic shift in funding toward carceral interventions (e.g., criminal legal responses to domestic violence, three-strikes laws) (Kim, 2013; Mehrotra et al., 2016), we may now be witnessing the beginnings of a new era of social work and law enforcement partnerships. While Mehrotra et al. (2016) discuss criminalization in the context of carceral solutions to violence against women, it can be more broadly understood as the process by which behaviors and individuals are transformed into crimes and criminals (Michalowski, 1985). Through criminalization, marginalized groups—Black, Indigenous and people of color (BIpoc), poor people, queer people, disabled people—are socially defined as criminal and targeted by legislation and state actors intent on restricting their rights and activities (Kaba, 2016).

Social work's ties to criminalization date back to the inception of the National Conference of Charities and Corrections, professionally aligning social work with correctional organizations from the very birth of the (U.S.) profession. Progressive era social workers spent decades campaigning for gender- and age-responsive prisons, eventually building the juvenile punishment system (Tanenhaus, 2004) and running reformatories (i.e., child prisons) (Waters, 1922). Yet social welfare scholars tend to ignore or minimize the field's relationship to policing and corrections. Prior to the modern welfare system, police were the only state providers responsible for many social welfare functions such as shelter, food, and employment services (Higgins, 1951; Walker & Katz, 1992). Throughout the late nineteenth century, police and social workers reconfigured private conflicts as public matters (Garland, 2001). In New York, Charity Organization Society members acted as "lay-police," arresting beggars and working alongside officers to offer their "special knowledge" of the "[beggar] class and their schemes" (Chapman & Withers, 2019, pp. 36–37). For social work's earliest pioneers, care and control, or welfare and the judicial, were mutually reinforcing agendas (Jackson, 2003).

Throughout the twentieth century, social work continued to push and rely upon criminalization. Social workers helped facilitate the forced removal, incarceration, and resettlement of Japanese immigrants and descendants in the 1940s (Park, 2019), displaced thousands of Native American children into non-Indigenous communities in the 1950s and 1960s (Thibeault & Spencer, 2019), and prompted the disproportionate arrest and further mass imprisonment of communities of color in the name of "anti-violence" (Kim, 2018).

In the twenty-first century, social work's role in criminalization spans countless fields of practice. Dorothy Roberts (2002) has written extensively about the role of the "family regulation system" in the monitoring, regulation, punishment, and devaluing of Black mothers. Most children in the U.S. foster care system have been removed from their homes due to alleged child neglect (poverty) (Kim et al., 2017) with Black children being both reported to child services (Krase, 2013) and placed into foster care at disproportionate rates (NCJFCJ, 2017; Fluke et al., 2003) based on significantly less risk (Roberts, 2019). These types of statistics are not unique to the U.S. but are reproduced in other countries including but not limited to Canada (Cénat et al., 2021), the U.K. (Morris et al., 2018), and New Zealand (Keddell, 2019). We heed Schenwar and Law's (2020) warning that "not all prisons have steel bars, not all police carry guns, and not all punishments are called punishments. Sometimes the police are called 'social workers' and the punishment is called 'care'" (p. 140).

Critical feminist theory can help social work research resist post-positivist epistemologies that obfuscate and deny the lived experiences of criminalized communities. By centering a structural analysis of power, critical feminism illuminates both macro forces (e.g., poverty) and micro practices (e.g., individual case decisions) contributing to the stigmatization, ostracization, and disappearance of criminalized peoples. A long history of anti-carceral and abolitionist praxis in critical feminism can help social work research problematize constructs like *crime* and *harm* and defer to the people most impacted by criminalization for community-led strategies for safety and well-being.

Impacts of the Braid on Social Work Research

The neoliberal context is anti-feminist—is not egalitarian, renders power invisible, values marketable productivity—and clarifies how and why feminist research is marginalized, particularly within social work. Feminist research does not appear on the "call for proposals" of any major federal or private funding source. This absence is a commentary on the value of feminist knowledge production within hegemonic institutions where critical feminist contributions are erased, marginalized, and even attacked. It also means that to engage feminist social work research is often to have one's scholarship dismissed and devalued.

This critical look at how the braid is manifest in social work practice and research offers a praxis of resistance that makes way for more emancipatory forms of knowledge production and practice. Feminist research rooted in a complex, historicized, and dynamic power analysis offers an interruption to the constraining implications of the braid on social work research, practice, and education. Feminist research principles offer an opportunity to center social change, trouble and expand the politics of evidence by moving against epistemological unconsciousness (Steinmetz, 2005) in research and practice, and break with carceral and policing logics toward a more liberatory and just praxis.

Case Examples from a Praxis of Resistance

What follow are our praxis of resistance to the impacts of the braid on social work research and our praxis toward rewiring social work scholarship toward anti-carceral, social justice, and emancipatory dreams. We apply the concept of the braid to concrete social work practice examples centered on social change, the politics of evidence, and a break with carceral logics from lead author Sam Harrell's practice experiences. Prior to entering a doctoral program, Sam worked in a range of social work settings rife with illustrations of the braid and its constricting power. We close each illustration with a discussion about how principles of feminist research can contribute to interrupting and disrupting the restrictions of the braid on social work practice, scholarship, and/or education.

Centering Social Change

Sam: During my BSW program, I began organizing with social workers, students, and people experiencing homelessness to address gentrification and policing in our college town. The city did not have enough shelter beds to respond to an increasing dearth of affordable housing. Shelters had barriers to entry that especially targeted transgender people and people with active addictions, "untreated mental health disorders," sex offense charges, and open warrants. The interfaith community organized a seasonal, volunteer-run, low-barrier shelter where I worked as a site director.

With gentrification driving up commercial rent and no organization with capacity to run a year-round, low-barrier shelter, community activists drafted a proposal for a city-funded shelter operated by a team of social workers and students. After the city met our proposal with resistance, we organized community forums, marches, council testimonies, and sleep-ins on city property. As the city continued to resist calls for emergency shelter services, activists erected an unofficial, outdoor tent shelter on private property. This operated for weeks until the city shut it down for zoning violations.

While people experiencing homelessness and housing insecurity had clearly articulated their needs for years (e.g., emergency housing, affordable rent), the city and local foundations paid a private consulting firm tens of thousands of dollars to study homelessness through a series of "charrettes." These fishbowl conversations between "community stakeholders" included representatives from the department of corrections, police, parole, and local court system. Only one member of the steering committee was homeless, and no homeless persons participated in the charrette process that informed the final recommendations.

> These recommendations largely echoed the (uncompensated) insights of people experiencing homelessness and housing insecurity. They did not include a year-round, low-barrier, city-funded shelter. When the community pushed back against the city for spending money on redundant research that could have gone to direct services, the city put pressure on nonprofits to support and legitimize the findings. And I found myself one month out of my BSW program being asked by two potential employers to sign contracts promising to discontinue shelter organizing.

In this example, we see the powerful impacts of neoliberalism, professionalization, and the city's carceral approach to homeless persons. The city's choice to pay for "measurable" and "accountable" results through a consultant, utilizing a process that favored professionals in the areas of police, corrections, parole, and court personnel, made a city-funded emergency shelter an impossibility. The process silenced community authority, voices, and needs and clearly positioned homeless individuals as potential offenders, rather than as individuals in need of housing who could clearly speak to their needs. The city's contract became an entity unto itself such that what the city as a community needed became secondary, if not completely irrelevant, to the needs and interests of the city as a corporation.

Feminist research could disrupt the braid in the example above. By calling attention to power and voice, feminist research can center the voices of those stakeholders experiencing homelessness and challenge the city's process emphasizing "professional" stakeholders. Bringing an orientation to power as structural, dynamic, and discursive, feminist research can offer new language to explore limits on participation and representation in decision making. In doing so, feminist research principles can help call attention to structural limitations on participation, voice, and power of those who are homeless and showcase their contributions to resolving their situations; in participating in this process, individuals become visible as contributors who help define the problem rather than being defined as the problem. Feminist research can also help the city identify who and what has been excluded from knowledge production to support a more inclusive process in service of the viability of results. Attention to praxis in feminist research can help open spaces for inclusive conversations regarding the political nature of the problem of and solutions to homelessness.

The Politics of Evidence

Sam: I spent an excruciating four weeks working at an outpatient site of a tri-state mental health organization. At that time, providers were expected to meet 85% productivity, meaning that 85% of work hours had to be spent on billable tasks and services. Staff meetings began with the passing out of a list of employee names and productivity percentages, with those below 85% designated in red ink. We were reminded that staff who repeatedly failed to meet productivity were terminated. This method of "accountability" presented contact hours with clients as an objective measurement of our efficacy as social workers and our value to the company. I quickly resigned. Less than a decade later, a BSW student told me they had raised productivity expectations to 100% and had no other choice but to "adapt." I continued to watch students and colleagues question or lose sight of the purpose and quality of their work while trying to meet impossible standards. Neoliberalism's roots in objectivist thinking and knowledge production foreclose ways of knowing that prioritize serving people rather than institutions and the State.

We build on Steinmetz's (2005) critique of *epistemological unconsciousness* that privileges objectivist epistemologies to argue that feminist research offers social work opportunities to deepen *what* and consequently *how* we know when it comes to services, interventions, and practice. While we've taken the position that a specifically feminist epistemology doesn't exist, feminist researchers and scholars share certain commitments and engagements that shape knowledge production. As we've noted elsewhere (Gringeri et al., 2010), an overall challenge to androcentric bias, attention to power, authority, ethics, reflexivity, praxis, and difference inform many types of feminist research, that is, ways of knowing that move beyond notions of objectivity and measurable markers. Feminist ways of knowing, particularly critical and post-structural paradigms can nudge researchers, administrators, and practitioners to examine the role of macro forces, such as the braid, power, and authority when evaluating programs, initiatives, and practice.

Break with Carceral Logics

Sam: While coordinating anti-violence prevention services for a midwestern shelter and rape crisis center, I was invited to a "feminist" anti-pornography conference organized by a state anti-violence coalition. Keynote speakers included radical, trans- and sex worker-exclusionary feminists who advocated for the criminalization of sex work. They preached the same radical feminism behind initiatives like Project Reaching Out to the Sexually Exploited (ROSE), a collaboration between police and social workers that "offers" eligible people in the sex trade the option of prostitution diversion programs as an alternative to arrest (Wahab & Panichelli, 2013). These programs are a prime example of social work interventions considered "progressive" when framed as an "alternative" to traditional policing, when in fact they make social work knowledge, resources, and relationships available to police, thereby expanding the scope and power of the carceral state (Walker & Katz, 1992).

Immediately, social workers and sex workers across the state began contacting the coalition to demand they drastically alter or discontinue the conference. Many providers were silenced because their rape crisis centers relied on funding controlled by the coalition. In a lengthy email exchange with me, conference organizers argued that all sex work was exploitative and that transphobic platforms held by their keynotes were irrelevant because the conference was not about trans people. When they learned that local sex workers were planning to picket the conference and hold an alternative event, they promptly emailed attendees with a statement from their (now former) Director of Campus Initiatives. The letter patronized sex workers; positioned conservative, non-sex worker academics as the "experts" whose evidence-based research "pornography survivors" needed to learn from; and asked sex worker activists to engage professionally to ensure a "safe learning environment" [for social workers] (Irvine, 2020).

While picketing the conference, several sex workers were heckled by social workers and some were kicked out. Protestors invited attendees to an alternative event, *Rights Not Rescue: Resisting a Single Narrative*, hosted at a nearby university. I continued to exchange emails with the coalition about carceral feminism, transmisogyny, and accountability. The following Spring, they named me "Prevention Educator" of the year. I declined the award with a detailed message about how their feminist theory and praxis conflicted with the feminism grounding the prevention programs I delivered. Immediately, my boss scolded me for my lack of "professionalism" and "diplomacy" with an important coalition in charge of our rape crisis center certification.

Feminist research, rooted in anti-carceral logics, holds the potential to shape understanding and knowledge about sex work and the people engaged in the sex trades in ways that interrupt social work's carceral legacy. As the braid has systematically legitimated the voices and authority of non-sex workers (rescue professionals and non-sex working academics) as the experts on the sex trades, dangerous social work interventions such as Project ROSE have further entrenched social work practice within the carceral state. In other words, carceral feminisms have legitimated a movement seeking to "protect" women and girls through collusion with the prison industrial complex. Critical and post-structural feminist research, committed to praxis, nuanced power analyses, and counter-narratives can lead scholars to look for multiple voices and experiences within the sex trades as a means of shaping practice away from policing and incarceration.

Conclusion

We've used the metaphor of the braid to present and discuss elemental forces that shape social work research and practice away from some of its key commitments, namely social justice and equity. We've also proposed that feminist research grounded in principles of social justice, praxis, voice, and attention to power can help us resist the constraining forces of the braid by shaping the knowledge we produce through research that informs practice. A more robust engagement with feminist research might help us recognize the knowledge-building capacity of social activism and the astonishing capacity of human solidarity. Feminist research can also help us find and assert our voices to dismantle the silence and complicity of social work in the face of systems of oppression by centering the margins, and those who are quietly doing the work of rejecting institutional norms that oppress and erase. Feminist research might help us see the centrality of social and economic inequality and the degradation of the environment as radical causes of physical and mental illness. This could spur us to develop new and rich collaborations with far-reaching transformative impacts. Perhaps more than ever, social work can lean on feminist research to dream a new social work, one that rejects complicity with systems of oppression once and for all.

References

Abrams, L. S., & Dettlaff, A. J. (2020, June 17). *An open letter to NASW and allied organizations on social work's relationship with law enforcement*. https://medium.com/@alandettlaff/an-open-letter-to-nasw-and-allied-organizations-on-social-works-relationship-with-law-enforcement-1a1926c71b28

Acker, S., & Wagner, A. (2017). Feminist scholars working around the neoliberal university. *Gender and Education, 31*(1), 62–81. https://doi.org/10.1080/09540253.2017.1296117

Bacchi, C. (2012). Why study problematizations? Making politics visible. *Open Journal of Political Science, 2*(1), 1–18. https://doi.org/10.4236/ojps.2012.21001

Baines, D., & Clark, N. (2020). [Under review]. "Ongoing Dysfunction?" Mandatory social work licensure. [Journal article].

Cénat, J. M., McIntee, S. E., Mukunzi, J., & N., & Noorishad, P. G. (2021). Overrepresentation of Black children in the child welfare system: A systematic review to understand and better act. *Children and Youth Services Review, 120*(C). https://doi.org/10.1016/j.childyouth.2020.105714

Chapman, C., & Withers, A. J. (2019). *A violent history of benevolence: Interlocking oppression in the moral economies of social working*. University of Toronto Press.

Dominelli, L. (1996). De-professionalization social work: Anti-oppressive practice, competencies and postmodernism. *British Journal of Social Work, 26*(2), 153–175. https://doi.org/10.1093/oxfordjournals.bjsw.a011077

Exec. Order No. 13929, 3 C.F.R. 37235-37328. (2020). Reprinted in *Federal Register, 85*(119). https://www.govinfo.gov/content/pkg/FR-2020-06-19/pdf/2020-13449.pdf

Fluke, J. D., Yuan, Y. Y. T., Hedderson, J., & Curtis, P. A. (2003). Disproportionate representation of race and ethnicity in child maltreatment: Investigation and victimization. *Children and Youth Services Review*. https://doi.org/10.1016/S0190-7409(03)00026-4

Garland, D. (2001). Modern criminal justice and the penal-welfare state. In *The culture of control: Crime and social order in contemporary society* (pp. 27–51). University of Chicago Press.

Gray, M., Dean, M., Agllias, K., Howard, A., & Schubert, L. (2015). Perspectives on neoliberalism for human service professionals. *Social Service Review, 89*(2), 368–392. https://doi.org/10.1086/681644

Gringeri, C., Wahab, S., & Anderson-Nathe, B. (2010). What makes it feminist? Mapping the landscape of social work feminist research. *Affilia, 25*(4), 390–405. https://doi.org/10.1177/0886109910384072

Heite, C. (2012). Setting and crossing boundaries: Professionalization of social work and social work professionalism. *Social Work and Society, 10*(2), online. https://doi.org/10.5167/uzh-85442

Higgins, L. (1951). Historical background of policewomen's service. *Journal of Criminal Law and Criminology (1931–1951), 41*(6), 822. https://doi.org/10.2307/3491292

Irvine, M. (2020). *ICESA statement in response to harms of pornography protest*. http://icesaht.org/wp-content/uploads/2017/06/ICESA-Statement-in-Response-to-Harms-of-Pornography-Protest-October-2017.pdf

Jackson, L. A. (2003). Care or control? The metropolitan women police and child welfare, 1919–1969. *The Historical Journal, 46*(3), 623–648. https://doi.org/10.1017/S0018246X03003182

Kaba, M. (2016). Criminalizing Bresha Meadows: A teach-in. *No Selves 2 Defend.* https://noselves2defend.files.wordpress.com/2016/09/criminalizing-bresha-teach-in.pdf

Kanuha, V. (1998). Professional social work and the battered women's movement. *Professional Development: The International Journal of Continuing Education in Social Work, 1*(2), 4–18.

Keddell, E. (2019). Harm, care and babies: An inequalities and policy discourse perspective on recent child protection trends in Aotearoa New Zealand. *Aotearoa New Zealand Social Work, 31*(4), 18–34.

Kim, H., Wildeman, C., Jonson-Reid, M., & Drake, B. (2017). Lifetime prevalence of investigating child maltreatment among US children. *American Journal of Public Health, 107*(2), 274–280. https://doi.org/10.2105/AJPH.2016.303545

Kim, M. E. (2013). Challenging the pursuit of criminalisation in an era of mass incarceration: The limitations of social work responses to domestic violence in the USA. *British Journal of Social Work, 43*(7), 1276–1293. https://doi.org/10.1093/bjsw/bcs060

Kim, M. E. (2018). From carceral feminism to transformative justice: Women-of-color feminism and alternatives to incarceration. *Journal of Ethnic & Cultural Diversity in Social Work, 27*(3), 219–233. https://doi.org/10.1080/15313204.2018.1474827

Krase, K. S. (2013). Differences in racially disproportionate reporting of child maltreatment across report sources. *Journal of Public Child Welfare, 7*(4), 351–369. https://doi.org/10.1177/0002716220980329

Kunzel, R. G. (1993). *Fallen women, problem girls: Unmarried mothers and the professionalization of social work, 1890–1945.* Yale University Press.

Lauve-Moon, K. R., Enman, S., & Hentz, V. (2020). Mainstreaming gender: An examination of feminist methodology in social work research. *Social Work, 65*(4), 317–324. https://doi.org/10.1093/sw/swaa039. PMID: 33064825.

McClain, A. (2020, June 15). Social workers cooperate with police forces. *Wall Street Journal.* https://www.wsj.com/articles/social-workers-cooperate-with-police-forces-11592255480

McPhail, B. A. (2004). Setting the record straight: Social work is not a female-dominated profession. *Social Work, 49*(2), 323–326. https://doi.org/10.1093/sw/49.2.323

Mehrotra, G. R., Kimball, E., & Wahab, S. (2016). The braid that binds us: The impact of neoliberalism, criminalization, and professionalization on domestic violence work. *Affilia, 31*(2), 153–163. https://doi.org/10.1177/0886109916643871

Michalowski, R. (1985). *Order, law and crime*. Random House.

Morris, K., Mason, W., Bywaters, P., et al. (2018). Social work poverty, and child welfare interventions. *Child & Family Social Work, 23*, 364–372. https://doi.org/10.1111/cfs.12423

Mupedziswa, R., & Sinkamba, R. P. (2014). Social work education and training in southern and East Africa: Yesterday and tomorrow. In C. Noble, H. Strauss, & B. Littlechild (Eds.), *Global social work: Crossing borders, blurring boundaries* (pp. 141–154). Sydney University Press.

Nadkarni, V. V., & Joseph, S. (2014). Envisioning a professional identity: Charting pathways through social work education. In C. Noble, H. Strauss, & B. Littlechild (Eds.), *Global social work: Crossing borders, blurring boundaries* (pp. 71–84). Sydney University Press.

National Council of Juvenile and Family Court Judges. (2017). *Disproportionality rates for children in foster care: Fiscal year 2015*. https://www.ncjfcj.org/wp-content/uploads/2017/09/NCJFCJ-Disproportionality-TAB-2015_0.pdf

Osgood, J. (2006). Professionalism and performativity: The feminist challenge facing early years practitioners. *Early Years, 26*(2), 187–199. https://doi.org/10.1080/09575140600759997

Park, Y. (2019). *Facilitating injustice: The complicity of social workers in the forced removal and incarceration of Japanese Americans, 1941–1946*. Oxford University Press.

Pollack, S., & Rossiter, A. (2010). Neoliberalism and the entrepreneurial subject: Implications for feminism and social work. *Canadian Social Work Review, 27*(2), 155–169. https://doi.org/10.2307/41669933

Reisch, M. (2011, June 10). *Being a radical social work in reactionary times* [Keynote Address]. 25th Anniversary Conference of the Social Welfare Action Alliance, Washington, DC. https://www.socialwelfareactionalliance.org/wp-content/uploads/2013/05/Reisch.Being-a-Radical-Social-worker-in-reactionary-times.pdf

Reisch, M. (2013). Social work education in the neo-liberal challenge: The US response to increasing global inequality. *Social Work Education, 32*(5), 715–733. https://doi.org/10.1080/02615479.2013.809200

Reisch, M., & Andrews, J. S. (2002). *The road not taken: A history of radical social work in the United States*. Brunner-Routledge.

Roberts, D. E. (2002). *Shattered bonds: The color of child welfare*. Basic Books.

Roberts, D. E. (2019). How the child welfare system policies Black mothers *S&F Online*. http://sfonline.barnard.edu/unraveling-criminalizing-webs-building-police-free-futures/how-the-child-welfare-system-policies-black-mothers/

Schenwar, M., & Law, V. (2020). *Prison by any other name: The harmful consequences of popular reforms*. The New Press.

Schram, S. (2012). Welfare professionals and street-level bureaucrats. In M. Gray, J. Midgley, & S. A. Webb (Eds.), *The Sage handbook of social work* (pp. 67–80). Sage.

Snyder, K. A., & Green, A. I. (2008). Revisiting the glass escalator: The case of gender segregation in a female dominated occupation. *Social Problems, 55*(2), 271–299. https://doi.org/10.1525/sp.2008.55.2.271

Soss, J., Fording, R. C., & Schram, S. F. (2011). *Disciplining the poor: Neoliberal paternalism and the persistent power of race*. University of Chicago Press.

Spolander, G., Engelbrecht, L., Martin, L., Strydom, M., Pervova, I., Marjanen, P., Tani, P., Sicora, A., & Adaikalam, F. (2014). The implications of neoliberalism for social work: Reflections from a six-country international research collaboration. *International Social Work, 57*(4), 301–312.

Staller, K. (2013). Epistemological boot camp: The politics of science and what every qualitative researcher needs to know to survive in the academy. *Qualitative Social Work, 12*(4), 395–413. https://doi.org/10.1177/1473325012450483

Steinmetz, G. (2005). The epistemological unconsciousness of US sociology and the transition to post-Fordism: The case of historical sociology. In J. Adams, E. S. Clemens, & A. S. Orloff (Eds.), *Remaking modernity: Politics, history, and sociology* (pp. 109–157). Duke University Press.

Tanenhaus, D. S. (2004). *Juvenile justice in the making*. Oxford University Press.

Thibeault, D., & Spencer, M. (2019). The Indian Adoption Project and the profession of social work. *Social Service Review, 93*(4), 804–832. https://doi.org/10.1086/706771

Trump, D. (2020, June 16). *Remarks by President Trump at signing of an executive order on safe policing for safe communities*. [Briefing]. https://www.whitehouse.gov/briefings-statements/remarks-president-trump-signing-executive-order-safe-policing-safe-communities/

Wahab, S., Anderson-Nathe, B., & Gringeri, C. (Eds.). (2014). *Feminisms in social work research: Possibilities for justice-based knowledge*. Routledge.

Wahab, S., & Panichelli, M. (2013). Ethical and human rights issues in coercive interventions with sex workers. *Affilia, 28*(4), 344–349. https://doi.org/10.1177/0886109913505043

Walker, S., & Katz, C. M. (1992). *The police in America: An introduction*. McGraw Hill.

Waters, M. V. (1922). Where girls go right: Some dynamic aspects of state correctional schools for girls and young women. *Theatre Survey, 48*(9), 361–376.

Weiss-Gal, I., & Welbourne, P. (2008). The professionalisation of social work: A cross-national exploration. *International Journal of Social Welfare, 17*(4), 281–290.

Wenocur, S., & Reisch, M. (1989). *From charity to enterprise: The development of American social work in a market economy*. University of Illinois Press.

Wies, J. (2008). Professionalizing human services: A case of domestic violence shelter advocates. *Human Organization, 67*(2), 221–233. http://www.jstor.org/stable/44127099.

CHAPTER 5

A Pedagogy of Our Own: Feminist Social Work in the Academy

Sarah Epstein, Norah Hosken, and Sevi Vassos

Introduction

The practice and teaching of western social work are shaped within a predominately managerial higher education sector underpinned by neoliberal principles that valorises the individual. Consequently, feminist social work educators face constraints and challenges when trying to imagine, co-construct and enact socially just pedagogy, which is invariably concerned with the interrelationships between individual, cultural and structural domains. In response, we developed a feminist collaborative, research, teaching and advocacy centre called Critical Edge Women (CrEW). CrEW enables us to interrogate the processes and tensions experienced in trying to embody feminist social work pedagogy. In this chapter, we draw upon feminist social work and pedagogical literature, and our own experiences in CrEW, to share our development of a feminist social work pedagogy as a

S. Epstein (✉) • N. Hosken • S. Vassos
Deakin University, Melbourne, VIC, Australia
e-mail: sarah.epstein@deakin.edu.au; Norah.hosken@deakin.edu.au;
Sevi.vassos@deakin.edu.au

C. Cocker, T. Hafford-Letchfield (eds.), *Rethinking Feminist Theories for Social Work Practice*,
https://doi.org/10.1007/978-3-030-94241-0_5

distinct critical (explanatory) theory and practice model. First, we describe the current context in which we teach and practice social work in the academy. A brief review of feminist social work principles and core feminist pedagogical practices are then identified to illustrate how the application of feminist social work in the academy is distinguished from other social work and pedagogical perspectives. A rationale for why social work needs a pedagogy of its own is then presented. The chapter concludes with a discussion of how we have integrated feminist social work and feminist pedagogy to fulfil our obligations to uphold social work values and ethics.

Feminist Social Work in the Academy

Feminism has a transformative goal that is grounded in a philosophy that critiques social injustice and demands its practitioners uphold human rights. In so doing, social work practices are responsive to service users' human rights entitlements. This aligns with the social work profession where practitioners are constituted as human rights advocates (Lynch et al., 2019). While the social work profession was slow to take up feminist theory (Allan et al., 2009), from the 1970s onwards feminism has steadily shaped the articulation of social work practice (Morley, 2009; Shepard & Dziengel, 2016). Importantly, without feminist effort and influence and without accountability to the lived experiences of female-identifying service users, many existing practice environments would not exist (Laing & Humphreys, 2013). Specifically, these include women's refuges, family violence support services and systems, pregnancy advice services, women's counselling and response services around sexual violence, women's health services, women's legal services, women's housing services as well as women's information and telephone support services.

As social work academics, the contemporary neoliberal academy is the location of our social work practice. The academy is where we work to uphold social work values and enact our ethical responsibilities to social justice. In order to try and achieve this, we engage a feminist social work pedagogical practice because without gender equality, there is no social justice.

As social work educators in a critical social work teaching programme, our intent is bound by a critical analysis of power and the ethical responsibilities that stem from a commitment to the core values of the social work profession: human rights, social justice and professional integrity (International Federation of Social Workers, 2014). However, as feminists, a gendered lens is always our starting point in understanding the shape these responsibilities take. We have arrived at a juncture where we realise that it is both 'natural' for us, and inescapable, that our pedagogical framework reflects our feminist activism.

Feminist social work pedagogy is context-dependent; therefore, the following ideas reflect both our social positioning and our workplace. It is not possible, nor desirable, to generate a universal pedagogy (Orme, 2003). Instead, it is our intent to contribute to ongoing feminist efforts and to be part of continuing discussions about feminist social work practices in the academy. We hope to do this by sharing our experiences and analysis of power relations as they shape female-identifying student and academic lives within the academy. Our experience, and response, is contingent on our social locations as three university-educated, non-Indigenous, female-identifying tenured academics who are attempting to speak aloud our pedagogical practice. The aim of this chapter is to invite collective conversation and reflection with feminist social work colleagues reading it.

Critical Edge Women (CrEW)

In 2018, we established Critical Edge Women (CrEW) as a direct response to the sense of alienation and distress about the ability to contribute to social change within the academy. CrEW draws on our individual and collective contributions and reflections, and, in turn, these are shaped by similar and different social locations and social processes. Our experiences of gender, race, culture, religion, ability, sexuality and class often generate different, situated and partial knowledges (Harding, 2004). Through coming together and collaborating, we were able to more easily identify neoliberal logics that lead to our feeling demoralised, thus helping identify opportunities to be subversive. Resisting the dominant paradigm of academic success as individually driven and expanding opportunities to share accolades for our work was key. CrEW is now consolidating our understanding of how feminist social work pedagogy can sustain us in these neoliberal times.

Neoliberal Contexts for Social Work Education

Neo-liberalisation of the modern university constructs the academy in economic rather than education terms (Brooks & Waters, 2011) at the same time as responding to and impacting diverse social locations and processes. This diversity also includes particular local, regional and national histories and contexts 'defined by the legacies of inherited institutional frameworks, policy regimes, regulatory practices, and political struggles' (Brenner & Theodore, 2002, p. 349). While recognising the importance of context, there are, however, some agreed core elements of neoliberalism (Harvey, 2005), referred to as 'neoliberal logics' (Strier, 2019). The core elements of neoliberalism comprise the principles of freedom from constraints to maximise self-interest, competition, accumulation and profit in a market not constrained by state action (Abramovitz, 2012, p. 33). In Australia, this involves 'transforming universities into a business by mobilizing processes, discourses, and practices of marketization, managerialism, and privatization' (Blackmore, 2015, p. 285).

Social work practice and education across many parts of the world has been widely documented as being affected by neoliberalism (Lavalette, 2011; Strier, 2019). Social work education is located in higher education sectors where most countries are responding to neoliberalism (Strier, 2019; Spolander et al., 2014). This positions many social work educators as potential conspirators in the neoliberal machine, aiding and abetting the preference for production of market-driven competencies over critical thinking and transformative learning (MacKinnon, 2009; Shahjahan, 2014). There is also a risk of being complicit in competitive, individual self-marketing and valorisation of career advancement (Hill, 2012). Further, there are ongoing invitations to engage in punitive administrative processes that closely resemble social control functions around student surveillance (Prinsloo & Slade, 2017). In addition, the alienation, exhaustion and atomisation of self that is a function and effect of neoliberalism (Morley, 2016; Rogowski, 2010) is not unfamiliar to the authors.

The neoliberal logics structuring much of contemporary higher education (Shahjahan, 2014) of homogeneity, marketisation, gender neutrality, individualisation and self-surveillance are explored in the next section.

Homogeneity

Madaus and O'Dwyer (1999, p. 692) argue that with the advent of industrial capitalism, universities became influenced by the concepts of 'standardisation, uniformity, precision, clarity, quantification and rational tactics'. The marketisation of the teaching and learning space draws attention away from the lived experiences, and diverse motivations, of student cohorts because the focus is on measurable (often non-qualitative) outputs. This is problematic because the massification of higher education

(Ross-Gordon, 2011) has resulted in greater representation of 'non-traditional' and 're-entry' students within our social work programmes. The contemporary cohort of students has diverse approaches to learning and study as well as significant differences in learning needs. The heterogeneity and often complicated lives of female-identifying students and academics are obscured, and made vulnerable, 'in higher education contexts that prioritise homogeneity' (Raaper, 2016, p. 185).

Also pertinent is the observation that female-identifying academics are less likely than their male counterparts to occupy positions of institutional power and authority (e.g. Ballenger, 2010; Morley & Crossouard, 2016). This further invisibilises the diversity of women's experiences and needs within the academy.

Marketisation

The marketisation of neoliberal university structures draws energy and attention towards compliance, monitoring and accountability procedures (Shahjahan, 2014). These procedures are designed to measure competencies against pre-set learning outcomes instead of the critical thinking and analysis that underpins transformative learning. Failure to meet these outputs is seen as failure on the part of the individual (student and/or academic). However, women are more likely to be adversely affected by market-driven strategies due to overall economic status and their relational responsibilities for caretaking (Evans, 2015). Women's social location particularly at the intersections of race, age, class, ability and sexuality position them further away from the material resources (Mehrotra, 2010) needed to maximise self-interest (Blackmore, 2019). Competitive individualism, where one acts based on self-maximising interest regardless of others, is central to the fiction of the market as a natural order and not socially constituted (Blackmore, 2019, p. 178).

Gender Neutrality

The massification of tertiary education has seen the proliferation of 'equity and diversity' measures ostensibly designed to support non-traditional student cohorts (Hosken et al., 2020). Within social work education, there has also been a 'strong push … to prepare students for twenty-first century practice that embraces diversity and culturally competent practice' (Shepard & Dziengel, 2016, p. 28). However, as a result, this has meant there is 'less importance being placed on specifically addressing the needs and concerns of women' (Shepard & Dziengel, 2016, p. 28).

The value of feminist frameworks for making sense of women's lived experiences is at risk of being displaced in order to raise the visibility of class, race, ability, sexuality and age. This is problematic because women's lives often intersect with oppressed social locations which diminishes their access to resources and increases risk of marginalisation. Connell (cited in Poggio, 2018, p. 173) has identified this practice as the 'postfeminist ideology of gender neutrality'.

Individualism

Neoliberal discourse advocates individual freedom on the proviso that individuals take personal responsibility for the pursuit of self-advancement (Evans, 2015). It is important to emphasise that this is regardless of the individual's circumstances (Blackmore, 2019). Neoliberalism colonises our way of being such that 'what seems thinkable and practicable in the neoliberal context, structures and limits our lives and social relations' (Shahjahan, 2014, p. 222). Further, the colonisation of ideas and terms such as freedom, choice and empowerment makes it difficult to articulate resistance (Evans, 2015). This poses specific challenges for feminism because historically, cultural and structural change for women has relied on mobilising *collective* political identities. Feminism has explicated the way that patriarchy produces and reproduces structural gender inequality constraining women's freedom and empowerment. However, within a neoliberal discourse, the individual unit is co-opted in such a way that to speak out about the problems with individualism is synonymous with a denial of women's right to make individual choices. The key here is that the feminist focus on the personal as political is the theoretical and material anchor. Feminist analysis draws attention to structural inequality through making sense of constructs such as the individual. The atomising of the individual draws attention away from collective advancement for women as a group, let alone diverse groups of women at risk of intersecting oppressive factors such as class, cultural, age, disability, sexuality and religion.

In this way, neoliberalism constructs both academic and student as context free, independent human beings who make decisions in order to exercise their free will. According to Blackmore (2019) feminists have long asserted that this type of human being bears the closest resemblance to the middle-class, white, heterosexual male—traditionally unencumbered by caring responsibilities and socialised to exercise independent decision-making that centres their own best interests. Those who do not do so effectively are set up as 'failures' and positioned as 'other than' the 'good academic' or the 'good student' (Blackmore, 2019; Hosken, 2017; Hosken et al., 2020).

The normative academic and student in the neoliberal university does not exist in relation to others. For female-identifying students and academics this poses specific obstacles especially when the measure of success is determined via market-based outcomes (Morley, 2015). For students, this is measured by the completion of a degree at a pass or above level. For the academic, the metric is how many articles produced in specific 'Q1' journals (Blackmore, 2019) and/or via student evaluations dominated by numerical measures of teachers' abilities (Morley et al., 2017). The context within which work is produced, or indeed despite the context, is largely ignored. This is anathema to the feminist practice of the personal is political.

Within the neoliberal paradigm gendered structural inequality is a feature and practice wherever the specificities of women's lives (their sexuality, caring status, age, class, culture, ability and religion) distance their access to resources required for students to complete their social work degree. Sexism and misogyny occur at the societal, cultural and organisation level of the university (Connell, 2011). This results in positioning women as 'problematic' should they not act, behave, appear or fit with accepted norms and stereotypes of the 'good' student (Hosken et al., 2020). Further, the 'good student' is able and willing to exercise independence and take individual responsibility for the completion of their studies (Hosken, 2017; Hosken et al., 2020). However, women are socialised and structurally positioned in such a way that they are more often than not relationally connected and restricted by those they care for and/or by those who seek to control them.

Self-surveillance

The neoliberal environment within the academy invites us to compare our academic work in relation to others who are deemed to have achieved neoliberal 'success' such as external grants, promotions and publications in 'Q1' journals. We are encouraged to measure our worth alongside groups of people we do not know, who we may not aspire to be, who we have no intention of learning from and who may not in any way even resemble our own discipline's values and ethos. Opportunities for self-surveillance abound (Gill, 2017). Under the guise of 'self-enhancement', neoliberal logics attempt to motivate us to glow with success, win popularity and level up to high scores around consumer-based satisfaction surveys. Accolades abound within the academy when this is achieved. The diverse lived experiences of academics have little impact on how success is measured (Mountz et al., 2015). The irony is not lost on us as social work academics who are charged with teaching future social workers to care about service users' stories and points of view. Cumulatively, the neoliberal higher education teaching, research and administration experience has caused sufficient alarm that we felt it time to document some of the practices we are engaging in as a form of our own resistance to the world we work in.

Feminism, Feminist Social Work and Feminist Social Work Pedagogy

The foundation for the development of feminist social work pedagogies in the neoliberal university relies on the synergy between principles of feminist social work and the central tenets of feminist pedagogy. Dominelli (2002, p. 7) defines feminist social work as a practice

> that takes women's experience of the world as the starting point of its analysis and by focusing on the links between a woman's position in society and her individual predicament, responds to her specific needs, creates egalitarian relationships 'client'-'worker' interactions and addresses structural inequalities.

Seeking to enact socially just practice within the neoliberal university often places feminist social work pedagogues at odds with dominant and powerful structures and practices (Epstein et al., 2018). Therefore, feminist social work in the academy is a subversive pedagogy because it is an attempt at non-compliance.

Why We Need a Pedagogy of Our Own

Feminist social work pedagogy operates from a feminist standpoint underpinned by two main assumptions: first, knowledge is situated and perspectival; and second, there are multiple standpoints from which knowledge is produced (Hekman, 2004). This means the social location of the subject constitutes the knowledge that is produced. This allows for both operationalising the category of woman as the starting analytic entry point, at the same time as accounting for the diverse and heterogeneous social locations that women (service users, practitioners and academics) occupy including class, age, race, sexuality, ability and culture. Bringing women in from the margins is necessary when a review of pedagogical literature (including critical pedagogy) indicates that the bulk of the knowledge produced about pedagogy does not arise from the standpoint of women (as students, as academics and certainly not from the standpoint of the female service user). Traditional accounts of pedagogy privilege the male standpoint and are similar to the way general power relations privilege men.

Operating from a feminist standpoint, feminist social work pedagogy is constituted from within female-identifying academic, student and service users' experiences. There is, therefore, more chance that the pedagogy is about women and discussed by women. A feminist standpoint re-qualifies women's experiences placing them at the centre of the pedagogical process where her experiences provide the starting point from which to build knowledge. Feminist standpoint theorists have asserted that women's vantage point from the margin of patriarchal structures and society can provide visual clarity, of the social structures and relations that they are both marginalised from and oppressed by (Frankenberg, 1993).

If the broader social work mission is social justice and human rights, then feminist social work pedagogy has an ethical responsibility to find ways to enact pedagogical practices that identify and address gendered injustice. Any dominant pedagogical practice that omits gender as a priority analytic category is at odds with the profession given numeric dominance of women in social work, as practitioners and service users as well as among social work student cohorts globally (Payne, 2014; Tower et al., 2019; Wendt & Moulding, 2016). Within the academy the majority of social work educators are also women. Recent research conducted in the United States into gender discrimination within the social work disciplines found the following: men's base salary is higher than women's; men who do not have administrative responsibilities earn almost the same as women faculty with administrative responsibilities; female social work faculty are more likely to have caretaking responsibilities for dependents and more women report exhaustion compared to men (Tower et al., 2019).

Feminist Social Work Pedagogy

Feminist social work pedagogical activism engages across the personal, cultural and structural levels of the university. In working with existing feminist social work (Dominelli, 2002; Wendt & Moulding, 2016) and general feminist pedagogical (Webb et al., 2002) principles, we have found five areas of practice relevant to our contemporary neoliberal context. These include:

- Foregrounding diverse lived experiences of women as fundamental to sources of knowledge
- Recognising and informing relations of power
- Working in collaboration
- Using the neoliberal university as a site for social change
- The personal is political

We now turn to providing a brief explanation of each principle interspersed with illustrative examples from our practice as feminist social work pedagogues.

Foregrounding Diverse, Lived Experiences of Women as Fundamental to Sources of Knowledge

Feminist social work pedagogy foregrounds women's diverse experiences of personal, cultural and structural injustice (Epstein et al., 2018) and includes the feminist practice of representing the voices and experiences of our diverse client base (Carrington, 2016). As it is connected in principle to feminist struggle, feminist social work pedagogical practice is concerned with contributing to change in gender relations. This is supported by the practice of consciously bringing women's experiences into the teaching, research and administrative domains of the university. We do this in order to re-qualify women's knowledge, to imbue women's knowledge with authority and ultimately to construct a better picture of women's lives.

Gender justice, a key goal of feminist social work pedagogy, requires resistance and activism across the personal, cultural and structural domains of the neoliberal university. The starting point is to reveal women's diverse lived experiences. At any given moment in time, multiple discourses and perspectives are at play, yet only a few are heard and listened to (Foucault, 1978). Feminist social work pedagogy is interested in working with women's lives and concerns as key knowledge sources for making visible their diverse experiences and revealing the power relations that have subjugated their social realities (Ackerly & True, 2010; DeVault, 1996). At the same time, feminist social work pedagogy needs to be accountable to women's lives by ensuring that descriptions of her experience do not reproduce problematic categories of identity (DeVault, 1996). Feminist social work pedagogy invites students' lived experience, to take up space in the classroom as a means for decentring the dominant narrative and repositioning subjugated knowledges as authoritative and meaningful. This serves as the fundamental platform for this knowledge development.

In addition to creating space for the voices and experiences of social work students in the classroom, feminist social work pedagogy uses a feminist lens to shape students' understandings of female service users in the teaching and learning space. This includes how service users might be spoken with and about. The service user's own gendered position is also considered. In sum, foregrounding diverse lived experiences of women as fundamental to sources of knowledge is inclusive of female-identifying students, female-identifying service users and female-identifying academics.

Some Ways to Think About and Do Feminist Social Work Pedagogy
Practically, we (CrEW) have found there are individual or combinations of pedagogical approaches across teaching and learning that facilitates the environment where many students bring their own experience and knowledge into the classroom and into assessment. These include instructions for, and content of, assessments that invite diverse life experience and knowledge of students; adjusting deadlines for assessment in ways that recognises diverse living conditions; marking feedback that validates diversity and structured inequalities; ongoing regular communication with students during the teaching semester that models feminist approaches; engagement with student critique and questions in online and face-to-face discussions that demonstrate commitment to the feminist goals of reducing injustice; normalising student questions and encouraging critique of learning materials; teaching that starts with examination of students' own experiences; and positioning of teacher as a collaborator and companion explorer in a learning journey. CrEW offers us space and time to share the development of feminist pedagogy At times, we have found this challenging where students have, rightly, critiqued our blind spots or areas of ignorance. It is, however, an exciting and authentic feminist process.

Recognising and Informing Relations of Power

Recognising power as key to exploring social injustice is part of the critical pedagogical tradition (McArthur, 2010). However, for feminist pedagogues gendered power imbalances and the way this constrains women's lives becomes the primary entry point for recognising power in the academy. Also, as a member of the critical pedagogy tradition, feminist pedagogy understands that the goal is not only to observe society but also to change it (McArthur, 2010).

Feminist social work principles (Dominelli, 2002) and feminist pedagogy (Webb et al., 2002) highlight identifying individual and collective opportunities for agency. For example, there are intersections of privilege where it is possible for the academic to reconfigure relationships between student and teacher. It is possible to disrupt the reproduction of domination and oppression through teacher and student collaboration around knowledge production, thereby resisting expert/novice dichotomies.

Domination and oppression are also reproduced at the structural level especially in regard to institutional administrative processes. Therefore, a feminist social work pedagogy identifies and critically demonstrates the potential for agency (of both the feminist social work pedagogue and her students) through advocacy such as assessment policy and special consideration policy change (Hosken et al., 2020).

Challenging traditional views and practices of power (Webb et al., 2002) also means making deliberate choices about who we research and what we seek to learn. Gathering collective accounts of the storied lives of female-identifying students and academics has the potential to disrupt neoliberal hegemony for a number of reasons. First, it challenges the ideology of individual personal failure and unjust constructs of individual responsibility for self-determination. When groups of people, women, speak to obstacles to self-determination, it is much harder to uphold individual neoliberal logic. Second, publishing accounts of women's storied and diverse lives reveals subjugated knowledge. It becomes harder to conceive of women as 'Other' and the concerns and struggles of women can be harder to dismiss. In turn, this knowledge can be used to advocate for cultural and structural change. Third, it is important to be strategic in terms of the focus of researching women's lives. Feminist social work pedagogy aims not only to reveal relations of power but also to inform these relations.

Gender Shapes Our Decisions About Who We Research and What We Seek to Learn
CrEW is currently researching female-identifying social work students' experiences of field-based placements. The aim is to gather evidence designed to advocate with governing bodies around policies that ignore caring responsibilities that female-identifying students are over-burdened with. Conducting this type of research, as a way of drawing attention to 'embodied struggles' (Shahjahan, 2014, p. 223) under neoliberal conditions, is an important component of feminist social work pedagogy. Once the stories are out in the world they can be used to agitate for social and material change.

Working in Collaboration

One of the six principles developed out of Webb, Allen and Walker's (2002) review of the literature of 1990s feminist pedagogy writing is the restructuring of the relationships between teachers and students, a focus on empowerment, and the commitment to building community. Similarly, looking for collective solutions to individual problems (Dominelli, 2002) is a core principle of feminist social work practice. Yet, as previously discussed collective interest has become harder to galvanise (MacKinnon, 2009) due to the impact of performance metrics, output focused measurements of individual value and university wide discourse around careerism.

In response, the goal is to utilise feminist approaches to develop collaborative partnerships and challenge underlying patriarchal cultural and structural causes of women's distress, marginalisation and subjugation in the academy. Aligning with women also means developing curriculum content that draws attention to patriarchal structures that cause the over-representation of female social work service users. Further, working out collective ways to reveal subjugated knowledges is more effective than repeated individual efforts.

Feminist social work pedagogues seek to work in collaboration with other academics and with students. It is important, however, to acknowledge relations of power and the limitations of the potential for egalitarian relationships between teacher and student, colleagues, social workers and clients. In addition to the structural role that teachers occupy, at play is also race, gender, class, age and so on of teachers and students.

Who We Collaborate with
As mentioned above, CrEW is currently researching female-identifying social work students' experiences of field work placements. While the research is designed to reveal lived experiences of women, CrEW has also chosen to conduct this research alongside social work student researchers. Through sharing feminist research goals and processes with undergraduate students CrEW has had the opportunity to take part in consciousness raising and collaborative learning experiences. While the research question is a feminist one, we have committed to establishing a research practice that also reflects collectivity and collaboration with students.

Using the Neoliberal University as a Site for Social Change and Social Justice

Identifying ways that particular policies attempt to normalise practices that reinforce individualising, alienating and vulnerable academic selfhoods enables us to develop a discursive distance. This distance helps us to resist being totalised by neoliberal logic. Understanding that it is not because we are not working hard enough or not working as efficiently as we think we should be provides relief from increasing self-blame and disillusionment. Discursive distance also creates ways to talk with each other about what we are experiencing, why we are experiencing it and to strategise ways to cope. Further, discursive distance provides opportunity to identify how to resist and attempt change. While there may be real limitations to our ability to create material change, naming the problematic practice gives us a language to speak about it. This is an important step towards working out where, and if, we can impact change in any substantially material way. Through recognising and naming these relations of power we are able to better exercise autonomy and disrupt individualising and alienating discourse.

Feminist pedagogy also questions the role and authority of the academic as expert (Weiler, 1991). This is a subversive resistance that refuses to reproduce dominant narratives and disrupts universalising truths. The teaching and learning space is a terrific location for this pedagogical practice. Using the teaching and learning space to openly critically analyse can be an important step towards transformative pedagogy (Giroux, 2002).

Critically Analysing Social Work

We have found that deliberately sharing our own critical analysis of the regulation of social work by professional bodies, and the university, elicits relief and validation amongst our students. As academics, we occupy positions of authority and when the authority is willing to expose problems that students feel they are struggling with then this has real significance. This type of subversive practice can make it more possible for students (and academics) to resist self-surveillance and self-critique around personal failure. We have found being overt about understanding that social work students face obstacles to accessing study because of their gendered location is an important way to resist reproducing shame, doubt and self-blame

The Personal Is Political

Many social work students have caring responsibilities that prohibit access to study spaces at certain times of the day and week. This is an issue experienced personally by our students but has cultural and structural precedence and cause. Many of our students have been, or are currently, dealing with gendered discrimination, oppression and violence. This creates obstacles for their learning and undertaking of assessment tasks. We have found that acknowledging this is not enough. Consequently, we have had to create assessment options, be flexible about access to study spaces and tailor assignment deadlines in ways that adapt and respond to female-identifying students' lives. More so, in order to work across the structural domain (and effect material change) we have had to advocate via our administrative roles with line management, and beyond, for accommodation requirements that adapt to students facing hardships as a result of their gendered social location.

The following example is a collective feedback to students about their first assignment. Attention is drawn to how feminist social work pedagogy can use combined feedback to lessen the focus on individualised notions of merit and normalise students' diverse capacity and opportunities.

In the example above, there is acknowledgement of the personal as political and that diversity of student's social locations impact their abilities to write and submit assignments. The academic (unit chair and marker) also validates students' sharing of their personal experiences in their

Collective Feedback on Assignment

Hi Everyone,

I have just finished marking and released the grades for those students who submitted on the due date for this unit. Thank you to all those who submitted, and those who are still working on their papers. There are a range of marks reflecting different levels of understanding, reflective and written abilities and, I suspect, social, cultural, economic, political and time resources. I was privileged to read the work from you where the literature and ideas were used to explore personal experiences. In response to some student critiques in their assignments about the LGBTQI resources, I have adapted the section about intersectionality in Topic Three and added some readings in Topics Four and Eight. I will attach the readings here in case you are interested. One of the primary incentives to change and update curriculum is in response to your feedback, thank you.

assignments as a source of knowledge. Relations of power are addressed where the academic feeds back to all students that she made immediate changes to the curriculum in response to the critiques of some students about the lack of LGBTQI resources. The collective feedback and changes to curriculum based on student critique are also examples of working in collaboration with students to foster social change and privileging students' knowledges as authoritative.

Conclusion

Feminist social work pedagogues have an ethical responsibility to gender equality in order to enact socially just, human rights-based pedagogy. In this chapter, we have engaged with existing feminist social work (Dominelli, 2002; Wendt & Moulding, 2016) and feminist pedagogical (Webb et al., 2002) principles, to suggest five areas of feminist social work pedagogy relevant to the contemporary neoliberal academy. These areas of practice include foregrounding diverse lived experiences of women as fundamental to sources of knowledge; recognising and informing relations of power; working in collaboration; using the neoliberal university as a site for social change and the personal is political. We provided brief explanations of each principle interspersed with illustrative examples from our practice as feminist social work pedagogues. The principles and examples inform our present conceptions of what constitutes a feminist social work pedagogy, contributes to ongoing conversations about this topic and has relevance for others who may be able to adapt for different locations and contexts. Finally, this chapter represents our attempt to make visible feminist social work pedagogical practice as a theory and practice model that is distinct from other pedagogies.

References

Ackerly, B., & True, J. (2010). *Doing feminist research in political and social science*. Palgrave Macmillan.

Allan, J., Briskman, L., & Pease, B. (Eds.). (2009). Critical social work: Theories for a socially just world (2nd ed.). Allen & Unwin.

Abramovitz, M. (2012). Theorising the neoliberal welfare state for social work. In M. GrayJ. Midgley, & S. A. Webb (Eds.), *The SAGE handbook of social work* (pp. 33–50). SAGE Publications Ltd. https://doi.org/10.4135/9781446247648.n3

Ballenger, J. (2010). Women's access to higher education leadership: Cultural and structural barriers. *Forum on Public Policy Online, 2010*(5), 1–20.

Blackmore, J. (2015). Disciplining academic women: Gender restructuring and the labour of research in entrepreneurial universities. In M. Thornton (Ed.), *Through a glass darkly: The social sciences look at the neoliberal university* (pp. 179–194). Australian National University.

Blackmore, J. (2019). Feminism and neo/liberalism: Contesting education's possibilities. *Discourse: Studies in the Cultural Politics of Education, 40*(2), 176–190.

Brenner, N., & Theodore, N. (2002). Cities and the geographies of "actually existing neoliberalism". *Antipode, 34*(3), 349–379.

Brooks, R., & Waters, J. (2011). *Student mobilities, migration and the internationalization of higher education*. Palgrave Macmillan.

Carrington, A. (2016). Feminism under siege: Critical reflection on the impact of neoliberalism and managerialism on feminist practice. In B. Pease, S. Goldingay, N. Hosken, & S. Nipperess (Eds.), *Doing critical social work: Transformative practices for social justice* (pp. 226–240). Allen & Unwin.

Connell, R. (2011). *Confronting equality: Gender, knowledge and global change*. Allen & Unwin.

DeVault, M. L. (1996). Talking back to sociology: Distinctive contributions of feminist methodology. *Annual Review of Sociology, 22*(1), 29–50.

Dominelli, L. (2002). *Feminist social work theory and practice*. Palgrave.

Epstein, S. B., Hosken, N., & Vassos, S. (2018). Theoretical research: Creating space for a critical feminist social work pedagogy. *Aotearoa New Zealand Social Work, 30*(3), 8.

Evans, E. (2015). *The politics of third wave feminisms: Neoliberalism, intersectionality, and the state in Britain and the US (gender and politics)*. Palgrave Macmillan.

Foucault, M. (1978). *The history of sexuality, vol. 1, an introduction* (Vol. 43). Vintage Books.

Frankenberg, R. (1993). *White women, race matters: The social construction of whiteness*. University of Minnesota Press.

Gill, R. (2017). Beyond individualism: The psychosocial life of the neoliberal university. In M. Spooner (Ed.), *A critical guide to higher education & the politics of evidence: Resisting colonialism, neoliberalism, & audit culture*. University of Regina Press.

Giroux, H. (2002). Neoliberalism, corporate culture, and the promise of higher education: The university as a democratic public sphere. *Harvard Educational Review, 72*(4), 425–464.

Harding, S. (Ed.). (2004). The feminist standpoint theory reader: Intellectual and political controversies. Routledge.

Harvey, D. (2005). *A brief history of neoliberalism*. Oxford University Press.

Hekman, S. J. (2004). Feminist identity politics: Transforming the political. In *Feminism and the final Foucault* (pp. 197–213). University of Illinois Press.

Hill, R. (2012). *Whackademia: An insider's account of the troubled university*. NewSouth.

Hosken, N. (2017). *Exploring the organisation of social injustice in Australian social work education*. University of Tasmania.

Hosken, N., Vassos, S., & Epstein, S. B. (2020). Feminist social work pedagogy: Personal, cultural and structural advocacy in the academy. *Social Work Education: The International Journal, 40*, 302–314.

International Federation of Social Workers. (2014). *Global definition of social work*. IFSW. http://ifsw.org/policies/definition-of-social-work/

Laing, L., & Humphreys, C. (2013). *Social work and domestic violence: Developing critical and reflective practice*. Sage.

Lavalette, M. (2011). 'Social work in crisis during crisis: Whose side are we on?'. *Canadian Social Work Review, 28*(1), 7–24.

Lynch, M., Bengtsson, A. R., & Hollertz, K. (2019). Applying a 'signature pedagogy' in the teaching of critical social work theory and practice. *Social Work Education, 38*(3), 289–301.

MacKinnon, S. T. (2009). Social work intellectuals in the twenty-first century: Critical social theory, critical social work and public engagement. *Social Work Education, 28*(5), 512–527.

Madaus, G. F., & O'Dwyer, L. M. (1999). A short history of performance assessment: Lessons learned. *Phi Delta Kappan, 80*(9), 688.

McArthur, J. (2010). Time to look anew: Critical pedagogy and disciplines within higher education. *Studies in Higher Education, 35*(3), 301–315.

Mehrotra, G. (2010). Toward a continuum of intersectionality theorizing for feminist social work scholarship. *Affilia, 25*(4), 417–430.

Morley, C. (2009). Using critical reflection to improve feminist practice. In J. Allen, L. Briskman, & B. Pease (Eds.), *Critical social work: Theories and practices for a socially just world* (pp. 149–159). Allen & Unwin.

Morley, C., Macfarlane, S., & Ablett, P. (2017). The neoliberal colonisation of social work education: A critical analysis and practices for resistance. *Advances in Social Work and Welfare Education, 19*(2), 25.

Morley, L. (2016). Troubling intra-actions: Gender, neo-liberalism and research in the global academy. *Journal of Education Policy, 31*(1), 28–45.

Morley, L., & Crossouard, B. (2016). Women's leadership in the Asian century: Does expansion mean inclusion? *Studies in Higher Education, 41*(5), 801–814.

Mountz, A., et al. (2015). For slow scholarship: A feminist politics of resistance through collective action in the neoliberal university. *ACME, 14*(4), 1235–1259.

Orme, J. (2003). It's feminist because i say so!' feminism, social work and critical practice in the UK. *Qualitative Social Work, 2*(2), 131–153.

Payne, M. (2014). *Modern social work theory* (4th ed.). Oxford University Press.

Poggio, B. (2018). Gender politics in academia in the neoliberal age. In J. W. Messerschmidt, M. A. Messner, R. Connell, & P. Y. Martin (Eds.), *Gender reckonings: New social theory and research* (pp. 173–192). New York University Press.

Prinsloo, P., & Slade, S. (2017). Big data, higher education and learning analytics: Beyond justice, towards an ethics of care. In *Big data and learning analytics in higher education* (pp. 109–124). Springer.

Raaper, R. (2016). Academic perceptions of higher education assessment processes in neoliberal academia. *Critical Studies in Education, 57*(2), 175–190.

Rogowski, S. (2010). *Social work: The rise and fall of a profession?* Policy Press.

Ross-Gordon, J. (2011). *Research on adult learners: Supporting the needs of a student population that is no longer traditional* (pp. 26–30). AAC&U Peer Review, Winter.

Shahjahan, R. A. (2014). From 'no' to 'yes': Postcolonial perspectives on resistance to neoliberal higher education. *Discourse: Studies in the Cultural Politics of Education, 35*(2), 219–232.

Shepard, M., & Dziengel, L. (2016). Feminist social work practice: Implications for the twenty-first century. In *Contemporary feminisms in social work practice* (pp. 24–39). Routledge.

Spolander, G., Engelbrecht, L., Martin, L., Strydom, M., Pervova, I., Marjanen, P., Tani, P., Sicora, A., & Adaikalam, F. (2014). The implications of neoliberalism for social work: Reflections from a six-country international research collaboration. *International Social Work, 57*(4), 301–312.

Strier, R. (2019). Resisting neoliberal social work fragmentation: The wall-to-wall alliance. *Social Work, 64*(4), 339–345.

Tower, L. E., Faul, A. C., Chiarelli-Helminiak, C., & Hodge, D. M. (2019). The status of women in social work education: A follow-up study. *Affilia, 34*(3), 346–368.

Webb, L. M., Allen, M. W., & Walker, K. L. (2002). Feminist pedagogy: Identifying basic principles. *Academic Exchange, 6*(1), 67–72.

Weiler, K. (1991). Freire and a feminist pedagogy of difference. *Harvard Educational Review, 61*(4), 449–475.

Wendt, S., & Moulding, N. (Eds.). (2016). Contemporary feminisms in social work practice. Routledge.

CHAPTER 6

Collaborative Autoethnography for Feminist Research

Jennifer Dyer, Sarah Pickett, Jennifer Davis, Kathleen Hackett, Cindy Holmes, Julie James, Daze Jefferies, Kimberley Manning, Annie Pullen Sansfaçon, and Julie Temple Newhook

Introduction

Recent research on transgender children and youth suggests that strong parental support helps achieve greater quality of life and positive wellbeing, and helps reduce risk of self-harm, depression, and suicide. However,

J. Dyer (✉) • S. Pickett • K. Hackett • D. Jefferies • J. T. Newhook
Memorial University of Newfoundland, St. John's, NL, Canada
e-mail: jdyer@mun.ca; spickett@mun.ca; kh4345@mun.ca; dsj272@mun.ca

J. Davis
Lethbridge College, Lethbridge, AB, Canada
e-mail: jennifer.davis@lethbridgecollege.ca

C. Holmes
University of Victoria, Victoria, BC, Canada
e-mail: cindyholmes@uvic.ca; jtemple@mun.ca

C. Cocker, T. Hafford-Letchfield (eds.), *Rethinking Feminist Theories for Social Work Practice*,
https://doi.org/10.1007/978-3-030-94241-0_6

the challenges that some parents face in supporting and advocating for their youth are immense, and only a fraction of parents will end up engaging in ally-ship and activism with, or on behalf of, their children.

As critical scholars deeply interested in feminist practices of disrupting, exposing, and critically analysing social and structural power, making room for the unspoken, unseen, invisible aspects of labour and taking action towards a more just society, collaborative autoethnography (CAE) appealed to us. As a team of critical scholars, we are driven ontologically towards pushing the boundaries of traditional epistemologies. We place value on interpreting the self as an artefact, whose voice is but a creation and a reckoning of the spaces in which we reside. The vulnerability of showing oneself in autoethnography (Lorde, 1984) is contrary to the traditional research ethos of participant confidentiality. In choosing collaborative autoethnographic as a form of advocacy, our stories do not end at the production of this research; rather, we hope they live on in our appeal for social justice (Bochner & Ellis, 2016b).

Authors of this chapter are parents of trans and gender diverse children or their allies and have engaged in advocacy work. This chapter presents how CAE was deployed through both face-to-face and online meetings between five Canadian scholars who are deeply engaged in advocating for trans children and youth and offers a reflection about their experiences in participating in such a process.

The purpose of this chapter is to demonstrate how collaborative autoethnography as a feminist research methodology and practice may be actioned in social work. We intentionally include personal and historical narratives, as both researcher and participant, in constructing parent-advocate selves (Chang et al., 2012), realizing we cannot write these representations of ourselves in a vacuum (Bochner & Ellis, 2016a, b,

J. James
Ryerson University, Toronto, ON, Canada
e-mail: julie.james@ryerson.ca

K. Manning
Concordia University, Montreal, QC, Canada
e-mail: kimberley.manning@concordia.ca

A. P. Sansfaçon
University of Montreal, Montreal, QC, Canada
e-mail: a.pullen.sansfacon@umontreal.ca

p. 131). Aligned with the tenets of feminist practice, collaborative autoethnography incorporates self-examination within a team of cooperating researchers (Chang et al., 2012). This process increases accountability and may participate in shaping our anti-oppressive lenses and development of ethical consciousness (Caron et al., 2020) in a research process aimed at highlighting the invisible in society and disrupting dominate narratives through an interrogation of the self.

This chapter is methodological in nature, describing how a group of academic mothers and allies of trans and gender diverse children used feminist methodologies to unpack our engagement with advocacy work. As we detail our process, we also keep close in mind the paper we wish we had as an interdisciplinary team to guide us through our use of collaborative autoethnography (CAE) and arts-based methods. Importantly, this inquiry is part of a larger research project exploring parental advocacy for gender diverse, creative, and trans children and youth. This case study aims to evoke an 'experience near' or insider view of the ways in which parental advocacy is performed by the researcher-participants and is limited by their social locations.

Background and Rationale

Until recently, the experience of parenting a gender non-conforming child has been "culturally unintelligible" (Rahilly, 2015, p. 342), and transgender children and their families were publicly invisible (Manning et al., 2015). Similar to transgender adults, they were subject to systemic processes of "erasure" (Namaste, 2000), scarcely mentioned in mainstream media. For decades, North American gender diverse children were pathologized and considered subjects for therapeutic intervention, much like LGBTQ+ children (Bryant, 2006). Professional support for gender nonconforming children advised universally reinforcing heteronormative, binary gender expression and repressing signs of gender diversity (e.g. Zucker, 2008); this led to the "non apparent" status of childhood gender nonconformity (Hellen, 2009). When researchers or clinicians focused on parents, it was through the framework of parental psycho-pathology (Pyne, 2014, citing Owen-Anderson et al., 2010; Zucker & Bradley, 1995), and specifically through a lens of "mother blame" (Caplan, 2000). This pathological perspective remains powerful in society (Elischberger et al., 2016; Grossman et al., 2005), and women remain vulnerable to what Johnson and Benson (2014, p. 124) call the "secondary stigma of mothering a transgender child".

The establishment of Gender Creative Kids Canada in 2013, the first Canadian organization dedicated to advocating for and supporting transgender and gender diverse children and families (Sansfaçon & Manning, 2015), coincided with rights challenges prompted by Canadian parents. Since 2013, parents have become active in provincial rights-based advocacy (British Columbia, Alberta, Saskatchewan, and Quebec), as well as in efforts to include gender identity in the Canadian Charter of Human Rights. Canadian parents sought out online advocacy communities. Parents began blogging or writing about their experiences, attending and running support groups, attending conferences, consulting with therapists and professionals with experience in children's gender diversity. Parents began educating and negotiating with their children's school communities and healthcare professionals (Manning, 2017; Travers, 2018). Strong parental support is shown to reduce youth suicide attempts by as much as 93% (Travers et al., 2012), yet parent advocates' needs are rarely examined. Given that parents who support their children's gender diversity face transphobic oppression and discrimination (Cowden & Pullen-Sansfaçon, 2012; Pullen-Sansfaçon et al., 2015; Riley et al., 2011), and that this transphobia is aggravated by other forms of oppression linked to sexual orientation, race, class, and ability (Mullaly, 2010; Saketopoulou, 2011), this is a glaring omission. Parental advocacy encompasses ethical dilemmas of safety, privacy, and personal risk. It demands significant time, energy, resources, and emotional labour (Hochschild, 1985; Meadow, 2011) and is also strongly gendered work (Kuvalanka et al., 2014; Rahilly, 2015): in an era of "intensive mothering" (Hays, 1996, pp. 6–9) mothers end up doing the labour to create safe worlds around their and other transgender and gender diverse children. In their study of mothers of transgender girls, Kuvalanka et al. (2014) found that while the mothers initially lacked education on transgender issues, they rapidly became 'expert' advocates. This public advocacy often extends from the mothers' personal processes of convincing others that their children's identities were valid, corresponding with feminist theories about intersectional thinking and privilege-checking that challenge 'everyday sexism' and exclusionary forms of representations (Robinson & Ross, 2013; Schuster, 2016).

Autoethnography in Research

Autoethnography is a suitable methodology for researchers writing and exploring effusive issues (Tetnowski & Damico, 2014). Autoethnographic writing moves away from traditional disengaged reason and distanced analysis towards an intimate and vulnerable narrative that aims to emotionally connect people in the name of social justice (Bochner & Ellis, 2016a, p. 62). Collaborative autoethnography (CAE) approach was used to document their experience of advocacy. CAE is a qualitative methodology that promotes community-building between participants and can, by its process, become transformative for participants. Where autoethnography focuses on the researcher self, collaborative autoethnography incorporates self-examination within a team of cooperating researchers CAE increases autoethnography's accountability as a research process born out of understanding society by interrogating the self and is defined as "a qualitative research method in which researchers work in community to collect their autobiographical materials and to analyze and interpret their data collectively to gain a meaningful understanding of sociocultural phenomena reflected in their autobiographical data" (Chang, et al., 2012, pp. 23–24).

Using Art in Research

Leading scholar of arts education Eisner argues that "science can be regarded as a species of research; so too can the arts" (2006, p. 9). As Cahnmann-Taylor suggests, "the literary, visual, and performing arts offer ways to stretch a researcher's capacities for creativity and knowing, creating a healthy synthesis of approaches to collect, analyze, and represent data in ways that paint a full picture of a heterogeneous movement to improve education" (2013 p. 4). Arts-based research crosses disciplinary boundaries to illustrate lines of significant connection shared between creation, learning, and knowing (Finley, 2011; Piantanida et al., 2003).

Following decades of decolonial, feminist, and poststructural thought, critical creative inquiry challenges the notion of art and science as mutually exclusive intellectual domains (Finley & Knowles, 1995; Finley, 2011). Arts-based scholars contend that social scientific inquiry in the qualitative paradigm has been limited by narrow understandings of legitimacy, reliability, and validity (Eisner, 1981; Seale, 1999; Siegesmund, 2014). For many qualitative researchers, doing arts-based research is a way of learning what can take form at the intersections of creative and scientific scholarship (Barry, 1996; Leavy, 2009). It is also a way to challenge what counts as scholarly knowledge, even within the qualitative realm of scholarship that does not always see meaning in aesthetic ways of knowing social and institutional worlds (Barone et al., 1997; Piantanida et al., 2003).

Methodology

CAE as a method allows the research team to approach parental advocacy as a type of gendered 'work' that recognizes a large commitment of time, energy, resources, and emotional labour. CAE and arts-based methods open up the many affective and ethically fraught challenges that some parents face as advocates. Further, it allows us to critically examine our own situations, how gender, heteronormativity, whiteness, class, and geography shape collective advocacy strategies, especially in the context of pre-existing transgender rights organizations.

Situating the Participant-Researchers

The researcher-participants are all white or white-passing, cisgender, academic mothers. In many ways, the whiteness of our team is not surprising: the Canadian academy has remained stubbornly white despite the near universal adoption of equity policies (Henry et al., 2017). At the same time, in the context of a settler-colonial state, racialized parents of transgender children are less visible in public advocacy work than white parents (Rahilly, 2015). Importantly, the researcher-participants, as described above, do not represent the social locations of the team members of the larger project or all authors on this chapter. In our efforts to disrupt the homogeneous researcher-participant team, the additional authors (graduate students and research assistants) include trans, non-binary, and Indigenous folks.

Establishing a Relational Framework for the Data Collection Process

We chose CAE as a method of exploration and analysis precisely for its profoundly enquiring, revelatory, and transformative aspects to better understand the experiences, challenges, strategies, and practices of advocating for trans kids. We also chose CAE because community-building is integral to the process, and not only did our research team require a method of information exchange that could develop trust, but we needed one that allowed us to explore experiences that disrupted the normative academic modes of investigation and authorship (Ellis et al., 2011). Our in-person meetings involved five or six team members meeting over three days. We also intentionally built in downtime (communally self-funded) given that both the cognitive and emotional toll of collaborative inquiry can be exhausting. This downtime was not meant to be part of the formal research exploration but intended to be self-care. We quickly observed that there is no hard line between interactive explorative research engagement; our enquiries were all productive of hitherto unrecognized aspects of advocating for trans youth, sometimes painful aspects of our advocacy, and still revelatory in ways impossible to anticipate. CAE doesn't occur in strictly work-defined spaces, it is performed when the group focuses together on the issue at hand (parent advocacy of trans kids), and this can be at the meeting table or at the dinner table. At the same time, throughout our CAE explorations, we were both mindful and often reminded of a gendered aspect to our academic research. As London et al. (2012) argue, the experience of working in competitive institutions, such as universities and colleges, is gendered and one potent aspect of this gendering is the "detection of gender-rejection threat and, to protect against rejection, … self-silencing" (p. 962). Our CAE incorporated work-spaces, arts-based spaces, and downtime or calming spaces.

Beginning, Re-entering, and Transformative Conversations

While some of us had used autoethnography as a method, and some had engaged with arts-based methods, few had ever combined the two into research projects. Our first face-to-face meeting in our Montreal hotel involved seven researchers; we met the first night and got to know one another; some people had worked together before in person, and others had collaborated on research remotely. We planned meetings the next morning in a university meeting room and subsequently held meetings at a local Art Hive, in one of the larger of our hotel rooms, and in local restaurants. The agenda included strategic planning, social action research and CAE training (led by individuals from our team), ethical considerations (primarily devoted to examining our critical whiteness), and the involvement of graduate students in our research, team roles, and decision-making processes.

While CAE was a major emphasis of our project planning as a methodological learning outcome, it quickly blended with our impromptu arts-based session at the local Art Hive into a fundamental, but not fully planned, part of the research output of this meeting. Art Hives "create multiple opportunities for dialogue, skill sharing, and art making between people of differing socio-economic backgrounds, ages, cultures and abilities" (ART HIVES n.d.) (Image 6.1).

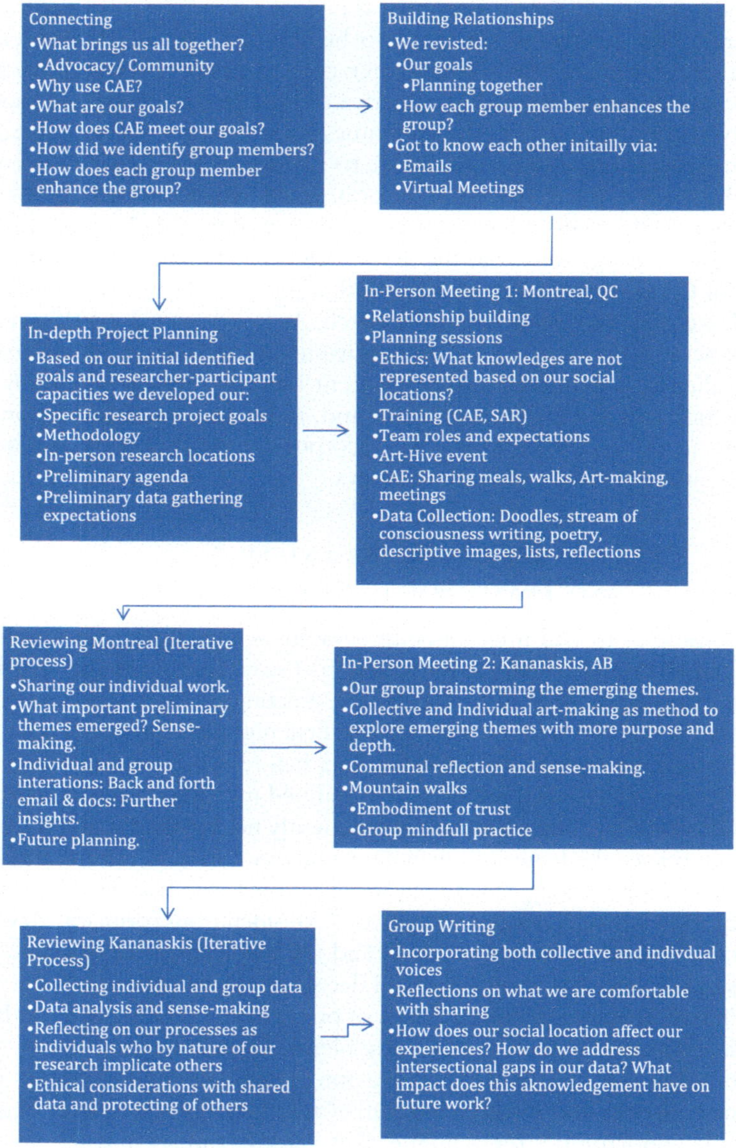

Image 6.1 Research team at Art Hive

CAE melded into urban walks around Montreal that precipitated conversations about shared experiences, shared fears, and recognitions of intersecting skills and experiential backgrounds with a resonant, recurrent question, "What do we already know collectively about parental advocacy, individually and systemically?"

The three-day meeting ended with a spontaneous social outing; we utilized this 'downtime' by noting that CAE can continue in less-focused and more free-flowing interpretive spaces. Our outing made palpable even more realizations of the stress, the physical burden, the inability to otherwise 'turn off', and the stark situational power differentials between group members that most of us felt differently but unconsciously, all key insights that informed our research. Such self-reflection and self-reflexivity leads to a self-respect where proximity in the CAE process works to open our concerns, emotions, and experiences generally. In this respect, time to emotively reflect reveals the gendering and racializing of emotion, what Sullivan (2014) explains as "how race and class intersect with gendered emotional expectations and demands" (p. 136), which also works to situate ourselves as white or white-passing, and to examine the racialized identities clearly missing from this aspect of our research. Here we reflected on "what we believe doesn't work, our confidence in these knowings, and who we need to engage help us unpack what we don't know". Further, following our meeting we each emailed to the others a document capturing the key themes that stood out individually of our experience, documents which took the form of large-scale doodles, stream of consciousness writing, poems, word-and-image pieces, lists, and expository reflections.

Art, Craft, and Embodiment

Our second meeting was more organized methodologically, because it involved producing content based on our training in CAE from our face-to-face meeting in Montreal. For this reason, this second meeting held in Kananaskis, Alberta, included a heavy component of arts-based materials that we brought with us to the meeting. In addition to paper, pens, cell phones with cameras, and computers, we brought coloured pencils, crayons, plasticine, glue, yarn, beads, coloured paper, figurines, craft supplies, pixel art, paint, markers, colouring pages, and used the hotel projector.

We used group brainstorming to observe thematic outcomes of our last meeting. Next, we narrowed our focus to certain themes and concepts of our CAE research. We reflected on "how our interconnection between the personal and professional influences all of our questions?" We communally participated in dedicated art-making directed by our early themes to develop expressions, trajectories, manifestations, and explorations of those themes emerging from documents that were not primarily word-based or expository. We utilized art-making as a way to refocus our engagement, to consider differently—less consciously or less intentionally—the themes, and to shift our emotional focus from immersion in those themes to a more balanced reflection that would take off some of the pressure, but still open hitherto unrecognized avenues of focus.

The mountain setting of this meeting was awe-inspiring. The aesthetic experience of walking mountain paths or simply viewing the mountainscape itself instigated a communal experience of the sublime: a cathartic, liberatory, emotional release with one another of the pride we feel, the fear, the exhaustion, and the protection over the children for whom we advocate. This was an experience that both made us recognize the firmly embodied experience of advocating for trans youth, and also how we carry on largely by putting that experience on the back-burner: the mountainscape played a role in our CAE practice of pulling our interiorized, embodied selves out and into the light of mutual self-recognition (Fig. 6.1).

Insights

Through our CAE process shown in Fig. 6.1, encapsulating reflexivity and repeated iteration of emergent themes we found five (5) key insights. First and not surprisingly, all members prioritized health. Existing hospital policies and healthcare experiences can expose trans youth to potentially harmful practices, either by omission of or by restrictions to affirmative services. Our voices are needed to educate and negotiate the physical and mental health of trans youth at the service and policy level of care.

Secondly, through exercises of writing five-minute reflections on our work and, later, on our workshop, through morning activities of 'checking in' with one another, and through mapping out aesthetically in the Art Hive the forms, patterns, and images that emerge in reflecting on advocacy work materially, we found new ways of considering and connecting what we do. Patterns and structures—whether in paintings, yarn-work, beading, drawing, or sculpting with found objects—manifested as a combination of bold symbols of gratitude and maze-like maps of survival.

The third insight, stimulated by and evolved through art-making, was the recognition of issues not often considered 'parenting' issues, obvious

Fig. 6.1 Developing a collaborative autoethnography process

forms of advocacy, or lived realities of resilience amongst parent advocates and trans children. Our art-making induced silent contemplations that led to riskier conversations about the advocacy experience: fears, frustrations, and the recognition that gender, sexuality, and a history of care- or trauma-work plays a larger role in advocacy than previously considered. Art-making put a spotlight on the mundane or repetitive activities of everyday life that comprised much of our advocacy practices, whether these are extraordinary practices of public activism or private acts of parental support. Creativity unlocked key features of CAE, namely the effort needed to trust and believe in not only others but oneself, leading to gambles in self-study. We trusted each other—people in many ways we barely knew—to reveal that we all live in various states of urgency, a theme that took on new meaning in our considerations.

Fourth, our research examined the intersections of our identities as scholar, advocate, parent as self, parent as protector. Our methods sought to address our responsibility as affirmative parents, described by Pullen Sansfaçon, Robichaud, and Dumais-Michaud (2015) as "the parents' desire to protect their children and place their safety above all else, including social norms and expectations". Through CAE we were able to disrupt traditional notions of consent, challenge how identity development may change over time in our collective and how to protect the identities of those possibly implicated in our stories and social action movements.

Our fifth important insight is that CAE and art-making opened up the theme of isolation, revealing intersections of self-censorship and of the isolating experience of a child coming out as gender diverse. In the first instance, we censor ourselves as mothers professionally, as if our experiences of advocating for trans youth engender protection and hypervigilance of our children. We found that friendships and allegiances change to the extent that we censor who we share with and how we bring this challenge into our research. In the second case, we foreground our privileged places as academic researchers and put behind us our own experienced oppressions and needs. Here, visibility became a key concept: we either invisibilize ourselves as parents/advocates of trans kids or we hyper-visibilize ourselves as advocates in order to protect the very children for whom we advocate. In each instance, our own experiences and needs are pushed to the background. The self-abnegating and care-focussed practices of advocacy opened onto the intersecting themes of the gendered nature of care work, the invisibilization of parenting, and the isolation of advocacy work from our professional lives and the trans affirming groups (primarily focussed on adults) around us. Together we recognized that risk-taking in our personal and professional lives, concerns about publicness, and the ongoing experience of urgency are key features of the parent/advocate engagement in fighting for the rights and wellbeing of trans and gender diverse youth.

Discussion

CAE as a method brought us to the aforementioned findings by promoting space to be informed by one's life experiences and as a collective, exploring researcher subjectivity through continuous reflexivity. Collaboration facilitated a pooling of resources, interdisciplinary insights, and access to a wider range of data sources (team members experiences) that were then collected through our two meetings and ongoing conversations. In this way, the interdisciplinary nature of our team was an asset as we engage in knowledge construction, mobilization, and dissemination. We are also critically aware of how CAE as a methodology could not render visible the stories and knowledge outside of the social location of our group members.

CAE opened space for our team to intentionally include personal and historical narratives, as both researcher and participant (Chang et al., 2012). However, the vulnerability of showing oneself in autoethnography (Lorde, 1984) is contrary to the traditional research ethos of participant confidentiality inherent in our inquiry. With this in mind collaborative autoethnography made way for us as researcher-participants to share ideas, generate new research, refine identities (Moore et al., 2013), and privilege our stories, parents of gender diverse kids, and advocate allies all as authentic sources of data (Shay & Wickes, 2017). CAE helped us to push the boundaries of traditional epistemologies by placing value on interpreting the self as an artefact, whose experience and voice is but a creation and a reckoning of the spaces in which we reside.

Meaning-making was captured through our researcher-participant co-collaborator stories to render visible the assumptions and/or events critical to the phenomena of parental advocacy with and for gender creative trans children and youth. This phenomenon has been previously skimmed over by the autoethnographic lens. In our context this revealed the stress, urgency, and labour inherent within our simultaneously shared and nuanced forms of parental advocacy.

Researcher-participant commitment to the vulnerability required in CAE allowed for deep probing and interrogation, permitting connection to wider issues such as guilt, isolation, self-negation, and the dialectic of hyper-visible/invisible parental advocate coalitional identities. This vulnerability in the research process fostered trust when researching complex and sensitive topics such as ours. Through the process of interrogation and probing these private communal stories were analysed so that great care has gone into protecting the confidentiality of the researchers themselves but also those implicated in the stories (Chang et al., 2012). Notably the influence of hyper-visibility and invisibility melded into ongoing conversation, which extended far beyond the researchers themselves often centring those implicit to and implicated by our stories.

In our team of feminist scholars, doing arts-based research matters most at the level of methodological innovation. It is experiential and experimental (Finley, 2003), which is appropriate and relevant to research that is unprecedented and that is unrecognized. Recognizing the significant limitations of traditional spoken language, arts-based research permitted us to explore how diverse voices and narratives can emerge through creative inquiry. Mobilizing artforms to showcase diversity in thought, language, and worldview was a way of encouraging aesthetic expression as a tool for social and institutional change.

The reporting of our work is timely as growing bodies of scholarship shows that researchers working with trans and gender non-conforming communities are using a variety of arts-based methods, including educational participative arts workshops (Rooke, 2010), photovoice (Holtby et al., 2015; Hussey, 2006), collaborative photography (Davidmann, 2014), video (Rhoades, 2012; Taylor & Bryson, 2016), as well as community discussions about art (Marshall et al., 2014). These all grapple with matters of embodiment, disability, intimacy, and resilience. Given the limitations of many research projects to effect meaningful change in the lives of trans people, particularly trans women (Namaste, 2000, 2009), doing arts-based research with trans communities is a way to distribute power relations between researchers and participants (Furman et al., 2019). Our bold engagement with arts-based methods as academics, parents, advocates, and allies of trans children and youth work to disrupt disciplinary power through the use of feminist methodologies and the promotion of collaborative identities. Further, mobilizing knowledge through art can be beneficial to trans communities because it encourages an interrogation of power and subjectivity on accessible, creative, and therapeutic terms (Zappa, 2017).

Some scholars suggest that arts-based methods can be particularly useful to explore issues of advocacy and resilience among trans communities (Asakura et al., 2019). We situated our inquiry outside the pathological violence of clinical literature by using arts-based research to encourage new ways of understanding axes of difference (Addison, 2005; Barbee, 2002; McNiff, 1998) through expressive communication as parent advocates.

In the context of our research, the combination of CAE and arts-based research employs abductive reasoning, making a probable conclusion from what one knows, to open the often surprising observations about the experience of advocating for trans youth that do not fit into the schemes by which we usually consider it. We approached the CAE and arts-based workshops with agenda and themes that demanded consideration in light of recent research; we left these workshops with surprising realizations about what counts as advocacy, the embodied, phenomenological effects of advocacy, and the social or inter-subjective role of advocacy work in our lives, professional and personal.

Conclusions

The combination of arts-based and CAE is a powerful feminist research methodology that allows participant-researchers to grapple with the subjective realities of one another. Participant-researchers can also develop stronger ties between them, which may contribute to their feeling of personal and collective empowerment. Applied to parents of trans children and youth, who are often experiencing social isolation and discrimination in their advocacy, it offers a powerful way to document experiences while answer some of their needs towards a greater social justice.

Our story of advocacy and ally-ship with trans and gender diverse children and youth does not end at the production of this research. Rather, we hope it lives on in our appeal for social justice (Bochner & Ellis, 2016b) and makes space for the investigation of intersectionalities, such as race, class, religion, or geography, not captured in this phase of our study.

Key Ideas for Social Work Practice

1. When engaging with families, it is important for social work practitioners to be aware of the stress, urgency, and intense labour parent advocates of gender diverse and trans children and youth may face.
2. The weight and intensity of the advocacy described by researcher-participants was immense. Social work practitioners can learn from advocacy initiatives that are already taking place thereby validating the skills and knowledge possessed by parent advocates.
3. CAE/Art revealed themes of anonymity, self-negation, guilt about self-care or self-focus, isolation, and invisibility/hyper-visibility, which may act as points of exploration for social work practitioners when engaging with parents of trans or gender diverse children. Importantly, social location may both intersect and mediate these experiences.
4. Social work practitioners can incorporate arts-based methods to enhance relationships and build trust.

5. Social work research that incorporates CAE and arts-based methods with individuals facing stigma and oppression may help build collective and shared understandings of self and social phenomena that may be inhibited or hidden due to societal stigma for both researchers and practitioners.

References

Addison, N. (2005). Expressing the not-said: Art and design and the formation of sexual identities. *International Journal of Art & Design Education, 24*(1), 20–30.

Art Hives. (n.d.). *What is the art Hove Network*. Art Hives. https://arthives.org/about

Asakura, K., Lundy, J., Black, D., & Tierney, C. (2019). Art as a transformative practice: A participatory action research project with trans* youth. *Qualitative Social Work, 19*(5–6), 1061–1077. https://doi.org/10.1177/1473325019881226

Barbee, M. (2002). A visual-narrative approach to understanding transsexual identity. *Art Therapy, 19*(2), 53–62.

Barone, T., Eisner, E., & Jaeger, R. M. (1997). *Complementary methods for research in education* (Vol. 2, pp. 75–116). American Educational Research Association.

Barry, D. (1996). Artful inquiry: A symbolic constructivist approach to social science research. *Qualitative Inquiry, 2*(4), 411–438.

Bochner, A., & Ellis, C. (2016a). *Evocative autoethnography: Writing lives and telling stories*. Left Coast Press.

Bochner, A. P., & Ellis, C. (2016b). The ICQI and the rise of autoethnography: Solidarity through community. *International review of qualitative research, 9*(2), 208–217.

Bryant, K. (2006). Making gender identity disorder of childhood: Historical lessons for contemporary debates. *Sexuality Research & Social Policy, 3*(3), 23–39.

Cahnmann-Taylor, M. (2013). Arts-based research: Histories and new directions. In *Arts-based research in education* (pp. 21–33). Routledge.

Caplan, P. (2000). *Don't blame mother: Mending the mother-daughter relationship*. Routledge.

Caron, R. E., Lee, O. J., & Pullen Sansfaçon, A. (2020). Transformative disruptions and collective knowledge building: Social work professors building anti-oppressive ethical frameworks for research. *Teaching, Practice and Activism, Ethics and Social Welfare*. https://doi.org/10.1080/17496535.2020.1749690

Chang, H., Ngunjiri, F., & Hernandez, K. C. (2012). *Collaborative autoethnography*. Routledge.

Cowden, S., & Pullen-Sansfacon, A. (2012). *The ethical foundations of social work*. Pearson.

Davidmann, S. (2014). Imag (in) ing trans partnerships: Collaborative photography and intimacy. *Journal of Homosexuality, 61*(5), 636–653.

Eisner, E. (2006). Does arts-based research have a future? Inaugural lecture for the first European conference on arts-based research Belfast, Northern Ireland, June 2005. *Studies in Art Education, 48*(1), 9–18.

Eisner, E. W. (1981). On the differences between scientific and artistic approaches to qualitative research. *Educational Researcher, 10*(4), 5–9.

Elischberger, H. B., Glazier, J. J., Hill, E. D., & Verduzco-Baker, L. (2016). "Boys don't cry"—Or do they? Adult attitudes toward and beliefs about transgender youth. *Sex Roles, 75*(5–6), 197–214.

Ellis, C., Adams, T. E., & Bochner, A. P. (2011). *Autoethnography: An overview* (pp. 273–290). Historical social research/Historische sozialforschung.

Finley, S. (2003). Arts-based inquiry in QI: Seven years from crisis to guerrilla warfare. *Qualitative Inquiry, 9*(2), 281–296.

Finley, S. (2011). Critical arts-based inquiry: The pedagogy and performance of a radical ethical aesthetic. In N. Denzin & Y. Lincoln (Eds.), *Handbook of qualitative research* (4th ed., pp. 435–450). Sage.

Finley, S., & Knowles, J. G. (1995). Researcher as artist/artist as researcher. *Qualitative Inquiry, 1*(1), 110–142.

Furman, E., Singh, A. K., Wilson, C., D'Alessandro, F., & Miller, Z. (2019). "A space where people get it": A methodological reflection of arts-informed community-based participatory research with nonbinary youth. *International Journal of Qualitative Methods*. https://doi.org/10.1177/1609406919858530

Grossman, A. H., D'Augelli, A. R., Howell, T. J., & Hubbard, S. (2005). Parent 'reactions to transgender youth' gender nonconforming expression and identity. *Journal of Gay & Lesbian Social Services, 18*(1), 3–16.

Hays, S. (1996). *The cultural contradictions of motherhood*. Yale University Press.

Hellen, M. (2009). Transgender children in schools. *Liminalis: Journal for sex/gender emancipation and resistance, 3*, 81–99.

Henry, F., Dua, E., James, C. E., Kobayashi, A., Li, P., Ramos, H., & Smith, M. S. (2017). *The equity myth: Racialization and indigeneity at Canadian universities*. UBC Press.

Hochschild, A. R. (1985). *The managed heart: Commercialization of human feeling*. University of California Press.

Holtby, A., Klein, K., Cook, K., & Travers, R. (2015). To be seen or not to be seen: Photovoice, queer and trans youth, and the dilemma of representation. *Action Research, 13*(4), 317–335.

Hussey, W. (2006). Slivers of the journey: The use of Photovoice and storytelling to examine female to male transsexuals' experience of health care access. *Journal of Homosexuality, 51*(1), 129–158.

Johnson, S. L., & Benson, K. E. (2014). "It's always the mother's fault": Secondary stigma of mothering a transgender child. *Journal of GLBT Family Studies, 10*(1–2), 124–144.

Kuvalanka, K. A., Weiner, J. L., & Mahan, D. (2014). Child, family, and community transformations: Findings from interviews with mothers of transgender girls. *Journal of GLBT Family Studies, 10*(4), 354–379.

Leavy, P. (2009). *Method meets art: Arts-based research practice*. Guilford Press.

London, B., Downey, G., Romero-Canyas, R., Rattan, A., & Tyson, D. (2012). Gender-based rejection sensitivity and academic self-silencing in women. *Journal of Personality and Social Psychology, 102*(5), 961.

Lorde, A. (1984). *Sister outsider*. Crossing Press.

Manning, K., Holmes, C., Sansfacon, A. P., Newhook, J. T., & Travers, A. (2015). Fighting for trans* kids: Academic parent activism in the 21st century. *Studies in Social Justice, 9*(1), 118–135.

Manning, K. E. (2017). Attached advocacy and the rights of the trans child. *Canadian Journal of Political Science/Revue Canadienne de science politique, 50*(2), 579–595.

Marshall, Z., Burnette, M., Lowton, S., Rainbow, R. D. T. S., Tiamo, J., Udegbe, O., & Vo, T. (2014). A conversation about art and activism with trans and genderqueer people labelled with intellectual disabilities. *Trans activism in Canada: A reader* (p. 125).

McNiff, S. (1998). *Art-based research*. Jessica Kingsley Publishers.

Meadow, T. (2011). 'Deep down where the music plays': How parents account for childhood gender variance. *Sexualities, 14*(6), 725–747.

Moore, J., Scarduzio, J. A., Plump, B., & Geist-Martin, P. (2013). The light and shadow of feminist research mentorship: A collaborative autoethnography of faculty-student research. *Journal of Research and Practice, 9*(2), Article M8. Retrieved from http://jrp.icaap.org/index.php/jrp/article/view/343/309

Mullaly, B. (2010). *Challenging oppression and confronting privilege* (2nd ed.). Oxford University Press.

Namaste, V. (2000). *Invisible lives: The erasure of transsexual and transgendered people*. University of Chicago Press.

Namaste, V. (2009). Undoing theory: The "transgender question" and the epistemic violence of Anglo-American feminist theory. *Hypatia, 24*(3), 11–32.

Owen-Anderson, A. F., Bradley, S. J., & Zucker, K. J. (2010). Expressed emotion in mothers of boys with gender identity disorder. *Journal of Sex & Marital Therapy, 36*(4), 327–345.

Piantanida, M., McMahon, P. L., & Garman, N. B. (2003). Sculpting the contours of arts-based educational research within a discourse community. *Qualitative Inquiry, 9*(2), 182–191.

Pullen-Sansfaçon, A., Robichaud, M. J., & Dumais-Michaud, A. A. (2015). The experience of parents who support their children's gender variance. *Journal of LGBT Youth, 12*(1), 39–63.

Pyne, J. (2014). Health and well-being among gender-independent children and their families: A review of the literature. In E. J. Meyer & A. Pullen Sansfacon (Eds.), *Supporting transgender and gender creative youth: Schools, families, and communities in action* (pp. 25–40). Peter Lang.

Rahilly, E. P. (2015). The gender binary meets the gender-variant child: Parents' negotiations with childhood gender variance. *Gender & Society, 29*(3), 338–361.

Rhoades, M. (2012). LGBTQ youth+ video artivism: Arts-based critical civic praxis. *Studies in Art Education, 53*(4), 317–329.

Riley, E. A., Sitharthan, G., Clemson, L., & Diamond, M. (2011). The needs of gender-variant children and their parents: A parent survey. *International Journal of Sexual Health, 23*(3), 181–195.

Robinson, M., & Ross, L. E. (2013). Gender and sexual minorities: Intersecting inequalities and health. *Ethnicity and Inequalities in Health and Social Care, 6*(4), 91–96.

Rooke, A. (2010). Trans youth, science and art: Creating (trans) gendered space. *Gender, Place & Culture, 17*(5), 655–672.

Saketopoulou, A. (2011). Minding the gap: Intersections between gender, race, and class in work with gender variant children. *Psychoanalytic Dialogues, 21*(2), 192–209.

Sansfaçon, A. P., & Manning, K. E. (2015). Maximising research outcomes for trans children and their families in Canada: Using social action and other participatory methods of inquiry. In *LGBT health inequalities* (1st ed., p 223). Policy Press.

Schuster, J. (2016). Intersectional expectations: Young feminists' perceived failure at dealing with differences and their retreat to individualism. In *Women's studies international forum* (Vol. 58, pp. 1–8). Pergamon.

Seale, C. (1999). Quality in qualitative research. *Qualitative Inquiry, 5*(4), 465–478.

Shay, M., & Wickes, J. (2017). Aboriginal identity in education settings: Privileging our stories as a way of deconstructing the past and re-imagining the future. *The Australian Educational Researcher, 44*(1), 107–122.

Siegesmund, R. (2014). The N of 1 in arts-based research: Reliability and validity. *International Journal of Education & the Arts, 15.*

Sullivan, S. (2014). *Good white people: The problem with middle-class white anti-racism.* State University of New York Press.

Taylor, E. T., & Bryson, M. K. (2016). Cancer's margins: Trans* and gender non-conforming people's access to knowledge, experiences of cancer health, and decision-making. *LGBT health, 3*(1), 79–89.

Tetnowski, J., & Damico, J. (2014). Autoethnography. In C. Forsyth & H. Copes (Eds.), *Encyclopedia of social deviance* (Vol. 2, pp. 49–50). SAGE.

Travers, A. (2018). *The trans generation: How trans kids (and their parents) are creating a gender revolution*. NYU Press.

Travers, R., Bauer, G., Pyne, J., Bradley, K., Gale, L., & Papadimitriou, M. (2012). Impacts of strong parental support for trans youth: A report prepared for Children's Aid Society of Toronto and Delisle Youth Services. TransPULSE.

Zappa, A. (2017). Beyond erasure: The ethics of art therapy research with trans and gender-independent people. *Art Therapy, 34*(3), 129–134.

Zucker, K. J. (2008). Children with gender identity disorder: Is there a best practice? *Neuropsychiatrie de l'Enfance et de l'Adolescence, 56*(6), 358–364.

Zucker, K. J., & Bradley, S. J. (1995). *Gender identity disorder and psychosexual problems in children and adolescents*. Guilford Press.

PART II

Feminisms and Intersectionalities

Trish Hafford-Letchfield and Christine Cocker

The term intersectionality was originally coined in 1989 by Kimberlé Crenshaw (1989), a Black law professor in the USA who described intersectional feminism as a prism for seeing the ways in which various forms of inequality and the relationships between them often operate together and exacerbate each other to compound and deepen concurrent experiences of discrimination. Crenshaw (2011) had used the concept within her own discipline (critical legal studies) to address the specific issues of Black women's experiences. The sociologist, Patricia Hill Collins, developed and explained the concept of intersectionality in her groundbreaking book 'Black Feminist Thought, Knowledge, Consciousness, and the Politics of Empowerment' published in 1990. However, intense debate has since raged (see Davis, 2020) on the misappropriation of intersectionality from its radical beginnings (Salem, 2018) and the displacement of Black feminists, not properly credited. Some voices have asserted that intersectionality has been whitened, depoliticized and transformed into a product of the neoliberal academy rather than being/becoming what it was supposed to be for (Davis, 2020). This state of affairs not only raises the issue of what should be done about intersectionality, but, more generally, how we should view the transnational circulation of ideas and theories in a globalizing world and what this means for how critical feminist scholars ought to think about the ownership and uses of the knowledge we produce and disseminate.

Taking a third way, Carastathis (2014) explains and responds to some critiques of intersectionality and cites as evidence four analytic benefits of intersectionality as a research paradigm: simultaneity, complexity, irreducibility and inclusivity. A subsequent blog hosted by the UN (2020) based on the voices of Black feminists also reflects on these historical contexts. Long histories of violence and systematic discrimination and the deep inequalities that disadvantage some from the outset have never been more visible than during the COVID-19 pandemic, #BlackLivesMatter and #MeToo movements, and their different impact on various parts of the world. The intersection of inequalities such as poverty, caste systems, racism and sexism has denied people their rights and opportunities across all generations. The contributors remind us that those most affected by gender-based violence, and by gender inequalities, are black and brown women, indigenous women, women in rural areas, young girls, disabled girls, trans youth and gender non-conforming youth. These are often from communities also most impacted on by natural disasters and the devastating effects of climate change (see Sontay Herrera, UN contributor).

The six chapters in this section cover a range of issues building upon the theoretical contributions of Black feminisms in the development of their epistemic, theoretical, analytic and/or substantive focus of inquiry. Kathomi Gatwiri and Sharlotte Tusasiirw in Chap. 7 address Afrocentric feminism and Ubuntu led social work practice in an African context. This is through the exposure and challenge to colonial education, which alienates Africans from their indigenous knowledges, beliefs and the philosophies of African people thus ontologically separating them from who they are. Colonial ideologies of domination which sit outside the principles of *Obuntu/Ubuntu* were advanced through formal education and western social work introduced in Africa by colonialists (missionaries). These were dominated by Eurocentric knowledges, theories and models of practice, as opposed to orality, storytelling and the power of experience. Gatwiri and Tusasiirw put African women's knowledges, experiences, agency and collective resistance at the centre of decolonizing feminism and social work practice and education in Africa, thus positioning African women as support for each other in addressing their struggles. Decolonized feminist social work in Africa promotes African ways of being, knowing and doing to prioritize interconnectedness/collectivity and interdependence, rather than individualism.

Fiona Kumari Campbell in Chap. 8 also follows this theme in her examination of configurations of shame (*lajja-baya*) as it is understood within a Sri Lankan Buddhist framework and its impact on mothers and partners with disabled girls. Campbell also refers to how feminist scholarship has fallen foul of the preoccupations emanating from the 'civil war' (1983–2009), focusing primarily on sex-based violence towards women and girls and, to a lesser extent, on feminist theory (to the neglect of issues such as those concerning disability and the intersections of sex, caste and poverty with some exceptions). Her highly original review of research on mothering and disability in the Sri Lankan context enables a complex analyses of the cultural context of relations, namely, the *parvula* (family-kin) in communities, the notable *absence of fathers* in actively parenting and also the place of *lajja-bhaya* (fear-shame) ethics which acts as the interpretative paradigm of familial and social relationships. Going beyond hierarchies, she questions the narrow scope of research in relation to the role of women as a sex class and its influence on parenting and requisites of support required beyond simply noting statistical sex differences. Campbell provides yet another example of the need for interdisciplinary research that draws on diverse epistemological, intersectional and methodological approaches.

In Chap. 9, Dina Pervez Sidhva explores the lived experiences of 20 Syrian refugees during the COVID-19 pandemic and the resultant lockdown in Jordan. Over the past decade, Jordan experienced a huge influx of Syrian refugees fleeing the civil war in their country, most of whom were women. Since the beginning of the Syrian conflict in 2011, an estimated 1.4 million Syrian refugees have fled to Jordan. Given the traditional conservative values and gender norms prevalent among the Syrian refugees, women and girls are disproportionally affected in terms of their ability to access education and meaningful employment. This resulted in serious problems—women losing the ability or right to pursue education, develop skills or improve personal qualities, which often exposed them to violence and abuse. Sidhva provides us with a valuable understanding of the feminist political and economic theory of gender-based violence through the narratives of the women in her research. She documents the use of power and violence stemming from inequality within society at both personal and structural levels. Violence and war were the fundamental reasons for fleeing Syria as a refugee; all the women in her study had experienced gender-based violence and personal trauma in their homes. Additionally, as refugees living in Jordan, structural issues such as poverty, immigration status, social isolation, persistent oppression, problems accessing health and humanitarian services all served to further marginalize them. Thus, the feminist political-economic understanding of gender-based violence unequivocally necessitates that as social workers we focus on the inextricable link between personal and structural vulnerabilities and inequalities in all our assessments and interventions in this area.

Chapter 10 by Jama Shelton, Maggie Dunleavy and Kel Krohele addresses transfeminism. Transfeminism is a critical approach to feminism rooted in the efforts of trans and cis women of colour's challenge of the omission of white supremacy in historical feminist conceptualizations of the oppression of women of colour. Utilizing their professional values, experiences and lived expertise as individuals with nonbinary identities, Shelton, Dunleavy and Krohele advocate for intentional confrontation of all systems of oppression that surveil and control individuals and their bodies, including colonialism, capitalism and white supremacy, and argue that this is essential for social work's justice-orientated practice.

Jennifer (JD) Drummond and Shari Brotman in Chap. 11 also incorporate an intersectional life course perspective in their analysis of a life story narrative, that of Josie, a gender non-conforming, queer woman living with chronic illness and disability. Drummond and Brotman use a feminist-narrative methodology to elicit Josie's description of her life experiences, and through attention to critical moments and turning points in her lived experience of disability, sexuality and gender, this chapter examines how intersecting identities both shape and are shaped by context and experiences of the body as both complex and fluid. Josie's narrative provides a rich account of her experiences embodying multiple marginalized and intersecting social locations and offers important insights on how to engage in a social work practice informed by a feminist-narrative approach, particularly when addressing the intersecting social locations of disability and queer identities.

Finally in this section, Trish Hafford-Letchfield, in Chap. 12, draws on multidimensional concepts comprising power, economic, social and symbolic relations operating simultaneously at intrapersonal, interpersonal, institutional and societal levels to apply an intersectional lens to gender and ageing. Through a review of the literature and key concepts in critical gerontology, Hafford-Letchfield outlines the contribution made by critical gerontology and feminist gerontology and its relevance for social work and social care. She describes how gendered ageism is a double jeopardy, where two interacting power systems lead to an increased vulnerability and emphasize the dominance of patriarchal norms. Theoretical work emphasizing gender as a relational process has also led to theoretical problematization of the concept to understand the processes of multiple marginalizations based on age, disability and race. Through empirical evidence on economic, health and sexual inequalities, professionals are urged to give more attention to strategies associated with lobbying, advocacy, coalition-building, increasing social awareness and supporting social movements that support women in later life.

This section illustrates many of the issues and debates on intersectionality and its relevance for social work. In order to engage with anti-racist practice, it is critical that practitioners and researchers embark on the journey of comparative travel, that is to shift our frame of reference, engage in cross-cultural debates and work at multiple levels. These will help us to engage in feminist social work that is cognizant of the micropolitics of context, subjectivity and struggle as well as the macropolitics of global economic and political systems and processes (Mohanty, 2013, p. 223). Looking through an intersectional feminist lens, we see how different communities are battling various, interconnected issues to question power structures and speak out against the root causes to build a better society for all.

References

Carastathis, A. (2014). The concept of intersectionality in feminist theory. *Philosophy Compass, 9*(5), 304–314.

Carastathis, A. (2019). *Intersectionality, origins, contestations and horizons.* University of Nebraska Press.

Crenshaw, K. (1989). Demarginalizing the intersection of race and sex: A Black feminist critique of antidiscrimination doctrine, feminist theory and antiracist politics. *The University of Chicago Legal Forum, 140*, 139–167.

Crenshaw, K. (2011). Postscript. In H. Lutz, M. T. Herrera Vivar, & L. Supik (Eds.), *Framing intersectionality: Debates on a multi-faceted concept in gender studies* (pp. 221–233). Ashgate.

Davis, K. (2020). Who owns intersectionality? Some reflections on feminist debates on how theories travel. *European Journal of Women's Studies, 27*(2), 113–127.

Hill Collins, P. (1990). *Black feminist thought: Knowledge, consciousness, and the politics of empowerment.* Polity Press.

Mohanty, C. P. (2013). Transnational feminist crossings: On neoliberalism and radical critique. *Signs, 38*(4), 967–991.

Salem, S. (2018). Intersectionality and its discontents: Intersectionality as traveling theory. *European Journal of Women's Studies, 25*(4), 403–418.

United Nations. (2020). Intersectional feminism: What it means and why it matters right now. https://www.unwomen.org/en/news/stories/2020/6/explainer-intersectional-feminism-what-it-means-and-why-it-matters

CHAPTER 7

Afrocentric Feminism and Ubuntu-Led Social Work Practice in an African Context

Kathomi Gatwiri and Sharlotte Tusasiirwe

Introduction

The roles of both feminism and social work in Africa *for* Africans remain contested to date. Feminism is often positioned as a western ideology, while social work is criticised for the underlying whitewashed knowledge it inculcates in both students and practitioners. Without proper contextualisation and theorisation, both feminism and social work can function as colonising agents which do more harm than good in the African context. Defining the needs of Africans in Africa through feminist lenses requires us to comprehend the pervasiveness of both colonialism and its ongoing impacts on African women. Feminist social work must critique global power constructs, situating both feminism and social work within their

K. Gatwiri (✉)
Faculty of Health, Southern Cross University, Gold Coast, QLD, Australia
e-mail: Kathomi.Gatwiri@scu.edu.au

S. Tusasiirwe
Western Sydney University, Penrith, NSW, Australia
e-mail: s.tusasiirwe@westernsydney.edu.au

C. Cocker, T. Hafford-Letchfield (eds.), *Rethinking Feminist Theories for Social Work Practice*,
https://doi.org/10.1007/978-3-030-94241-0_7

historical and political context. In this chapter, we theorise African feminism as an Afrocentric framework that centres African knowledge, theories, practices and ways of being, doing and knowing. It is within this context that we articulate its role in social work practice in Africa while also problematising the epistemological contradiction it presents. We argue that Afrocentric feminism is a collection of multiple knowledges that are focused on *returning* to our forgotten and often marginalised ways of being, doing and knowing and reclaiming a somehow lost or disfigured identity which was and is continually altered by colonialism and its ever-lingering presence in Africa.

Afrocentric Feminism

It is common knowledge now that "professional" social work emerged from the Anglosphere (Tascon & Gatwiri, 2020). As a result, the ways of *doing* social work across the world, including in the Global South, are dictated by Eurocentric ideologies (Gatwiri, 2020). Similarly, the role of feminism *in* Africa, and *for* Africans, remains contested. African feminists are concerned with the hegemony of western feminisms which are dominated by voices and experiences of white middle-class women (Chilisa & Ntseane, 2010; Sendi, 2017). The dominance of whiteness in social work contributes to the ideological imaginations that locate the African woman as being without agency and one who can only be liberated through white western saviourism. In writing this chapter, we seek to address the limitations of western feminisms that Sendi (2017) highlighted: that is the universalisation of western values and the imposition of western beliefs and values on African women. We contextualise African feminism as a theoretical position that privileges African experiences, which allow for the production of knowledge that is "contextually relevant, builds relationships, heals the self, the community and the larger socio-cultural context" (Chilisa & Ntseane, 2010, p. 619). Gatwiri (2020, p. 11) defined African feminism as

> an attempt to undo the multiple complexities and challenges presented by sexism—a derivative of patriarchy and colonisation—faced by the African woman. An African feminist approach is one that seeks to educate, empower, and elevate African women to positions of self and structural power while recognising the extra disadvantage laid upon their backs due the social constructions of their blackness and Africanness.

In this chapter, we adopt an African feminist, Indigenist and decolonising lens purposefully to name and locate African women's experiences as contextually and theoretically poignant in the feminist discourse. We are particularly seeking to disrupt the deficit discourse that omits and erases the agency, resistance and strengths of African women from dominant narratives. This interrupts the universalisation of western knowledges in social work theorisations and practices where African knowledges and experiences are constructed as *Other*. We do this by sharing stories of African women, who in the face of different struggles and hardships draw on their cultural and *Ubuntu/Obuntu* values and philosophies to ensure each other's collective survival. It is our argument that the self-organised initiatives that African women engage in to address their struggles should be validated and built on through cultural frames of reference that are relevant to their experiences. We position African women as custodians of indigenous African philosophies, values and knowledges, which they pass onto younger generations through orature. These indigenous knowledges are embedded within stories, proverbs, songs and folklore and are an important resource in rethinking and decolonising feminism and social work.

In understanding how colonisation and imperialism impact both the mind and the body, we consider Frantz Fanon (1961, p. 36) argument that colonisation "leaves behind germs of rot which we must clinically detect and remove from our land and from our minds as well." This clinical removal of the rots of colonisation is part of the decolonisation process that Tusasiirwe (2019) argues include mourning of what was lost and what is left. Decolonisation requires confronting ongoing epistemic colonisation with a view to understand, question and challenge how it devalues African knowledges and experiences. Engaging in this process is an important step that leads to recovering, re-centring and returning to forgotten indigenous African ways of knowing, being and doing. In the chapter, we use the terms *Obuntu* or *Ubuntu* to refer to an African philosophy that defines what being human is by emphasising that we are made human through the process of humanising others. This entails embracing "values of interconnectedness, collectivism, solidarity, caring for and *about* others, and the environment" (Tusasiirwe, 2019, p. 1). This philosophy is diverse and is expressed differently in different African communities and languages. For instance, *Obuntu* is the terminology utilised in the *Runyankole/Rukiiga* language in Western Uganda where the case studies used in this chapter were derived. *Ubuntu* is utilised in other countries such as South Africa, Burundi, Zambia, Zimbabwe while Kenya employs the terms *utu/*

mtu (Mugumbate & Chereni, 2019). These terms "all point to one thing, [that is], an authentic individual human being, is part of a larger and more significant relational, communal, societal, environmental and spiritual world" (Mugumbate & Chereni, 2019, p. v).

Our Place in the Resistance

We are African-born social work academics based at two separate Australian Universities. We see our role as African feminist-activist educators and researchers as one where we are obliged to critique the ongoing devaluation of knowledges that emerge from the Global South. As academics, we cannot continue to uphold an education system that offers a singular way of *knowing* and one that does little to recognise the epistemological diversity and richness embedded in different cultures and systems. We reflect on our experiences both as academics and as former social work students in Africa, Sweden and Australia where we encountered alienation from our African knowledges through uncritiqued consumption of curriculums that whitewashed *us*. In class, discussions where Africa women featured, a picture of poverty, illiteracy, lack of agency and disempowerment was displayed—often with a pitiful display of "*Oh poor women*." In those presentation, there was no room for us, as African women to be complex or multi-storied. Through the white, Eurocentric gaze, we were one thing only *oppressed*. At our most complex, we were oppressed, poor *and* illiterate. We have been brought up by our families and communities who have accumulated knowledge, wisdom and worked hard to support their families and communities through negotiating and bargaining with the patriarchy, so as to benefit from it, rather than be immobilised by it (Gatwiri & Mumbi, 2016; Tusasiirwe, 2020). While African women would be considered oppressed by western critiques, we see our mothers and grandmothers as having learnt a way to dismantle patriarchy from *within*.

To counter these deficit narratives, we present the missing voices that have been substantial to ourselves and our communities. We acknowledge how challenging it is for us to remember and re-validate experiences, knowledges, that we have previously been taught to devalue. Remembering is important for decolonisation especially when knowledges are shared and stored in orality. To remember, we spent many hours together, telling stories of our childhood, evoking with fondness, validating and affirming different knowledges that we had forgotten. This chapter, therefore, is about *returning home*, returning to ourselves and revisiting forgotten

African (feminist) knowledges. I (Kathomi) remembered and told stories of my father who used to teach me key life lessons through the invocations of African proverbs. As a parent, my father never made decisions *for* me; neither did he tell *why* anything I did or said was wrong. Instead, he would invoke a proverb and then invite me to think about the meaning and the implied lessons. Sometimes the *thinking* would take days, sometimes months. Other lessons are still unfolding now in my adulthood. Employing that parenting technique allowed me the freedom to think and come to my own conclusions about life and about the choices I was making. To me, this is the essence of feminism. My parents were custodians of stories and lessons hidden in proverbs, songs, tongue-twisters and metaphors—all of which were theoretical frameworks used to parent, guide, discipline and interpret personal and political experiences.

Colonialism, Indigenous Education and African Women

Colonial education alienates Africans from their indigenous knowledges, beliefs and philosophies of African people, thus ontologically separating them from who they are. The global and imperial oppression of African women, for example, is inextricably linked with the devaluation of the knowledges that they have historically been custodians of. As Okech (2020, p. 314) argues, "a key aspect of decolonial thought involves interrogating the coloniality of knowledge, which includes epistemological questions, the politics of knowledge generation, as well as questioning who generates which knowledge, for what purpose and from where." Colonial ideologies of domination which sit outside the principles of *Obuntu/Ubuntu* were advanced through formal education. As a profession, western social work was introduced in Africa by colonialists (missionaries) and to date remains dominated by Eurocentric knowledges, theories and models of practice (Tascon & Ife, 2020; Gatwiri, 2020; Tusasiirwe, 2020; Twikirize & Spitzer, 2019). Colonial systems of education determined that indigenous, local knowledge systems were inferior and needed to be repealed to meet the "acceptable" western standard (Chilisa & Ntseane, 2010). Historically, indigenous knowledges were passed on orally across generations through proverbs, wise sayings, storytelling, observations, rituals, tongue-twisters, plant biology and songs (Avoseh, 2012; Chilisa, 2012; Wa Thiong'o, 1986). Through orality and storytelling, the power of experience as a teaching tool was extolled (Mucina, 2011). Stories were told not only to entertain but also to teach about shared values, morals and other cultural ways of being (Mucina, 2011; Wa Thiong'o, 1986).

Historically, cultural knowledge and stories were developed across time through observation of how nature interacts with people, through experience and through familiarity with patterns of change (Njoki et al., 2015; Ogungbure, 2013). African stories were about relationships between humans, animals and the environment. This interconnectedness between everything that sustains humanity in a way that is dignified is what we defined earlier as *Obuntu/Ubuntu* (Gatwiri, 2020). Local and cultural knowledges are collectively and communally shared and valued, so they are not monopolised by a few expert individuals. While there are community elders who are custodians of *some* sacred stories or specific knowledge for and about the community, they had the responsibility to pass it on to younger generation at the appropriate time (Chinyowa, 2001; Wane, 2008). These stories were only shared at specific times to specific people for specific reasons. In other words, indigenous knowledges are not only of the past but rather are present through stories of experience and cultural observations across time (Wane, 2008). As Okech (2020, pp. 314–315) states, "indigenous forms of knowledge are not only a historical task, but a contemporary creative project to grasp how [resistances] emerge in response to the demands of the moment."

These systems of knowing were later swept aside, undermined and denigrated as superstitious, unscientific, unsophisticated and backward. The invalidation and sidelining of local, cultural African knowledges were all done under the guise of "advancing Africa and Africans" in the interest of colonial empire (Ogungbure, 2013). The intent to destroy indigenous African knowledge systems was also pursued through the lack of incorporation of such knowledges into formal western education imposed on most colonies (Njoki et al., 2015). Both Kenya and Uganda (where we come from) were colonised by the British who introduced "formal" education as part of the strategy to achieve colonial interests through material exploitation, cultural domination and European self-aggrandisement. The education system was established to train locals to serve the colonial empire's interests. With the introduction of this education system, English language was imposed as the "official" language while indigenous languages were banned in colonial missionary schools. When Kenya and Uganda countries obtained "independence" from colonial rule in 1963 and 1962 respectively, the local elites inherited the colonial systems and sought to govern and educate in the same way as the colonisers (Bulhan, 2015) rather than dismantle Eurocentric systems that exploited local people and denigrated African ways of being and knowing.

After decades of colonial indoctrination, those in power currently have established that what our communities "need" is a western education system to "save us." We refer to this colonisation of the mind as "the pedagogy of the colonised" where the colonised elites, who have been educated in colonial education systems, have internalised and adopted the coloniser's doctrine and way of thinking as superior. Unlearning this ideology is most difficult as it requires one to unlearn what has been presented as fact. However, despite the challenges, we also celebrate the fact that indigenous knowledges and education have survived the "oppression, hegemony and arrogance on the part of the former colonial powers, contemporary Eurocentric scholars, and some African elites" (Twikirize & Spitzer, 2019, p. 3).

Across Africa, the demand to rethink the nature of education is now gaining momentum. African women in particular remain some of the guardians of traditional knowledge and are leading the resistance of colonial education through their traditional teachings. Despite the perverse power of colonisation, women's roles as indigenous teachers have never ceased (Wane, 2008). Later in the chapter we present a case study of African women in a rural community in South Western Uganda who have remained at the centre of indigenous education but whose knowledges remain erased in the education system that rewards only western knowledge holders.

Afrocentric Feminism and Social Work

Afrocentric feminism is any form of African feminism that centres the experiences of Africans in its feminist critique and discourse. It recognises the African woman as *fully human*, and not necessarily only a *female* human being. This is similar to Kolawole's definition of African womanhood and womanism which is "the totality of feminine self-expression, self-retrieval, and self-assertion in positive cultural ways" (1997, p. 24). Afrocentric feminism prioritises the experiences of African women, their scholarship and "vibrant feminist knowledge" which have been under-theorised in current literary discourse (Okech, 2020, p. 314). Despite this, there is an apparent detachment from the use of the term feminism and the alignment to the feminist movement is discouraged. This is despite the fact that African discourses and literature espouse feminist values (Nkealah, 2006). In *Third World women and the politics of feminism*, Mohanty (1991, p. 7) argued that women in the Global South "have always engaged with feminism, even if the label has been rejected in a number of instances." In reflecting on the multiple meanings of feminism, Collen Dryden wrote that many African women have always been feminist and feminism for African women is not a "response of African women to their western feminist sisters, as the addition of the word 'African' before feminism implies … [rather] a struggle to free ourselves against male oppression and to actively assist women who find themselves in oppressive situations- in everyday lives" (Dryden et al., 2002, p. 117).

The connections between gender, patriarchy imperialism and race are important for African feminists too. In the 1980s and 1990s Gqola and Arnina Mama offered important critiques of African feminism. This included the obvious patterns of the "Africa/Diaspora divide, differences in naming ourselves as 'womanists', 'black feminists', 'African feminists' or 'postcolonial feminists', as well as varying forms of engagement with western feminism" (Lewis, 2001, p. 4). This means that the diversity therein, including the "essentialist evocations of geographical, national or racial criteria as decisive grounds for defining African feminism are especially untenable in our current context of intensified globalisation" (Lewis, 2001, p. 4). Caution is necessary not to homogenise Africa or Africans as there are diverse pluralities of women experiences everywhere. These pluralities are articulated in feminist scholarship across the continent. As Arndt (2002) argues, "the diversity of social realities on the African continent has had a lasting effect on conceptions of feminism, making it necessary to use the plural with respect to feminism in Africa as well" (p. 32). Undoubtedly, the location and the geography of the African continent is important—but "not in the sense that it gives us a common identity, but in the sense that it gave us a common starting point" (Dryden et al., 2002, p. 117). Our proposition of Afrocentric feminism recognises this complexity as it adopts the heterogeneity of multiple African feminisms, while still offering a sharp critique of the "problems" of gender in Africa and for Africans.

Afrocentric feminism is a tool with which to decolonise social work within an African context. An Afrocentric perspective locates the body, mind, spirit and the community as being interconnected. That is, the individual is reflected through the *wholeness* of community and vice versa.

While the western perspectives tend to put emphasis on the wellbeing of the "individual, autonomy and rights while traditional African societies tended to put more emphasis on the family and community welfare" (Ndungi Wa Mungai, 2015, p. 67). Afrocentric feminism reconnects social work with the importance of *holistic thinking*, focusing on the well-being of communities that African women live within.

Afrocentric Feminism in Context: Reflections from African Women in Uganda

Obuntu-Led Feminism

The second author (Sharlotte Tusasiirwe) in her recently completed PhD (2016–2019) had conversations with ten older women living and practicing *Obuntu-led* social work in a community in *Bwambara, Rukungiri* District, Uganda. Using oral storytelling methods to engage in conversations, the older women told stories of their lives, making sense of their lived experiences from their perspective. Story telling is an indigenous relational approach embedded within Obuntu framework (Tusasiirwe, 2019). The stories told by the women were of hardship and struggle in their care for themselves, families and communities. They reflected on the struggles to access basic health care in neoliberal dual systems where quality care was for those who could afford it. The women did not have access to clean water given their location in the rural geography and they constantly navigated the impacts of natural disasters, in particular drought, which made it hard for them to grow food. However, in these struggles, the women did not see themselves as passive victims but active agents who mobilised and supported each other through the grassroot mutual helping initiatives rooted in *Obuntu* philosophies, values and ways of being and doing.

The women established mutual helping groups which valued principles of working in solidarity, interdependence, sharing and reciprocity. These self-organised groups transcended kinships and were founded on strong and long-existing communal friendships and relationships. The mutual aid groups addressed different need areas of the community such as burial aid and also engaged in developmental activities including income-generating activities. As Okech (2020, p. 314) states, "communities and individuals-in-communities are complementary conceptualizations" of Afrocentric feminist realities. These grassroot women groups represent the Obuntu principle of caring about and for each other's wellbeing through mutual support and provided important pathways for collective survival. The survival of the collective community was seen as being just as important as individual survival. There was a shared understanding that individual wellbeing is intrinsically connected to communal wellbeing too. Indeed, this process of humanisation through collectivisation demonstrates how people are made human (*abantu)* as opposed to a thing (*ekintu)* in situations where relationships are grounded in the recognition of the humanity of others. The women put into practice a common African proverb that argues *Ageteriine nigo gaata igufa*, meaning *the teeth that work together can break a bone*. Through consistent concerted efforts and the mutual helping process, the women *shared power together* to support each other through significant hardships (Tusasiirwe, 2019, 2020). In centring these women as feminist agents in the community, their embodied transformative Afrocentric knowledge is validated.

Demonstrating the agency of the women in this study, one participant reflected,

> *Right now, we are constructing a building with four rooms and we would like to complete them and rent them out to get a communal income. In one room we are going to put in our group things like saucepans, and other things. If you do not collect money together the group will not grow, it will not develop. Because we even bought these big saucepans so that sometimes we can hire them to generate some money. That is the group we are in.*

Additionally, the women organised rotational farming groups where they worked in solidarity to help each other cultivate large gardens, weed and harvest their crops, hence ensuring that their families and communities had food across the year. These efforts are fundamental to the sustenance of their families and communities but not necessarily acknowledged by formal authorities. As Wane (2003, p 9) states "rural women in post-colonial African countries continue to be invisible due to the patriarchal landscape that dominates the social, economic, and political spheres. While these women labour over the planting, production,and harvesting of agricultural produce, their endless tasks and roles remain unrecognized, not only by the local governments, but by external agencies." Despite the challenges, The women also had a savings and credit group where they saved money in a pool fund. They used the money depending on what they defined as the priority needs for the members.

> *This group has also helped us so much. We meet and save money, but we save little; every month we save 1000 UGX [0.4 AUD]. So, after a certain time we share that money and sometimes some people may suggest that we buy things that help us in our homes. Now, in the first year, we forewent the 80,000 UGX [32 AUD] that we were supposed to get. Instead, we bought mattresses and each member got a mattress. Then the second year, we bought chairs, then the third one, we shared the money, and even now [fourth year] we shared the money. The group has benefited me because like those chairs I would not have been able to buy them (by myself).* (Older woman, Uganda)

Keeping the community cohesive in this manner reaffirms the *Obuntu* value of existing interdependently where a person is dignified through others. The philosophy of I *am because we are*, is embodied in the Kenyan philosophy of *Harambee*, which means "all pull together" and is consistent with feminist values collectivity and solidarity.

Older African Women as Custodians of Afrocentric Feminist Values and Knowledges

The older women interviewed in this project were also mothers and grandmothers. These are roles and statuses that are celebrated within the context of Afrocentric feminism. The women spoke to the importance of motherhood and the significant role it played for their identity, status and power in the community. African feminisms counter western forms of feminism that locate motherhood as disempowering or oppressive. As Chilisa & Ntseane (2010, p. 618) argue, some forms of western feminisms have denied "African women's power within the indigenous relational worlds that celebrates motherhood." In African households, understandings of how women's relational gender roles are seen "sites of resistance and sources of empowerment."

Power within parenthood and grand-motherhood is manifested through how the women used these roles to pass on their cultural knowledge and wisdom to their children. Their roles as mothers and grandmothers afforded an opportunity to reclaim that which was devalued by the formal systems of colonial education. Oral knowledges passed on through proverbs, and stories of courage and resilience, are used to instil cultural values of good discipline and collectivism. In these spaces, the women play pivotal roles in their homes and society as educators, carers, providers and custodians of cultural and social values. In addition to passing on *Obuntu* ways of doing and knowing, the women pass on livelihood and traditional knowledge in intercropping, rearing animals and organic agriculture which sustains their families and communities.

One of the women, Prisca, manifested her feminist agency by describing the traditional knowledge she holds in agriculture. She stated,

> *I did not go to school. I grew up using my inherited knowledge only, and knowledge from parents, mainly about digging/cultivation and it is what has sustained us; growing food like beans, cassava, sweet potatoes, most of the foods. If it was not for the bad sunshine which hits this area, there isn't anything that I do not grow. I divide the plots; on one plot, I put beans, on the other millet, and all the other foods, I make sure I grow all the foods, and even here* [pointing to one of the plots of land] *I had planted some coffee, but it is now not doing well, but all in all what has kept us moving is growing food.*

Prisca, like all the other nine older women, knew what crops to grow in which season, in which soils and how to rear animals. Many other women as well demonstrated agricultural know-how through the diversity of activities that have been their source of livelihood for themselves and their families. Using this knowledge, they have been able to lead a dignified life, sustained by organic and sustainable farming. Although these women did not complete formal education, we consider them literate. As Wane (2003, p. 6) suggests "women's knowledge of food processing is intertwined with their knowledge of the environment. By holding a maize cob or biting the grain, they know whether the moisture content is high or low. Standing in their doorways, they are able to determine whether the weather is good for winnowing, threshing, and/or drying food." We argue that the knowledges and accumulated wisdom embedded in the livelihood of these women carry inherent value and literacy that is devalued in formal education.

A colonial system that chooses to place more value on the ability to write or read to over the women's ability to *speak* is another form of erasing orality as a form of literacy. Even though the women interviewed are teachers of the community's children as well as custodians of culture, the formal systems only reward the *literate* teachers employed in formal schools. Retired teachers in formal education settings are paid a pension while the older women who are indigenous teachers age without any pension. That is, despite the fact that they do the work of teaching and passing on African ways of being that formal or colonial schools have continued to ignore and erase. As several African scholars have argued, it is this indigenous knowledge, orature, that has become a powerful source of resistance to ongoing colonisation (Chilisa & Ntseane, 2010; Chilisa, 2012; Wane, 2008). In line with the African proverb: *what elders can see sitting down, the young cannot see while standing on top of a tree*, the older women who were (are) relegated to margins as illiterate by past and present colonisers, were (are) able to see beyond us, beyond the erasure and maintain cultural dignity. This futuristic thinking, sense of agency and perseverance from our knowledge custodians is now the force behind our resistance and feminist empowerment. As Chilisa and Ntseane (2010) reiterate, "indigenous knowledge makes visible the spaces of agency ever so present in the life experiences of marginalised feminisms and yet so absent in the academic debate" (p. 621).

From our own experiences as academics, having these indigenous knowledges that we learnt from our varying custodians of cultural knowledge has been a strong force for us, a basis for resistance, emancipation and decolonisation of self and the work we do. I (Sharlotte), despite having had minimal teaching in formal class about indigenous methodologies or African *Obuntu* ways, was able to return to this knowledge during my doctoral research to draw from the *hidden* indigenous knowledges, practices and beliefs held by my numerous mothers in the rural community I grew up in (Tusasiirwe, 2019). I (Kathomi) also attach great value to the feminist teachings of my grandmothers, mothers, aunties and employed my doctoral research as an avenue to return to my village roots to learn from these women's experiences.

Implications of Afrocentric Feminism for Social Work Practice and Education

Afrocentric feminism points us to the urgent need for professional decolonisation in social work practice and education in African but also in the western contexts we currently work in. While *doing* social work with African women and people, African knowledges, philosophies, ways of being and doing must take centre stage in informing, imagining and constructing community-led practices and education. By African philosophies and experiences taking centre stage, we do not mean that they are *squeezed in* or just *added on* in the existing white western colonial theorisation of social work. This adding-on would be what Gaudry and Lorenz (2018) call indigenous inclusion especially in western settings where the status quo is maintained and indigenous peoples are adapted into the current and often alienating culture of the coloniser, with little, if any, emphasis on changing the very structures that have caused the exclusion of and hostility to indigenous people in the first place.

Instead we are advocating for decolonial indigenisation which goes beyond inclusion policies to an envisioning of radical transformation of the whole education system where dominant structures are reimagined and dismantled. We advocate for social work classrooms that actively "poke holes" in the systems' colonialism with the hope of unsettling them and finally dismantling them. A time has come where white, male worldviews currently held as the standard are being challenged by Black voices. Uprooting the rots of colonialism and returning humanity, connection, relationships and shared reciprocity to social work *in* Africa and *for* the African is what Gatwiri (2020) refers to Ubuntu consciousness.

Gatwiri (2020) has argued that *Ubuntu* consciousness which is imbedded in Afrocentric feminism

> enables us to view each other equally important in building the community and fostering social change. Fostering change is not about meeting the needs of the ego (unlike white saviour complex) but about communal growth and healing. This means that our humanity levels us equally regardless of our class, race, gender, sexual orientation, ethnicity, religion. Ubuntu seeks to affirm and honour the dignity of every human being and seeks to develop and maintain mutually affirming and respectful relationships that are grounded on deep respect of the other people's humanity.

African philosophies like *Obuntu/Ubuntu* which were never forgotten, just erased, play an important role in reaffirming the knowledge of the African as valid and important. Reimagining social work through a framework where different knowledges can co-exist alongside each other is necessary if effective decolonisation of social work is to be realised.

Conclusion

This chapter has highlighted the importance of putting African women's knowledges, experiences, agency and collective resistance at the centre of decolonising feminism and social work practice and education in Africa. In decolonised social work, the focus of social workers goes beyond viewing African women as passive victims to also recognising how they use African philosophies to support each other in addressing their struggles. Decolonised feminist social work in Africa must embrace *not* what is different but what *is* equal. This means promoting African ways of being, knowing and doing to prioritise interconnectedness/collectivity rather than individualism, interdependence rather than independence, community rather than self-aggrandisement, value for humanity and affirming humanness of others rather than domination or subjugation of others, value for relationships/community/people over material things/wealth.

References

Arndt, S. (2002, January 1). Perspectives on African feminism: defining and classifying African-feminist literatures. *Agenda*. https://doi.org/10.1080/10130950.2002.9676176.

Avoseh, M. B. M. (2012). Proverbs as theoretical frameworks for lifelong learning in indigenous African education. *Adult Education Quarterly, 63*(3), 236–250.

Bulhan, A. H. (2015). Stages of colonialism in Africa: From occupation of land to occupation of being. *Journal of Social and Political Psychology, 3*(1), 239–256.

Chilisa, B. (2012). *Indigenous research methodologies*. SAGE.

Chilisa, B., & Ntseane, G. (2010). Resisting dominant discourses: Implications of indigenous, African feminist theory and methods for gender and education research. *Gender and Education, 22*(6), 617–632.

Chinyowa, K. (2001). The Sarungano and Shona storytelling: An African theatrical paradigm. *Studies in Theatre and Performance, 21*(1), 18–30. https://doi.org/10.1386/stap.21.1.18

Dryden, C., Erlank, N., Haffejee, Y., Hardy, K., Nhlapo, S., Tonkin, S., & Tshamano, H. (2002, January 1). The many voices of feminism. *Agenda, 17*(54), 113–121.

Fanon, F. (1961). *The wretched of the earth* (R. Philcox, Trans.). New York: Grove Press.

Gatwiri, G. (2014). Silence as Power: Women Bargaining with patriarchy in Kenya. *Social Alternatives, 35*(1), 13–18.

Gatwiri, K. (2020). Afrocentric ways of doing social work. In S. Tascon & J. Ife (Eds.), *Disrupting whiteness in social work* (1st ed., pp. 44–60). Routledge.

Gaudry, A., & Lorenz, D. E. (2018). Indigenization as inclusion, reconciliation, and decolonization: Navigating the different visions for indigenizing the Canadian academy. *AlterNative: An International Journal of Indigenous Peoples, 14*(3), 218–227.

Kolawole, M. (1997). *Womanism and African consciousness.* Eritrea: Africa World Press.

Lewis, D. (2001, January 1). African feminisms. *Agenda, 16*(50), 4–10. https://doi.org/10.1080/10130950.2001.9675989

Mohanty, C. (1991). Cartographies of struggle: Third World women and the politics of feminism. In C. Mohanty, A. Russo, and L. Torres, (Eds.), Third World women and the politics of feminism, Bloomington and Indianapolis: Indiana University Press.

Mucina, D. D. (2011). Story as research methodology. *AlterNative: An International Journal of Indigenous Peoples, 7*(1), 1–14.

Mugumbate, J., & Chereni, A. (2019). Now, the theory of Ubuntu has its space in social work. *African Journal of Social Work, 10*(1), v–xvii.

Ndungi Wa Mungai. (2015). Afrocentric social work: Implications for practice issues. In V. Pulla & B. Mamidi (Eds.), *Some aspects of community empowerment* (1st ed., pp. 65–76). Allied Publishers Private Limited.

Njoki, A. M., Kinyua, P. L., & Muli, L. N. (2015). The practice of African indigenous education and its relevance to theory and practice of modern education in Africa. *International Journal of Innovative Research and Studies, 4*(12), 132–149.

Nkealah, N. N. (2006). Conceptualizing feminism(s) in Africa: The challenges facing African women writers and critics. *English Academy Review, 23*(1), 133–141. https://doi.org/10.1080/10131750608540431

Ogungbure, A. A. (2013). African indigenous knowledge: Scientific or unscientific? *Inkanyiso: Journal of Humanities and Social Sciences, 5*(1), 12–20.

Okech, A. (2020). African feminist epistemic communities and decoloniality. *Critical African Studies, 12*(3), 313–329. https://doi.org/10.1080/21681392.2020.1810086

Sendi, K. (2017). The limitations of U.S. white middle-class feminism in the Middle East. *International Journal of Gender and Women's Studies.* https://doi.org/10.15640/ijgws.v5n1a11

Tascon, S., & Gatwiri, K. (2020). Towards Cultural Humility: Theorising Cultural Competence as Institutionalised Whiteness Social Work & Policy Studies: Social Justice. *Practice and Theory, 3*(1), 1–16.

Tascon, S., & Ife, J. (2020). *Disrupting whiteness in social work.* Routledge.

Tusasiirwe, S. (2019). *Stories from the margins to the Centre: Decolonising social work based on experiences of older women and social workers in Uganda.* Unpublished Doctoral thesis, Western Sydney University.

Tusasiirwe, S. (2020). Decolonising social work through learning from experiences of older women and social policy makers in Uganda. In S. Tascon & J. Ife (Eds.), *Disrupting whiteness in social work*. Routledge.

Twikirize, M. J., & Spitzer, H. (2019). Indigenous and innovative social work practice: Evidence from East Africa. In M. J. Twikirize & H. Spitzer (Eds.), *Social work practice in Africa: Indigenous and innovative approaches*. Fountain publishers.

Wa Thiong'o, N. (1986). Decolonising the mind: The politics of language in African literature. *James Currey*. https://search.library.wisc.edu/catalog/999573905602121

Wane, N. (2003). Embu women: Food production and traditional knowledge. *Resources for Feminist Research, 30*(1), 137–148.

Wane, N. N. (2008). Mapping the field of indigenous knowledges in anti-colonial discourse: A transformative journey in education. *Race Ethnicity and Education, 11*(2), 183–197. https://doi.org/10.1080/13613320600807667

CHAPTER 8

Tears of Shame: Sri Lankan Mothers Negotiating Experiences of Caregiving and Disability

Fiona Kumari Campbell

Introduction

This chapter presents original research about configurations of shame (*lajja-baya*) as it is understood within a Sri Lankan Buddhist framework and its impact on mothers and partners where a family member experiences disability. Understandings of care ethics in Western societies are formulated through a prism of possessive individualism and a demarcation of the private and public spheres. In order to engage with anti-racist practice it is critical that practitioners and researchers embark on the journey of comparative travel, that is, to "shift our frames altogether so we see things

Dedicated to Mala (Malathi de Alwis, 1953–2021), a fervent anthropologist and feminist who assisted with the clarification of my ideas.

F. K. Campbell (✉)
University of Dundee, Dundee, UK
e-mail: f.k.campbell@dundee.ac.uk

C. Cocker, T. Hafford-Letchfield (eds.), *Rethinking Feminist Theories for Social Work Practice*,
https://doi.org/10.1007/978-3-030-94241-0_8

differently from another angle" (Butnor & McWeeny, 2014, p. 11). It behoves us engaged in cross-cultural feminist research to be cognisant of "the micropolitics of context, subjectivity, and struggle as well as the macropolitics of global economic and political systems and processes," including our own scholastic training, which not only effects research questions and design, but also what is seen and not noticed (Mohanty, 2006, p. 223).

Understandably, feminist scholarship has also fallen foul of the preoccupations emanating from the 'civil war' (1983–2009), focusing primarily on sex-based violence towards women and girls (Jayawardena & Pinto-Jayawardena, 2017; Wickramasinghe, 2003), and to a lesser extent on feminist theory (Jaysinghe, 2004; Kiribamune & Samarasinghe, 1990), on to the neglect of issues such as those concerning disability and the intersections of sex, caste and poverty; the exception being the work of Nelfoufer de Mel (2006, 2007). First, I critically outline the delimitations of the study. Second, I engage with the notion of mothering and child learning. I then turn to a discussion of family-kin relations (parvula), and finally move to an exploration of fear-shame through the concept of *lajja-baya* in women's and girls' lives and how this might impact on caregiving roles.

Delimitations

There are many elements that shape storying about Sri Lankan mothers, but not all these elements are able to be addressed in this chapter, as there is simply not enough space to do so adequately. Although Sri Lanka is a multi-ethnic and multi-religious society, my focus is on Buddhist communities and Buddhist beliefs. Even though I would have liked to explore the experiences of disabled mothers or disabled parenting, I do not. The hegemonic assumption, in the minuscule amount of literature, assumes that all parents are inherently abled, which should not come as a surprise given the dominance of ableist norms within the research enterprise (Campbell, 2019, 2014). This is as much a problem of imagination as an indication of the hierarchy around which cohorts in society are in need of study. A separate work is needed to explore the experiences of disabled mothers (who may or may not have disabled children).

Of course, family studies tend to focus on children, almost forgetting that these children eventually become adults and partners. Again, I have had to limit my focal concerns but have undertaken, for comparative purposes, some discussion of partners of disabled people. Finally, this chapter is a combination of desk research related to mothering, and more significantly, it is an analysis of the *conceptual building blocks* that may impact on thinking about mothers of disabled children and their experiences of living in a Buddhist family-kin polity. This chapter acts as a precursor and grounding for future empirical work of the lived experiences of mother with a disabled child or partner, and what those experiences suggest about the religious-cultural beliefs about disability and non-normative people.

Mothering as Pivotal

Despite the overwhelming patriarchal orientation of South Asian societies, the domestic economy and family-kin relations are central to the transmission of culture, nation building and community relations. Whilst Sri Lankan women have a subordinated status in society, with only 34% of women in the workforce, the lowest in the region (International Labour Organization, 2016, p. 68), mothering and motherhood play a central and pivotal role in South Asian communities, especially in relation to the socialisation of children and the shaping of communal citizenship As de Alwis (1997, p. 186) argues, beyond axiomatic biological reproduction, as mothers, "women's signification [is] as moral guardian, care givers and nurturers." Because of the *centrality of mothering* as a cultural practice in Sri Lanka, the act of mothering can be a source of empowerment. Derived from Buddhist scripture, mothering is a spontaneous act of *selfless care* for another. In this sense the sanctity of mothering and family life means that women's identity as mothers becomes intertwined with the needs of the family. *Obligation and duty* and a mother's subsequent role involves providing selfless services to the family, and hence, suppressing her own needs (Sangha, 2014).

Mothering is a form of social personhood and should a mother oscillate away from her pre-determined role, this action not only invokes a questioning of her own identity, but also impacts on her *standing* within her wider kinship network. As bearers of culture, Sangha argues that this ideal of being a good mother can be experienced as *overwhelming* (Sangha, 2014, p. 217). Keeping up appearances of being a 'good mother' and not appearing to challenge dominant interpretations of mothering and marriage, the work of Abeyasekara (2017) on failed or childless marriages documents the narration of life histories of women that engage a culturally coherent, non-threatening repertoire, subsumed within the tropes of 'long suffering wife' and 'devoted mother'. When exploring atypical mothering, conventional notions of resistance and agency drawn from Western research require a reconceptualisation. Abeyasekara (2017, p. 429) warns of interpreting narratives about resistance at face value, because it ignores how women are

> enmeshing themselves in a set of new power relations. [There is] the tendency to reproduce dichotomous narratives, [however, these are] entrenched in everyday conversations about 'being modern' … Assertions about agency in marriage do not necessarily mean that people are more willing to act out their inner desires than before; rather it signals a change in the narratives devises used in managing self-presentation.

Selflessly giving involves offers of love to a person who receives it *without request*. This in turn can generate an *empathetic imagination* and interrelational expansion of love. Agency is full of complexities and ambiguities, rather than being associated with a 'loud' defiance or resistance (Sangha, 2014). The agency of mothers is impacted by the way cultural practices shape or constrain, and has much to do with the intersection of power and subordination in a communal context. Abeyasekara's (2017) insights into how divorced and childless women narrate their life stories and recuperate what may initially appear to be a form of 'failed mothering' is not through a *critique* of patriarchal mothering; rather, women transform the *idea of self-sacrifice* in order to (re)situate themselves as 'good mothers'. This ensures any acts of defiance and self-assertion are culturally intelligible and would be useful for thinking about the narrations of mothers with disabled offspring.

Children: Learning Hierarchy and Deference

Most complex and large-scale societies, particularly those with high population levels, are organised around certain hierarchies. In the West there is a great deal of suspicion around the notion of hierarchy, to which it is assumed that hierarchies have connotations with abuses of power and asymmetry. Is it possible or desirable to do away with social hierarchies or reconfigure 'just' hierarchies? Hierarchy is a relationship characterised by difference and ranking according to some attribute, and sometimes these hierarchical roles are interchangeable (Dumont, 1980). Hence, social hierarchies have a normative dimension. This, in turn, extends to hierarchies in business and public life, where reference to more senior members has primacy.

At the level of family and kin networks, age-based hierarchies, called in Buddhism *filial piety*, between adults and children involve the *value of reciprocity*—parents cared for us when we were children and, therefore, adult children have an obligation to care for their elderly parents. We need to excise the Western idea of atomistic individualism. In age-based hierarchies there is the notion of a 'shared body', for example, in eating practices, ideas of privacy and shared spaces (bedding). This, in turn, becomes an engagement with working models of the world, of others, of relationships, with ourselves and is played out in the responses of stigma and organisation of social care. Indeed, Wickremesoorya argues, "children observe their mothers and become partners, they learn the art of self-advocacy." Maturity is judged by increasing sensitivity and deference to the preferences of parents and other senior members of the community, by way of self-control rather than increased self-assertion and independence. This does not mean that individuals lack autonomy; rather, to adhere to *parvula* (kin-family) relations, one requires an autonomy of a different kind that *attunes* to the wishes of others, than a discernment about those wishes in a deferent way (Chapin, 2010, 2014). Through practices of recognition and subtle shaming (*lajja-bhaya*), children come to recognise that while they may get what they desire, in doing so, "they may destroy affection from and inclusion with those they care about" (Chapin, 2010, p. 357)—and hence, they are on the road to cultivating practices of *lajja-bhaya*, that is, restraint, self-discipline and deference to hierarchical and inter-communal relations.

Parvula: *Kinship Relations*

One of the core sites for the socialisation and performance of deference occurs within the private sphere (the householder life), namely, fictive kinship groups (*parvula,* meaning 'kindred') where survival strategies are played out. Unlike in the West, where there is a decline in family norm groups by virtue of the idealisation of rampant individualism, the *parvula's* role in modern times has become *even more central* with kin-family groups acting as gatekeepers of social mores, bonded together by *lajja-bhaya*. Even in urban middle-class Colombo, where a Western-orientated nuclear family manifests, there still exists a network of interpenetrating *parvula* relations which impact on family life.

The parvula incorporates family friends as part of the family social network. Said (2014, p. 21) explains: "a 'dominant family network' is the nuclear family unit (the *gē*) that has the most far-reaching social network, encompassing both kin and fictive kin, and therefore the widest arena of support, influence and access to material resources." The *parvula's* propinquity is more exacting than the responsibility of familial relatives and is suggestive of an 'ambivalence of kin' due to an acquiesce towards norm conformity in dealing with family problems (*gedara prashna*) or physical illness *(asanipaya*) through instilling obligations, such as assisting an older sister's children. *Parvula* relations are complicated by the two forms of marriage patterns within Singhalese society, namely, *diga* and *binna*. *Diga* is patrilocal, whilst in *binna*, the husband takes up residence with his wife's family. Inter-sibling etiquette is intricately scaffolded according to degrees of seniority; hence, de Silva argues that elder brothers become the guardians of the 'honour' of the family unit, which could be at stake when ridiculed by outsiders (de Silva, 1998, p. 36). Mothers then are surveilled by not just husbands, but by older brothers as well as senior members in kin-family networks.

Any derogation from that obligation results in being severed from the support of the extended family network, a network that acts as an informal social welfare system. Of interest in terms of disability are the practices of negotiation, which are characterised by an ethos of '*sharing' without 'reckoning*'. This lack of reciprocity may produce unequal power relations, suggesting the depth of a moral impetus to relations which may strain family networks due to experience with disability. Indeed, the study by Landry et al. (2015, p. 154), examining mothers of disabled children, indicates that the major source of additional emotional pressure is from the influence of the husband's family, wherein the husband's family views the existence of a disabled child as a source of negative stigma for the whole kin-family network, "to the extent that some wanted to hide the child within their home and community."

A '*sharing without reckoning*' ethos and its durability is governed and maintained through discourses of shame and deference (de Silva, 2005). Despite pressures on the *parvula*, this form of family structure has grown in *strength* in Sri Lanka; hence, an examination of the techniques of containment and resistance are critical. de Silva does suggest some leakiness in the *parvula's* disciplinary practices in her conclusion that "while the fear of public ridicule also remained intense, the *parvula* itself seemed increasingly unable to gauge at which point its own members have contravened the norms of appropriate conduct" (de Silva, 2005, p. 149). What does an ethic of sharing without restraint mean for the ongoing support needs of disabled people and their capacity and agency to contribute to *parvula* life? And what is the basis of *prioritisation* of sharing in the *parvula* and how do these prioritisations induce strain in the lives of families with disability, especially adolescents and adult children?

Lajja can bring shame to the family, thereby lowering prestige. With a positive connotation, having *lajja* ensures that a person, especially a disabled person, will conform to and be included in social customs. A belief in *doing the right thing* by the *parvula* is buttressed by the Buddhist ethic of *filial piety*, that is, honouring one's parents. This ethic is inculcated in children from an early age, and failure to imitate these norms often involves social derision. Such piety becomes a powerful invective, making individuals "less aggrandising and concerned not only about their own welfare, because it realises that the well-being of others is equally important for establishing peace and harmony among co-inhabitants of society" (Sinha & Chauhan, 2013, p. 135). Hierarchy and stratification produce practices of never naming people older to one's self, but only deployed kin-relation names.

de Silva (2005, p. 106) uses an example of children being indulged by their parents as contributing to social status, thus establishing the basis for future *exchange relations*, a moral contractual pact, if you like, of interpenetrative hospitality and obligation. This is more than an ethic of *saving face*. The organisational structure of the home economy of the *parvula* does not directly control action, but "rather the *interpretation* [and permissibility] of the action. The primary interpretations of action, with continual adjustment for the basis for them, are the tenuous threads which link and reconcile action and ideas" (Southall, 2012, p. 115, emphasis added), thus reinforcing unquestioned hegemonic societal norms. In homes in Sri Lanka and the diaspora one hears 'bad' behaviour being admonished in the expression '*lajja nādda, mokada minissa kiyanne*', meaning, 'aren't you ashamed; what'll people say?' This signifying practice instils in the child a sense of deference from an early age and a heightened awareness of the reactions of others (de Silva, 2005; Obeyesekere, 1984). Resistance seems futile:

> [T]he failure to conform is associated with ridicule and laughter by the parent, especially the father… The father not only ridicules the child, but he does it before a 'public', as is inevitable with shame socialisation… Socialised

> adults as a consequence have an added 'fear' of authority figures; the figures, who like to 'pull up' subordinates in the presence of others. (Obeyesekere, 1984, p. 505)

What does this mean for advocacy around promoting change through intervening in the conditions that enact ableist norms? The key is to discern the interpretative drivers for action. The agents of socialisation then extend beyond the home to school and workplaces, wherein individuals introject the bearers of socialising practices and internalise their values. Shame can be a reaction to other people's criticism or the inference of revulsion through contamination.[1] Harm is produced by being openly ridiculed or rejected or anticipating situations wherein one has been made to feel ridiculous. While *lajja* is important to both women and men, it is women particularly who are expected to conform to the rules of modesty and right behaviour (Sinha & Chauhan, 2013).

As wife and mother, she is refracted as a 'Buddha of the household', bearing the remit for emulating and inculcating the *Eightfold Noble Path* to her children. Hence, de Alwis (1998) has translated and associated the concept of *lajja-baya* with 'respectability'; a respectability that insists on a government of oneself through practices of *restraint* to alleviate communal anxieties (Hewamanne, 2003). Nonetheless, in this dominion, households are normally headed by men unless there are exceptional circumstances such as the death or disappearance of her spouse. Disability in the family magnifies this compulsion of respectability for mothers or spouses of disabled people.

Defining Lajja-Bhaya

The literature indicates that the preoccupation of research to date has focused on *lajja,* almost exclusively as it relates to gender relations (Hewamanne, 2003; Wijewardena) in India and Sri Lanka, and suicide, Sri Lanka (de Alwis, 2012; Widger, 2012a, b). *Lajja-Baya* is a particularised cultural formation of ableist relations that impacts on disabled people, their families and those with variable bodies in South Asia. In contrast with many non-Western societies, in Sri Lanka, *social status* plays a heightened role as compared to honour in social relations, with the epitome of high status inhering in a hegemonic masculinity. Interestingly, persons of status do not perform menial tasks that are set aside for cronies, resulting in a tendency for those with status to have "a certain sedateness of bearing, [nor do they engage in] ... practices of risking the body" (de Silva, 2005, p. 27).

The inverse of status is *shaming* or *humiliation* (*lajja*), which endures as an undercurrent of social interactions and relations in the possibility and anxiety of being shamed *publically* (*lajja-bhaya*). Due to space requirements, this chapter has limited its focus to a Buddhist concept of *lajja*, even though the conceptualisation *lajja* is apparent in Hinduism. A rupture or biographical disruption by virtue of the event of disability (either at birth or later in life) impacts on the performance of kinship and *lajja* relations; hence, the personal experiences of the disabled person and their immediate family interfolds into this context; even when they 'do not have the words' (*mama vacana nā*) to communicate the depth and complexities of incomprehensibility of pain caused by *lajja-baya*. The subordination of persons deemed without status is articulated in performances of *deference* to those perceived to have power.

We can see that *lajja* can act as a productive emotional effect, having the potential of being transgressive. The cultural existence of *lajii-bhaya* in Sri Lankan society magnifies ableist, patriarchal and caste-based[2] relations, widening the gap between those understood to be 'abled' and those persons regarded as 'disabled' or 'deficient' by the community buttressed by sex-based discourses. Also, consideration needs to be given to the divide between disabled veterans deemed 'war heroes' and civilian disabled groups. Groups that might be considered as liable to act with deference include lower caste groups, sexual minorities, women and of course disabled people.

Lajja represents the self-management, self-respect and honour of a woman or disabled person who follows cultural values. In terms of identity in Sri Lankan society, whilst there are individual rights bearers, identity *is not* individualised in an atomistic sense. Identity is *intersubjective*, and is validated in the collectivity, the group's status in public and economic life. Male-dominated Sri Lankan culture suggests the most moral way to control one's power is through the emotion of *lajja*. To have a sense of *lajja* is to be conscious of one's duties and responsibilities, to persevere in the performance of social role obligations, to continue commitment to the maintenance of social harmony.

The UN *Convention of the Rights of People with Disabilities* is a catalyst for significant change in the role and participation of disabled people in society, inducing shifts in social relations and familial responses. As with changes to women's role in rural areas, shifts to different kinds of employment have produced gendered social and familial tensions. What are the implications of these changes to the social status and roles of people with disability within the dynamics of *parvula* life? Subtle changes to expectations can affect existing power dynamics and social hierarchies, leading to threats to abled-bodied (masculinist) dominance in the family and village structures. Changed roles can produce an 'ableist panic' and upset to normally peaceful existence:

> [G]irls [are] encouraged to achieve and make a positive contribution to the cultural and intellectual life of the community. But to chart out the kind of independent social identity which would allow them to do this, it [is] necessary to circumvent the processes of *lajja*; to avoid the controlling eye of the community. (de Silva, 2005, pp. 214–215)

Mothering Disabled Children

I could only locate three studies that specifically examine the relationship between mothering or parenting and disabled children in the Sri Lankan context. Champa Wijesinghe, Natasha Cunningham, Pushpa Fonseka, Chandanie Hewage, and Truls Østbye (2015) (hereafter Wijesinghe et al., 2015) is heavily biomedically focused on the idea of 'burden' (which is not critiqued) as understood by 375 'caregivers' (primarily mothers, but gender neutered) of children with cerebral palsy attending a hospital in Galle, southern Sri Lanka. Similarly, Michel Landry, Sudha Raman, Elise Harris, Layla Madison, Meera Parekh, Cecile Banks, Huda Bhatti and Champa Wijesinghe (2015) (hereafter Landry et al., 2015), medical and public health clinicians, also adopt a medical paradigm with a narrow focus on a rehabilitation service in Galle, Sri Lanka. Finally, there is a positivist study conducted by medical researchers Sudath Abeywickrama, Inoka, Jayasinghe, and Samanmali Sumanasena (2013) (hereafter, Abeywickrama et al., 2013) in Kandy, Central Province, about the experiences of mainstream education by disabled children, their families and teachers.

Absentee Fathers

All three studies (Abeywickrama et al., 2013; Landry et al., 2015; Wijesinghe et al., 2015) note the absence or minimal role that fathers as husbands play in decision-making and support of their disabled children. Wives as mothers are effectively left on their own in raising and supporting their disabled offspring. Abeywickrama et al. (2013, p. 88) identified 97% of mothers having principal caregiver roles, whereas fathers acted as a principal caregiver in only 1% of the children studied. As mothers are principal decision-makers regarding the disabled child, their understanding and perception of disability will have a critical influence on health and rehabilitation decisions (Landry et al., 2015, p. 151). In a study of Punjabi mothers with disabled children living in Canada, Bains (2013) also noted that fathers took a passive role in parenting and left it up to the mother to monitor the child's progress and deal with any difficulties. Interestingly, she observed that

> [t]he mothers did not show a critical response to the lack of father involvement; they had limited or no expectations of involvement by their partner ... Partner involvement was not an option and the mothers had *resigned themselves* to that fact. However, usually, mothers agreed that their own needs took a *secondary position* to all the other members of the family in the household. (Bains, 2013, p. 107, emphasis added)

Wijesinghe et al. (2015) support the conclusion that spousal support, or lack of it, impacts on how a mother responds, in terms of wellbeing, to stressors associated with caregiving. In response, Wijesinghe argues for increased communal or multidisciplinary approaches to caring for children, with the addition of routinised marital stress assessments at the time of conducting psychosocial assessments of a child and caregiver (Wijesinghe et al., 2015, p. 92). Landry et al.'s (2015) findings are more forthright. Whilst there was variation in the role that fathers played in the life of their disabled child, "a small minority of the participants, however, indicated their spouse was completely unsupportive of the child, and that this lack of support had an important and long lasting negative emotional effect on the mother" (Landry et al., 2015, p. 154). In this study we get a glimpse of 'protest' in the sense of disclosure to the researchers, for the researchers conclude that "lack of support of a spouse seemed to emerge as *insurmountable* for mothers who participated in this study" (Landry et al., 2015, p. 154, emphasis added). The implication of this critical finding, however, is left hanging.

Supports

Given that the (woman) caregiver actively sought support as a coping method, and the experiences of 'burden' were significantly reduced when the caregiver actively sought social support, it is surprising then that the research team did not examine gendered relations (Wijesinghe et al., 2015, p. 88). Mothers occupy the role of principal parent, and hence experience associated *lajja-bhaya*, but this critical dimension is not challenged by the researchers in terms of specific programmatic outcomes or sex-specific strategies required. Instead, the focus of recommendations is on child health delivery, with caregiving as an additive. We do not know about any enhancing strategies deployed by mothers, aside from engagement in religious or spiritual practices, within family-kin networks, or whether having a male child made any difference to the support strategies pursued. Because the study was based on caregivers who were recruited from hospitals, we do not know what inhibiting factors are associated with caregiving for mothers, not accessing hospital support services.

Landry et al.'s study (2015), despite exploring factors that influence the perception of disability among mothers of disabled children attending a community rehabilitation facility, does not engage with or test a gendered or sex-class epistemology. Indeed, there is limited benchmarking in which to analyse the thematic data. The interview questions focus on three areas of family profile: knowledge of disability changes in role since having a disabled child, religious impact and explorations as to what disability means to the mother as well as community views. There were no explicit questions about prejudice, social exclusion or combatting stigma. Only 13 women were interviewed, with the majority of their children being male. The study focused specifically on the experience and perceptions of mothers who had ***already decided to seek*** rehabilitative care for their children. Again, we see an absence of perspectives regarding perceptions of disability by mothers who did not avail themselves of rehabilitation services.

Three major themes emerged regarding factors which since seemed to influence perceptions: (1) level of family and community support, (2) spiritual and cultural interpretations of disability, and (3) outcomes of rehabilitation services (Landry et al., 2015, p. 154). What is clear from this study is that mothers who had emotional support and logistic help with the care of their disabled child from their immediate and extended family seemed to perceive disability more positively (Landry et al., 2015, p. 154). Buried in the study are indications as to the role of *lajja-bhaya* in community relationships. Although this area is underexplored, it does provide us with some insights. In general, mothers were critical of the lack of local community support of their children. However, their use of community support services was able to positively influence and transform the mother's perception of disability and then, in turn, change community perceptions:

> And then, there was a friend of mine … [he] also encouraged me to take the child out, saying that, if the child is exposed to this new experience 10 times, [the child] will learn at least one thing. So, that helped me a lot to bring the child out. Also, the child is now getting on very well with the neighbours, everybody loves [the child] and everybody is very supportive. (Respondent: Landry et al., 2015, p. 154)

Yet the tension for mothers and the disabled child endures, as some community members openly and consistently stated that disabled children should not be encouraged to go to school, play sports, get married, have a family or even use public transportation (Landry et al., 2015, p. 154). This attitude provides a nice segway into the place of education as a strategy to nullify the effects of *lajja-bhaya*, reported in the study by Abeywickrama et al. (2013).

Abeywickrama et al. (2013) completed a small study involving 20 disabled children, 18 parents and 8 teachers, drawn from the Hill Country, city of Kandy. Whilst there were different opinions as to why children required schooling, overall "most parents considered education as a *route to escape from the stigma* attached to the child's impairment" (Abeywickrama et al., 2013, p. 120, emphasis added). In line with this sentiment, the primary goal of parents and teachers in mainstream education was the child's impairment (in contrast with a barrier-free focus) and correction. Families pressured teachers to achieve 'normalcy' through school-based education (Abeywickrama et al., 2013, p. 124). Despite the ongoing issue of bullying, the drawings and interviews of disabled students strongly portrayed support and relationships as the most significant factor influencing their perceptions of participation in school (Abeywickrama et al., 2013, p. 120).

Despite training in inclusive education, teachers resisted the participation of disabled students through their undue focus on impairment as a problem, rather than attitude or a poor-built environment. Whilst acknowledging that attention should be towards parental needs, as the sustainability of all integration projects hinges on *parental satisfaction* with children's educational attainments and experience, the study did not explore cultural kin-family and community impediments to social inclusion. Instead, achieving normalcy within *parvula* and community relations remained unproblematised.

Conclusion

What I have attempted to do in this chapter is to outline some of the core challenges for mothers with disabled children or adult offspring. To my knowledge this is the first time, in the Sri Lankan context, that research on mothering and disability has been undertaken. I have outlined the cultural context of relations, namely, the *parvula* (family-kin) in communities, the notable *absence of fathers* in actively parenting and also the place of *lajja-bhaya* (fear-shame) ethics, which acts as the interpretative paradigm of familial and social relationships. One aspect, the existence of hierarchical relations in Sri Lanka, needs to be analysed in terms of creating 'just' hierarchies.

Finally, I reviewed three studies and found problematic the research design (positivist empiricism), the absence of an upfront epistemology, the narrow scope of research questions and the exclusion of mothers who did not use hospital or medical services. These studies did not really interrogate the role of women as a sex class and its influence on parenting and requisites of support required beyond simply noting statistical sex differences. Research in this area is required to be interdisciplinary and provide diverse epistemological and methodological approaches. There is much more to learn about mothers of disabled children—when they are young, adolescents and become adults—as to how they negotiate *lajja* and engage in practices of *fearlessness* that dispute the negative rendering of disability which fragments *parvula* relations.

Notes

1. "*Chi* is the typical Sinhala expression of revulsion, which prototypically refers to faces. In socialisation, the child is soon made to feel that all dirt is faeces … The idiom of faeces and foul smells are extended to other context also, for examples, to 'shame': a person who has lost face is someone who 'stank'; dirt, mud, and bad smells all appear in the idiom of shame" (Obeyesekere, 1985, p. 146). How is physical disability then countenanced in its association with leaky bodies, bodies that seize, dribble or are incontinent?
2. For more on caste and gender in the Indian context, see Chakravarti (2018).

References

Abeysekara, A. (2017). Living for others: The writing agency in the context of failed marriages and singleness in urban Sri Lanka. *Feminism and Psychology, 27*(4), 427–446.

Abeywickrama, S. P., Jayasinghe, I. K., & Sumanasena, S. P. (2013). Excluded in inclusive schools: Experiences of children with disabilities, their families and teachers in Sri Lanka. *Disability, CBR & Inclusive Development, 24*(1), 115–129.

Bains, S. (2013). South Asian mothers with special needs children. In J. Sangha & T. Gonsalves (Eds.), *South Asian mothering negotiating culture, family and self and selfhood* (pp. 101–110). Demeter Press.

Butnor, A., & McWeeny, J. (2014). Feminist comparative methodology performing philosophy differently. In A. Butnor & J. McWeeny (Eds.), *Asian and feminist philosophies in dialogue: Liberating traditions.* Columbia University Press.

Campbell, Fiona Kumari (2014). Geodisability Knowledge Production and International Norms: A Sri Lankan Case study. In Karen Soldatic and Helen Meekosha (eds.), *The Global Politics of Impairment and Disability: Processes and Embodiments* (pp. 87–106). London: Routledge.

Campbell, F. K. (2019). Precision ableism: A *studies in ableism* approach to developing histories of disability and Abledment. *Rethinking History, 23*(2), 135–156.

Chakravarti, U. (2018). *Gendering caste: Through a feminist lens, revised edition.* Sage.

Chapin, B. (2010). 'We have to give': Sinhala mothers' responses to Children's expression of desire. *Ethos, 38*(4), 354–368.

Chapin, B. (2014). *Childhood in a Sri Lankan village: Shaping hierarchy and desire.* Rutgers University Press.

de Alwis, M. (1997). Motherhood as a space of protest: Women's political participation in contemporary Sri Lanka. In A. Basu & P. Jeffrey (Eds.), *Appropriating gender: Women's activism and the politicization of religion in South Asia* (pp. 185–201). Routledge.

de Alwis, M. (1998). *Maternalist politics in Sri Lanka: A historical anthropology of its conditions of possibility.* PhD, University of Chicago.

de Alwis, M. (2012). 'Girl still burning inside my head': Reflections on suicide in Sri Lanka. *Contributions to Indian Sociology, 46*(1 & 2), 29–51.

de Mel, N. (2006). Playing disability, performing gender: Militarised masculinity and disability theatre in the Sri Lankan war and its aftermath. In S. Grech & S. K. Karen (Eds.), *Disability in the Global South. International perspectives on social policy, administration, and practice* (pp. 99–116). Springer.

de Mel, N. (2007). *Militarizing Sri Lanka, popular culture, memory and narrative in armed conflict.* Sage.

de Silva, J. (1998). Shifting frames of masculinity. In M. Tiruchelvam & C. S. Dattsthreya (Eds.), *Culture and politics of identity in Sri Lanka* (pp. 25–40). International Centre for Ethnic Studies.

de Silva, J. (2005). *Globalization, terror and the shaming of the nation. Constructing local masculinities in a Sri Lankan Village*. Trafford Publishing.

Dumont, L. (1980). *Homo Hierarchicus: The caste system and its implications*. The University of Chicago Press.

Hewamanne, S. (2003). Performing dis-respectability: New tastes, cultural practices and identity performance by Sri Lanka's free trade zone garment factory workers. *Cultural Dynamics, 15*(1), 71–101.

International Labour Organization. (2016). *Factors affecting women's labour force participation in Sri Lanka*. Colombo: International Labour Organization, ILO Country Office for Sri Lanka and the Maldives.

Jayasinghe, V. (2004). *A New Vision: A Feminist Perspective in Sri Lanka*. Colombo: Pace Printers.

Jayawardena, K., & Pinto–Jayawardena, K. (2017). *The Search for Justice–The Sri Lanka Papers*. New Delhi: Zubaan Publishers.

Kiribamune, S., & Samarasinghe, V. (1990). *Women at Crossroads: A Sri Lankan Perspective*. New Delhi: Vikas Publishing House Pvt Ltd.

Landry, M. D., Raman, S. R., Harris, E., Madison, L., Parekh, M., Banks, C., Bhatti, H., & Wijesinghe, C. (2015). Exploring the factors that influence the perceptions of disability: A qualitative study of mothers of children with disabilities at a community-based rehabilitation Centre in Sri Lanka. *WHO South-East Asia Journal of Public Health, 4*(2), 150–158.

Mohanty, C. T. (2006). *Feminism without borders decolonizing theory, practicing solidarity*. Zubaan/Kali Publishing.

Obeyesekere, G. (1984). *The cult of the goddess Pattini*. University of Chicago Press.

Obeyesekere, G. (1985). Depression, Buddhism, and the work of culture in Sri Lanka. In A. Kleinman & B. Good (Eds.), *Culture and depression: Studies in the anthropology and cross-cultural psychiatry of affect and disorder* (pp. 134–152). University of California Press.

Said, M. (2014). Suicide and shame in southern Sri Lanka: Networks of dependency. *South Asia Research, 34*(1), 19–30.

Sangha, J. (2014). Contextualizing South Asian motherhood. In A. O'Reilly (Ed.), *Mothers, mothering and motherhood across cultural differences – A reader* (pp. 413–427). Demeter Press.

Sinha, M., & Chauhan, V. (2013). Deconstructing Lajja as a marker of Indian womanhood. *Psychology and Developing Societies, 25*(1), 133–163.

Southall, A. (2012). A critique of the typology of states and political systems. In M. Banton (Ed.), *Political systems and the distribution of power* (pp. 113–125). Routledge.

Wickramasinghe, N. (2003). *Dressing the Colonised Body: Politics, Clothing and Identity in Colonial Sri Lanka*. Hyderabad: Orient Longman.

Widger, T. (2012a). Suffering, frustration, and anger: Class, gender, and history in Sri Lankan suicide stories! *Culture, Medicine and Psychiatry, 36*(2), 225–244.
Widger, T. (2012b). Suicide and the morality of kinship in Sri Lanka. *Contributions to Indian Sociology, 46*(1 & 2), 83–116.
Wijesinghe, C., Cunningham, N., Fonseka, P., Hewage, C. G., & Østbye, T. (2015). Factors associated with caregiver burden among caregivers of children with cerebral palsy in Sri Lanka. *Asia-Pacific Journal of Public Health, 27*(1), 85–95.

CHAPTER 9

Voices of Syrian Refugee Women in Jordan Living with Exacerbated Gender-Based Violence During COVID-19: Conceptualizing a Feminist Perspective for Social Work

Dina Pervez Sidhva

Introduction

In a relatively short time, the COVID-19 pandemic has wreaked destruction and chaos across the globe, with over 5.8 million deaths worldwide, by mid-February 2022 (Worldometers.Info, 2022).

This has pushed health services to capacity, triggered nationwide lockdowns, and disrupted international markets—transforming people's livelihoods and lives in a fundamentally adverse manner.

The unprecedented and rapid spread of COVID-19 has resulted in dramatic global political and social upheaval—especially on vulnerable

D. P. Sidhva (✉)
University of the West of Scotland, Paisley, UK
e-mail: Dina.Sidhva@uws.ac.uk

C. Cocker, T. Hafford-Letchfield (eds.), *Rethinking Feminist Theories for Social Work Practice*,
https://doi.org/10.1007/978-3-030-94241-0_9

populations, who are already disproportionately affected. This chapter is based on small-scale qualitative research that explored the lived experiences of 20 Syrian refugees during the COVID-19 pandemic and the resultant lockdown in Jordan. All respondents were service users of two Jordanian non-governmental organizations (NGOs) that support refugee women experiencing gender-based violence. Vignettes from these women's narratives reflect the exacerbated violence experienced by them, as they are forced to stay at home with their aggressors. The conclusion was that women's empowerment is crucial, as it increases employment opportunities for them, and reduces the risk of exposure to violence.

The chapter considers some feminist approaches that provide social work practitioners, academics and researchers with a framework to undertake a gendered analysis of power and injustice for refugees and other marginalized women—by providing a foundation for critical reflection, engaging in action, and challenging the gender-based violence perpetrated globally.

Context

Over the past decade, Jordan has experienced a huge influx of Syrian refugees who were fleeing the civil war in their country, most of whom were women. Since the beginning of the Syrian conflict in 2011, an estimated 1.4 million Syrian refugees have fled to Jordan (Morris, 2019). Women make up a significant part of the Syrian community in Jordan (Sidhva et al., 2021). According to the United Nations High Commissioner for Refugees (UNHCR, 2019), 50% of all Syrian refugees are women. It is understood that Jordan hosts approximatively 660,000 Syrian refugees, although Ghazal (2016) avers that the 2015 governmental census is nearly double that figure.

Given the traditional conservative values and gender norms prevalent among the Syrian refugees, women and girls are already disproportionally affected in terms of their ability to access education and meaningful employment. This resulted in serious problems—women losing the ability or right to pursue education, develop skills, or improve personal qualities, which often exposed them to violence and abuse (Spencer et al., 2015).

Violence against women is a global concern, and it is estimated that one in every three women experience one form of violence in their lifetime (WHO, 2021). Violence against women cuts across all religions, socio-economic groups and cultures, and touches fundamental issues of gender, sexuality, and power. The United Nations (1993, p. 1) defines violence against women as "any act of gender-based violence that results in, or is likely to result in, physical, sexual, or mental harm or suffering to women, including threats of such acts, coercion or arbitrary deprivation of liberty, whether occurring in public or private life." Moreover, it states that intimate partner violence and sexual violence are major public and clinical health problems. Both these issues result in violating women's human rights and are rooted in and perpetuate gender inequalities. The European Institute for Gender Equality (2021) defines gender-based violence "as a phenomenon that is deeply rooted in gender inequality and continues to be one of the most notable human rights violations within all societies. Gender-based violence is violence directed against a person because of their gender. Both women and men experience gender-based violence but the majority of victims are women and girls." Gender-based violence is a crucial public health problem because of its serious physical, emotional, and social consequences that impose threats at the local, regional, and international levels (James et al., 2002).

Gender-based violence against women is a concerning issue in the Arab world in general and in Jordan in particular, because of the ever-increasing numbers of women who experience it (Al-Badayneh, 2012; Ghanim, 2009; Oweis et al., 2009). In Jordan, gender-based violence significantly contributes to ill health and social problems, including family problems. Earlier studies in Jordan (Clark et al., 2009) found that nearly one out of three women were exposed to intimate partner violence during their marital life. Another study reported that one in every four Jordanian women experience physical violence, mostly by husbands, fathers, and brothers (Spencer et al., 2015).

The coronavirus pandemic has adversely affected the economy and the social status of communities globally. In Jordan too, Syrian refugees suffered considerably from the consequences of the crisis, resulting in extreme stress and tension that has caused an increase in gender-based violence. The impact of increased gender-based violence also deprives women of their agency and exacerbates vulnerability, thus rendering them highly dependent on their perpetrators for survival. Amid the unprecedented pandemic, the imposed lockdown measures have deepened pre-existing conditions of abuse and violence for refugee women and girls, particularly because of their limited access to resources. According to the United Nations News (2020a, p. 1), "the combination of economic and social stresses brought on by the pandemic, as well as restrictions on movement, have dramatically increased the numbers of women and girls facing abuse, in almost all countries." This prompted the UN's chief to appeal to all governments to "put women's safety first as they respond to the pandemic." Additionally, the United Nations News (2020b) also suggests that sometimes families may force women living in deteriorating socioeconomic conditions, without documentation, into survival-sex or child marriages. Moreover, reduced community interaction, because of restricted mobility, exacerbates ongoing intimate partner violence, as trapped women remain unable to access community workers or seek in-person support within camps.

Methods

The research aimed to develop an empirical understanding of the heightened impact of gender-based violence on Syrian refugee women and girls in Jordan, with aim of reducing the impact on them through the recommendation of policies.

The central purpose of the research was to understand the lived experience of Syrian refugees, with a special focus on their experience of gender-based violence, during the lockdown.

A constructivist paradigm (Denzin & Lincoln, 1994) provided the philosophical standpoint for my research, as it focused on life experiences and lived realities. "Adopting a constructivist epistemology of knowledge allowed for accepting that the respondents were the 'experts' and knowledgeable about their own lives" (Sidhva, 2003). As my research aimed to develop an understanding of the meaning of living with gender based-violence, applying elements of the phenomenological perspective stressed the perspective and voices of the women themselves because the experience of living with gender-based violence was an inseparable part of their consciousness. Moreover, I used a feminist research lens, as my study was fundamentally "connected in principle to feminist struggle" (Sprague & Zimmerman, 1993, p. 266) and focused on understanding and giving hitherto silenced, marginalized, and oppressed voices a chance to be heard, thus fostering social justice for them (Hesse-Biber et al., 1999). Utilizing the qualitative research paradigm facilitated data collection in naturalistic settings that highlighted the complexity and diversity of the research content. It involved asking simple questions and getting complex answers from the *emic* perspective or the perspective of the women themselves (Merriam, 1998).

The research tool consisted of an interview schedule with open-ended questions that covered women's experience of violence in general; their familial life and exposure to violence in Jordan; changes to their social and familial life, including the level of violence encountered during the lock-down resulted from the pandemic; reasons for violence; and what services and support could improve their lives. All due ethical consideration was applied to this study both at the level of university ethical approval and with the women themselves in terms of consent, confidentiality, and so on. My Jordanian research partner from the Higher Population Council (Amman, Jordan) undertook all the face-to-face interviews, in Arabic. The COVID-19 pandemic made travel to Jordan for me an impossibility from the UK, so all research meetings were undertaken remotely across the miles. Twenty Syrian refugee women, who were service users from two Jordanian NGOs, the Jordanian Women's Union and the Noor al Hussein Foundation, volunteered to participate in this research.

Experiences of Gender-Based Violence: Voices of Syrian Refugee Women

The narratives of all 20 Syrian refugee women in my research point to the lived experience of some form of gender-based violence. Sadly, violence was something that they had all experienced in Syria prior to seeking asylum in Jordan. The vast majority of them unquestioningly accepted this violence as part of their lived day-to-day lives. As one woman put it, "All the women in Syria are like me, they are facing violence, so we feel it is a part of our custom that the woman has to be hit and punished; it's like a cultural custom which we just accept."

As discussed in the background, the incidents of violence experienced by women during the pandemic increased significantly. All women reported experiencing some form of either verbal, physical, or sexual violence: "Lately, he started punching or beating or cursing me whenever he goes in and out the house." Their narratives palpably describe the anguish, angst, and agony experienced on account of the escalating degree of gender-based violence experienced by them during the time they are locked down with their aggressors.

Several women recounted routinely experiencing verbal violence if their husbands were "upset or disturbed in any way." They suggested that lack of food, no income, and "wanting to feel in control of the home and not lose his manhood" were the key reasons for their ill-treatment. The aspect of control was experienced as a key driver for violence; women repeatedly stated that when men lashed out either verbally or physically, it was with retorts like: "Do you want to be the man of the house?" or "This will teach you before you try to become the man in the house."

Describing her situation, one woman said, "He gets angry about everything. If I tell him that our child did something wrong, he starts to beat me. If I buy something small for myself, he gets angry and curses." Another woman added, "If water falls from a cup, he curses; if a child asks him for something, he curses, for any reason, he curses." Yet another refugee suggested that her husband's bad behaviour was because he drank. "Alcohol is the main reason why he was angry, now that he got married a second time, 'a bad thing has turned him good' (an Arabic proverb). He has stopped his bad behaviour," she emphasized.

One woman recollected an incident where her husband was physically violent with her for the first time: "During the corona, the pressure increased so much on him that he was constantly cursing, angry, angry all the time and then one day he just beat me so badly, it was so frightening and painful." Recalling the pain, she had endured, another said: "He beat me so badly; I thought I was going to die." Yet another refugee woman cried as she recounted what she called a painfully humiliating experience. "I didn't go to the hospital, although he beat me until my eye was blue and my bones hurt me from the stick he used to beat me. I stayed sick for three days in the bed, but I don't go to the hospital because I married him by choice and I don't want the community to make fun of me," she said.

In general, women felt that their relationships had suffered badly due to the exacerbated violence borne during the lockdown. One woman spoke about the physical pain and mental humiliation when she confronted her husband about having an affair. She recalled, "I fought with my husband a lot after I found him naked, talking with a lady on the phone and she was also naked … he beat me hard with a water pipe on my eye, as you can see, and in the end, he left me for that woman." Another said wistfully, "Lockdown has changed my husband's personality a lot; he is more agitated for any insignificant reason, cursing me, beating me. Because of his mental state, he got diseases, including a stroke, and diabetes that caused his big toe to be cut. He is in pain and is bothered a lot, so he just kicks and hits me." An older 38-year-old woman explained that she had endured beatings all her married life, but the current situation was even grimmer. "Umm, he beat me and beat me until the stick was broken on me - you can go and see there are three sticks at home under the table, tied together. Three sticks! Whenever he beats me, I curse his mother, yes I do, I don't beat him but I curse … even my little child said to me, you are 'bent haram' (i.e. daughter born from an illegal relationship). Where did he get that from … from his father who always calls me that, of course."

The violence experienced by women sometimes extended itself to their children. When recollecting these experiences, most women expressed real pain and sadness. One woman tearfully recalled, "I didn't go to the hospital when he hits me, just bear the bruises, but one time he hit our boy really hard and his mouth was full of blood, I could not bear to see my child suffer; he can beat me, but not my child." Another woman was clearly distressed: "He used to beat me, and then he started beating my girls. The pressure was so much that I tried to kill myself several times. I ate many medications, but it did not work. I cut my veins with a knife but it did not work. I want to die, I hate life." Another respondent echoed her thoughts: "Life has become unbearable since he has started hitting the children for no reason at all … I don't know what to do."

Several women spoke about their lived reality of being coerced into having sex with their husbands against their will, mainly because of the fear of negative repercussions: "I don't oppose that, because I don't want to be beaten, and all my neighbours hearing that, which is so shameful for me." Another woman spoke about feeling disgusted: "By God, he always compels me, I understand now in Jordan it is called marriage rape—always kicking and hurting me; I hate when he comes near me."

Some women spoke about violence experienced during sex: "first he beats me badly and then he forces himself on me." One woman elaborated, "He forces me to sleep with him even if I'm bleeding. If I do not, he will go out for another woman. If I cry, he sings with happiness, when I cry in pain, he always laughs, can you imagine?"

Women also spoke about the sexual harassment that they frequently experienced in public places. One woman commented, "Uh, in the street, a lot of harassment, even people who try to help and stop the harassment get into trouble as well. Although I'm not young and able to protect myself, I still get harassed." Another added, "every time we walk in the street, we hear dirty words a lot." However, many women commented on feeling much safer in Jordan than in Syria. "I was forbidden to leave home or go out in Syria. But here, we have to make living; our personality is strengthened here by God."

A number of women spoke about the link between the joblessness of their husbands, food shortages, and the violence experienced by them. One woman recalled, "By God, life has become very bad, getting food and being hungry is difficult, but more difficult are the beatings and humiliations." Another refugee spoke in desperation: "We ran out of bread and food, how could I get food? From where? I called different organizations asking for help...I have nothing in my freezer, zero ... he got very stressed and hungry, more hunger more anger, more anger more beatings and the cursing increased." Some women pointed out that the violence during the lockdown was linked to decreased food supplies due to lack of money and increased consumption of food because the entire family was always at home, with little else to occupy them. One woman recounted, "Of course, the kids were so hungry and asking for more food. I did not have enough money to store extra food, so I picked 'khobaizeh' (hibiscus flowers) from the ground and made some soup for them to eat so they would be quiet. But my husband kept cursing me and pushing me with anger, even though it was not my fault." Another woman said forcefully that the pain of violence was insignificant compared to the pain she experienced because of her child's illness caused by the famine-like conditions experienced by them: "It was a tragedy, I had no milk for my child and his fever was high ... the civil defence ambulance didn't come until four days had passed. His brain shrank because of the fluids that filled his head ... they had to take this fluid out, and he is so sick my child."

A pervasive acceptance of gender-based violence and discriminatory gender roles and social norms seemed to be reinforced by their socialization. This was particularly evident in the way that most women appeared to condone the exacerbated violence experienced by them. A 27-year-old woman said, "In Syria, women are all the same: the man beats his wife, it is normal; breaks her hand, it is normal; her head bleeds, it is normal. When I came to Jordan, I was surprised that this is not the custom here. I just believe the woman should be patient with her husband and not mind his beatings."

Additionally, many women seemed to disregard the violence experienced by them, because they compared the lockdown hardships with the hardships that they experienced during the war in Syria, before becoming refugees in Jordan. One woman said: "The hitting and cursing are okay, during the corona we were safe, at least I can look out of the window and forget about his insults. But during the war, with snipers around, it was very different, corona war and the beatings from him is difficult, but it's easier than during the war." Another one added, "I'm not scared of him or the disease, bombing is much worse. My bruises and aches will heal, people will get well with corona, but during the war, we saw blood, dead bodies in the street that you cannot remove because you will be killed, there was no water no electricity, nothing."

A significant number of women seemed to condone their partner's increased violent treatment towards them on the basis that their men were under additional pressure and stress. They believed that once "corona was over" their "madness would stop." One woman laughed and said, "We couldn't pay the rent, so the pressure increased a lot for three months and his madness became worse, but after corona, we will be finished with his angriness and madness."

Finally, most women alluded to keeping the violence they experienced before being interviewed for my research a secret. One woman said, "We hear about things, but no one shares these things, all houses have their secrets, which should remain in the house." Another added, "We don't say it, but frankly, there is a lot of violence … beating after being drunk -- like that kind of violence." As one woman poignantly reflected, "It feels like suffering in silence is the only thing that we can do."

This section throws light on the varied physical, verbal, and sexual forms of gender-based violence and the stark personal and structural inequalities experienced by them. The voices of these women uprooted from their homes and lives in Syria by war and persecution echo the persistent unrest and day-to-day challenges that they encounter. It also illumines how their socialization perpetuates their acceptance of unacceptable gender roles, leading to self-effacement and a general acceptance of male violence. In particular, it focuses on how the COVID-19 pandemic and the resultant lockdown have exacerbated the gender-based violence already a part of their lived experiences. Yet, what also comes through in my research is the resilience and strong persistence of these women—to educate their children and themselves further, to learn a skill that will make them self-sufficient, and ultimately to empower themselves.

Feminist Perspectives and Gender-Based Violence: Implications for Social Work

Feminist perspectives provide social workers with a lens through which a profound understanding of gender-based violence is stimulated. At the heart of feminism is the belief that equality is only possible when both sexes are economically, socially, and politically equal: where there is no space for either victim or perpetrator, that all are equal.

The perspectives of some feminists detailed in this section provide us with a framework to analyse the gender-based violence experienced by the Syrian refugee women in my research. However, the lived experiences of these Syrian refugees are not exclusive to Jordan. In fact, they mirror the lives and challenges refugee women in countries across the globe are experiencing. Thus, the feminist approaches discussed in this section provide social work with deeper insights into working with marginalized, vulnerable women and girls experiencing gender-based violence irrespective of whether it is in the global north or south, irrespective of whether the women and girls are refugees or not. Additionally, in the light of the reality that the COVID-19 pandemic and resultant lockdowns worldwide have compounded the challenges experienced by women facing gender-based violence worldwide (United Nations News, 2020a), social workers in varied fields of practice need to be ever vigilant. It must be noted that while this section touches upon several feminist theories that throw light on issues of gender-based violence (inter-personal, social, and structural), it does not claim, in any way, to be covering the whole range of feminist approaches to gender-based violence.

True (2012) provides us with a valuable understanding of the feminist political and economic theory of gender-based violence. She argues that the use of power and violence stems from inequality within society both at a personal and at a structural level. This is evidenced in the lives and stories of the Syrian refugees in my research. Violence and war were the fundamental reasons for fleeing Syria as a refugee; all the women in my study had experienced gender-based violence and personal trauma in their homes. Additionally, as refugees living in Jordan, structural issues such as poverty, immigration status, social isolation, persistent oppression, and problems accessing health and humanitarian services all served to aggravate the marginalization experienced by them. Thus, the feminist political-economic understanding of gender-based violence unequivocally necessitates that as social workers we focus on the inextricable link between personal and structural vulnerabilities and inequalities in all our assessments and interventions in this area.

The work of Heyes (2013), Cockburn (2004), and Pence and Paymar (1993) focuses on the violence that is embedded in "patriarchal structures of power." Their work provides a critical lens to understand how the refugee women in my research were socialized into accepting and condoning the pervasive violence carried out against them. Heyes argues that violence is intrinsically connected to power, and therefore, all acts of violence intersect with gender. Cockburn suggests that feminist approaches enable an analysis of how violence is both connected and embedded in patriarchal power structures. Moreover, because gender power "shapes the dynamic of every interaction" (p. 28), all gender norms form and are formed by power structures. She, therefore, argues that all forms of violence are fundamentally linked to power and intersect with gender. Pence and Paymar stress on power inequality and gender in opposite-sex relationships. Their focus on how societal messages condone male violence and perpetuate inadmissible gender roles that order how men can behave towards women in their intimate relationships is particularly helpful in explaining the lived experiences of the Syrian refugee women in my study.

The theory of femicide refers to gender-specific violence that is directed only to women (Russell, 1977). Corradi et al.'s (2016) suggest that the feminist theory of "femicide" is crucial in developing an understanding of how gender-based violence is perpetrated on women through the dominance and assertion of males within societal power structures. It focuses on how patriarchal power structures enforce male dominance over every aspect of a woman's life.

In the same vein, Renzetti and Campe (2021) throw light on liberal and radical feminist perspectives, which are particularly important for social workers working with women who are vulnerable at personal, social, or structural levels. They suggest that liberal feminists conceive that gender inequality is a product of differential gender socialization. This in turn ascribes diverse and inequitably valued roles to men and women and unequal legal rights. Thus, from birth, girls are socialized into disparate social roles, providing a fertile ground for gender inequality. They cite the work of Stanko (1986) and MacKinnon (1982), who suggest that all forms of physical, sexual, and psychological abuse such as beating, battering, and so on help to render women relatively powerless against powerful men.

Radical feminism, on the other hand, stresses that violence against women is much more than a manifestation of gender inequality and sexism: it is a means by which male power, domination, and control are both upheld and imposed.

Finally, a particularly crucial lens in understanding the multiple inequalities and the impact of their interaction on each other is provided by Crenshaw (1991), who coined the term intersectionality. Intersectionality refers to the harmful costs of treating multiple, intertwining identities as if they are separate or mutually exclusive. Intersectionality looks beyond the liberal and radical feminist theories or what is referred to by some feminists as "hegemonic feminism" (Burgess-Proctor, 2006). It stresses the importance of studying the power relations rooted in gender, race, and other socially created differences alongside the dynamic intersection of these differences. Intersectionality as a framework encourages us to connect inter-personal violence with structural violence and draws particular attention to the complexity of gender-based violence for marginalized, oppressed groups. Thus, it provides social workers and humanitarian players with a critical tool that enables a deeper understanding and empathic grasp of the intersectional susceptibilities experienced by refugee women.

Conclusion

According to Amnesty International (2021), there are 26 million refugees across the globe; half of these refugees are children and 85% of refugees are hosted in developing countries. In an increasingly globalized world, these staggering figures provide a stark reminder to us as social workers of the role we need to play towards the amelioration of the conditions of some of the most marginalized and oppressed people in our society.

I would like to conclude this chapter by putting forward several ways in which social work can make a difference in the lives of refugee women and children.

As twenty-first-century social work practitioners, researchers, and academics, we have a moral responsibility to commit to fostering a gendered analysis of power and injustice for refugee women and girls by providing a foundation for critical reflection in relation to the vulnerability, oppression, and marginalization experienced by them. Moreover, as a profession, social work needs to engage in actions that challenge the gender-based violence perpetrated globally by men against women and girls, particularly those living in fragile, protracted crises.

My research reinforces the understanding that the COVID-19 pandemic has had a profound impact on the lives of Syrian refugee women by intensifying the gender-based violence experienced by them. However, it is unable to comment definitively on what this profound impact will be as the effects of the pandemic continue to affect these vulnerable and marginalized refugees. Palattiyil et al. (2021) suggest that social workers should advocate for further research and resources to be put towards advancing a greater understanding of "the longer term effects of Covid-19 on forced migrants and measures to mitigate them. We need a 'build-back better' agenda for forced migrants to genuinely improve their situation from what they were pre-Covid-19" (p. 13).

Social work has both an opportunity and a responsibility to promote gender equality through the empowerment of women and girls in our society. Education and enterprise are two effective channels that lead to empowerment.

As social work educators, we need to create specific spaces in our curriculums to teach about refugees and asylum seekers. The curriculum should encompass the lived experiences, intersectional needs, and priorities of these vulnerable and marginalized individuals. This will enable us to build awareness and capacity for social workers to work locally and globally with refugees in an effective and ultimately meaningful way to reduce gender inequality.

Finally, as social workers, feminists, advocates for marginalized people, and members of the human race, we all have a duty to challenge structural and physical violence in all its forms and support gender-transformative actions wherever possible. Renzetti & Campe (2021) suggest that "[u]ntil we recognize the significance of structural violence in perpetuating interpersonal violence and develop ways to dismantle it, our individual-level responses will continue to be merely band-aids on an enlarging and festering wound" (p. 423). By following a feminist human rights-based approach, transformative change is possible and need not be a lofty aim or distant dream.

Acknowledgements The findings of this chapter are based on funding through the Scottish Funding Council's (SFC) Global Challenges Research Funding (GCRF), administered through the University of the West of Scotland.

I also owe thanks to the Syrian refugee women who so generously shared intimate details about their personal lives; my Jordanian partners H.E. Dr. Abla Amawi and Manal Al-Ghazawi from the Higher Population Council, Jordan, without whom this research would not have been possible; and to Dr. George Palattiyil, University of Edinburgh, for commenting critically on the draft.

References

Al-Badayneh, D. (2012). Violence against women in Jordan. *Journal of Family Violence, 27*, 369–379.

Amnesty International. (2021). *The World's Refugees in numbers: The Global Solidarity crisis*. UpToDate. Retrieved April 13, 2021. https://www.amnesty.org/en/what-we-do/refugees-asylum-seekers-and-migrants/global-refugee-crisis-statistics-and-facts/

Burgess-Proctor, A. (2006). Intersections of race, class, gender, and crime: Future directions for feminist criminology. *Feminist Criminology, 1*, 27–47.

Clark, C., Hill, A., Jabbar, K., & Silverman, J. (2009). Violence during pregnancy in Jordan: Its prevalence and associated risk and protective factors. *Violence Against Women, 15*(6), 720–735.

Cockburn, C. (2004). The continuum of violence: A gender perspective on war and peace. In W. Giles & J. Hyndman (Eds.), *Sites of violence: Gender and conflict zones* (pp. 24–44). University of California Press.

Corradi, C., Marcuello-Servós, C., Boira, S., & Weil, S. (2016). Theories of femicide and their significance for social research. *Current Sociology, 64*(7), 975–995.

Crenshaw, K. (1991). Mapping the margins: Intersectionality, identity politics, and violence against women of color. *Stanford Law Review, 43*, 1251–1299.

Denzin, N. K., & Lincoln, Y. S. (Eds.). (1994). *Handbook of qualitative research*. Sage Publications.

European Institute for Gender Equality (EIGE). (2021). *What is gender-based violence?* UpToDate. Retrieved April 13, 2021. https://eige.europa.eu/gender-based-violence/what-is-gender-based-violence

Ghanim, D. (2009). *Gender and violence in the Middle East*. Praeger.

Ghazal, M. (2016, January, 16). Population stands at around 9.5 million, including 2.9 million guests. *Jordanian Times.com*. UpToDate. Retrieved April 5, 2021. https://www.jordantimes.com/news/local/population-stands-around-95-million-including-29-million-guests

Hesse-Biber, S., Gilmartin, C. and Lydenberg, R. (Eds.). (1999). Feminist approaches to theory and methodology: An interdisciplinary reader. Oxford University Press.

Heyes, C. (2013). Feminist solidarity after queer theory: The case of transgender. In S. Stryker & A. Z. Aizura (Eds.), *The transgender studies reader 2* (pp. 201–213). Routledge.

James, M., Dahlberg, L., & Zwi, A. (2002). The world report on violence and health. *Lancet, 11*(1), 1083–1088.

MacKinnon, C. (1982). Feminism, Marxism, method, and the state: An agenda for theory. In C. Renzetti, & M. Campe (Eds.), (2021). *Feminist Praxis and gender violence. companion to feminist studies*. In N. A. Naples (Ed.), *Companion to feminist studies* (1st ed., pp. 4411–426). Wiley.

Merriam, S. (1998). *Qualitative research and case study applications in education.* Jossey-Bass.

Morris, J. (2019, October). *The politics of return from Jordan to Syria Forced Migration Review.* Retrieved April 14, 2021. https://www.fmreview.org/sites/fmr/files/FMRdownloads/en/return/return.pdf

Oweis, A., Gharaibeh, M., Al-Natour, A., & Froelicher, E. (2009). Violence against women: Unveiling the suffering of women with a low income in Jordan. *Journal of Transcultural Nursing, 20*(1), 69–76.

Palattiyil, G., Sidhva, D., Macgowan, M., & Derr, A. S. (2021). Global trends in forced migration: Policy, practice, and research imperatives for social work. *International Social Work, 00*(0), 1–19. https://journals.sagepub.com/doi/pdf/10.1177/00208728211022791

Pence, E., & Paymar, M. (1993). *Education groups for men who batter: The Duluth model.* Springer Publishing.

Renzetti, C., & Campe, M. (2021). Feminist praxis and gender violence. Companion to feminist studies. In N. A. Naples (Ed.), *Companion to feminist studies* (1st ed., pp. 4411–4426). Wiley.

Russell, D. (1977). Report on the international tribunal on crimes against women. *Frontiers: A Journal of Women Studies, 2*(1), 1–6.

Sidhva, D. (2003). *Living with HIV/AIDS: Turning points, transitions and transformations in the lives of women from Bombay (India) and Edinburgh (Scotland).* Unpublished Doctoral Dissertation, University of Edinburgh. http://hdl.handle.net/1842/25185

Sidhva, D., Zuntz, A.-C., Akash, R., Nashwan, A., & Al Majali, A. (2021). 'In exile, the woman became everything': Middle-aged Syrian women's contributions to family livelihoods during protracted displacement in Jordan. *Journal of Humanitarian Affairs, 3*(1), 4–15. https://doi.org/10.7227/JHA.054

Spencer, R., Usta, A., Essaid, S., Shukri, Y., El-Gharaibeh, H., Abu-Taleb, N., Awwad, N., Alianza por la Solidaridad, United Nations Population Fund-Lebanon, & Clark, C. (2015). *Gender based violence against women and girls displaced by the Syrian Conflict in South Lebanon and North Jordan: Scope of violence and health correlates.* Alianza por la Solidaridad. UpToDate. Retrieved April 10, 2021. https://data2.unhcr.org/en/documents/download/50015

Sprague, J., & Zimmerman, M. K. (1993). Overcoming dualisms: A feminist agenda for sociological methodology. In P. England (Ed.), *Theory on gender: Feminism on theory* (pp. 225–280). Aldine de Gruyter.

Stanko, E. (1986). Intimate intrusions. Routledge and Kegan Paul. In C. Renzetti, & M. Campe (Eds.), (2021). *Feminist Praxis and gender violence. Companion to feminist studies.* In N. A. Naples (Ed.), *Companion to feminist studies* (1st ed., pp. 4411–426). Wiley.

True, J. (2012). *The political economy of violence against women.* Oxford University Press.

United Nations, *Declaration on the elimination of violence against women*. (1993). Available at: https://www.un.org/en/genocideprevention/documents/atrocity-crimes/doc.21_declarationeliminationvaw.pdf. Accessed 11 Apr 2021.

United Nations High Commissioner for Refugees, (UNHCR). (2019). *Syria regional refugee response – Jordan*. https://data2.unhcr.org/en/situations/syria/location/36

United Nations News. (2020a). *Global perspective human stories: Women living in deteriorating socio-economic conditions and without documentation may be forced into survival sex or child marriages by their families*. UpToDate. Retrieved April 5, 2021. https://news.un.org/en/story/2020/04/1062132

United Nations News. (2020b). *Global perspective human stories: UN chief calls for domestic violence 'ceasefire' amid 'horrifying global surge'*. UpToDate. Retrieved April 5, 2021. https://news.un.org/en/story/2020/04/1061052

World Health Organization (WHO). (2021). *Violence against women 2021*. UpToDate. Retrieved April 5, 2021. https://www.who.int/news-room/fact-sheets/detail/violence-against-women

Worldometers.Info. (2022). *COVID live update: Deaths from the coronavirus-worldometer*. UpToDate. Retrieved February 18, 2022. https://www.worldometers.info/coronavirus/

CHAPTER 10

The Transformative Potential of Transfeminist Social Work Practice

Jama Shelton, Maggie Dunleavy, and Kel Kroehle

Introduction

Minimal literature exists regarding the application of transfeminist theory in social work practice. The absence of literature is of concern, given the high rates of discrimination and violence targeting trans people, specifically trans women of color, and the commitment of the social work profession to work toward social justice. Social workers are charged with the task of equitable service delivery, and research has demonstrated both the need for direct services to respond to the disparities faced by trans people and the need for structural change to address the underlying inequities that

J. Shelton (✉)
Hunter College, New York, NY, USA
e-mail: jshelton@hunter.cuny.edu

M. Dunleavy
Silberman School of Social Work, New York, NY, USA
e-mail: MAGGIE.DUNLEAVY61@myhunter.cuny.edu

K. Kroehle
University of Pennsylvania, Philadelphia, PA, USA

C. Cocker, T. Hafford-Letchfield (eds.), *Rethinking Feminist Theories for Social Work Practice*,
https://doi.org/10.1007/978-3-030-94241-0_10

result in the need for a direct service response (Kcomt et al., 2020; Shelton et al., 2019; James et al., 2016).

Transfeminism has the potential to expand social work practice and the delivery of social services beyond patterns of normative thinking that are rooted in white, cisgender womanhood (Pyne, 2015). Scholars have extensively examined transphobia and the exclusion of trans women in feminist movements (Salamon, 2008; Stone, 1991). It is not the aim of this chapter to re-state that history. Rather, this chapter provides an overview of transfeminism and documents the inherent potential in a partnership between transfeminism and the social work profession, particularly within the United States, as it is in this context that the authors are engaged in social work practice and education.

Language and Terminology

Language is dynamic and context dependent, often reflecting specific geographic, socio-cultural, and political differences (Shelton & Dodd, 2020; McInroy & Craig, 2012). The terms individuals use to describe their identities may evolve, just as the words that join groups of people for social or political purposes often do not reflect the diversity of the individual group members or of the community represented by the group members. In this way, our language is constrained (Shelton & Dodd, 2020). In this chapter, we intend the term trans to be inclusive of individuals whose gender differs from societal expectations associated with their assigned sex at birth, or who resist adherence to traditional gender categories in some way (Austin et al., 2016; Serano, 2016). Thus, the term trans as utilized in this chapter includes individuals who identify as transgender, genderqueer, nonbinary, gender fluid, transsexual, agender, autigender, and other categories and labels of which we are not yet aware. We use the term cis (cisgender) to refer to individuals for whom their gender identity and expression correspond with societal expectations associated with the sex they were assigned at birth. In practice, it is always best to ask people what words they use to describe themselves, and to respect the terms each individual uses.

Situating Ourselves

For all three authors (Jama Shelton, Maggie Dunleavy, and Kel Kroehle), the application of transfeminism to social work practice is in deep alignment with our professional values, our professional experiences, and our lived expertise. The act of positioning oneself is a foundational component of justice-oriented social work practice, and for bringing attention to power dynamics within practice and educational settings (Wagaman et al., 2018; Dentato, 2014). All three authors identify as nonbinary, and have also worked alongside trans individuals and communities in direct practice and educational settings, as well as in community organizing and advocacy efforts.

What Is Transfeminism? A Brief History

Transfeminism is a critical approach to feminism rooted in the efforts of trans and cis women of color's challenge of the omission of white supremacy in historical feminist conceptualizations of the oppression of women of color (Shelton et al., 2019; Simpkins, 2016; Stryker & Bettcher, 2016; Weerawardhana, 2018). Transfeminists (particularly transfeminists of color) assert that gender equity cannot be realized without an intentional confrontation of all systems of oppression that surveil and control individuals and their bodies, including colonialism, capitalism, and white supremacy (Silva & Ornat, 2016; Simpkins, 2016). Though the term may have been used prior to 2001, scholar Emi Koyama is responsible for introducing transfeminism to academia. Koyama published *The Transfeminist Manifesto* in 2001, providing the following definition:

> *Transfeminism* is primarily a movement by and for trans women who view their liberation to be intrinsically linked to the liberation of all women and beyond. It is also open to other queers, intersex people, trans men, non-trans women, non-trans men and others who are sympathetic toward needs of trans women and consider their alliance with trans women to be essential for their own liberation. (p. 244)

As such, transfeminism extends beyond the idea of mere inclusion of trans perspectives into feminism; the goals are not to simply assimilate trans perspectives in a kind of "add and stir" manner. Rather, it involves an awareness of shared needs, a recognition of and reckoning with intersecting forms of oppression, and includes projects of emancipation and coalition building (Arfini, 2020). Transfeminism reminds all feminists, especially women, to contemplate the ways in which they are all impacted by the internalization of patriarchal, heterosexist, and gender essentialism (Koyama, 2003).

The primary principles of transfeminism outlined by Koyama (2003) include:

1. The belief that every individual has the right to define their own identities and to have those identities respected by society. This includes the right to outwardly express gender without fear of violence.
2. Every individual has the right to make decisions about their own bodies. This means that medical, political, and religious institutions do not have the authority to impede decisions about what people do with their bodies, nor should these institutions or their representatives violate the integrity of an individual's body.

As a practice, transfeminism is grounded in an awareness of intersectionality and amplifies the voices and priorities of those most affected by overlapping vulnerability to patriarchal, racist, capitalist violence. The mobility of transfeminism is quite useful, possessing the ability to transcend borders and be taken up outside of an Anglo-American context. This transcendent mobility can be contrasted to the word "queer," which has been reclaimed by LGBTQ+ people and adopted theoretically by English-speaking academics. "Queer" often loses its utility, traveling outside of the United States as an untranslatable word, disconnected from its unique context and history, described by Arfini (2020) as "a disembodied, non-descript, and hyper-theoretical standpoint imposed by American cultural imperialism. Transfeminism in contrast is able to be taken up in local political movements and across languages, showing up as (transféminisme [French], transfeminismo [Spanish], transfemminismo [Italian])" (p. 162). The global reach of the transfeminist movement, then, suggests a release of a disembodied Anglo-American theoretical frame and a foregrounding of transfeminism as an anti-racist, grassroots, and decolonial project in order to successfully resist transmisogynist, cissesixt violence" (Espineira & Bourcier, 2016; Kroehle et al., 2020; Silva & Ornat, 2016; Weerawardhana, 2018).

Where Does Transfeminism "Fit" in Feminist Theories?

Transfeminism is a building-out of existing feminist scaffolding, which has historically been both radical in its aims of liberation from patriarchal oppression and limited in and by definitional constraints (Halberstam, 2018). Transfeminism is inherently intersectional. In fact, it is imperative to recognize the shared oppression of all people who exist outside of the heteronormative in order to effectively respond to the violence caused by the patriarchy (Sharma, 2009). Thus, the rights of all women—trans women and cis women alike—are challenged by the patriarchal order, and any attempt to address the rights of some women while ignoring the rights of other women will fail (Carrera-Fernández & DePalma, 2020). The goal of a global transfeminism "will not be simply the enhancement of opportunities for trans*women but the creation of a trans*feminism that works for all women" (Halberstam, 2018).

Transfeminism and Social Work

Prior to outlining the transformative potential of a partnership between transfeminism and social work, we must first acknowledge the ways in which the profession of social work has reproduced a binary gender system and contributed to the oppression of trans people, as well as other bodies deemed nonconforming. The social work profession writ large has, in the recent past, moved away from regarding trans identities as pathological. This move is indicated in updated practice guidelines issued by the National Association of Social Workers and the Council on Social Work Education in the United States (Austin et al., 2016; Social Work Speaks, 2009). However, the historical context of the profession as accomplice in trans oppression must be acknowledged because current practices of diagnosing and treatment are rooted in this history (Markman, 2011).

Social Work and "Nonconforming" Bodies: A Difficult Legacy

Broadly speaking, the profession of social work is populated with historical demands that nonconforming bodies be assimilated into the mainstream, or learn to "pass" within white dominant, cis/heteronormative culture and Eurocentric ideals. Examples include the profession's backing of

Native American Residential Schools, harmful therapies aimed at "curing" homosexuality and gender nonconformity, and a plethora of eligibility requirements connected to public assistance (Kroehle et al., 2020; Bowles & Hopps, 2014). In specific relation to the reinforcement of binary gender and the pathologization of trans people, social work has both directly and indirectly contributed to trans oppression. A brief account of this history follows.

Social Work and Trans Oppression

Much of the historical conflict between the social work profession and many marginalized communities, in particular trans communities, can be located in the profession's reliance on the Diagnostic and Statistical Manual of Mental Disorders (DSM) (Shelton et al., 2019; Markman, 2011). The DSM is used by healthcare professionals in the United States and much of the world as the authoritative guide regarding the diagnosis of mental disorders (5th ed.) (*DSM–5*; American Psychiatric Association, 2013). Given its use as a tool to create and enforce boundaries of normal and deviant gendered bodies, trans communities hold a particularly fraught relationship with the DSM. The 3rd edition of the DSM introduced Gender Identity Disorder (GID) in 1980, with two diagnoses: gender identity disorder of childhood (GIDC) and transsexualism. With the publication of the DSM III-R in 1987, an additional diagnostic category was added to the GID repertoire—gender identity disorder of adolescence and adulthood, nontranssexual type (Drescher, 2009). This diagnosis was removed in the subsequent edition of the DSM (DSM IV), replacing it with gender identity disorder (GID).

Most recently, the diagnostic verbiage has changed from GID to gender dysphoria (DSM-V 2013). Some have lauded this attempt to better reflect the incongruence between one's gender identity and societal expectations regarding gender. While this change in terminology does more accurately connect the diagnosis to the actual problem—the incongruence between societal expectations of gender that do not reflect people's lived experience of gender—it continues to pathologize individuals and to locate the source of dissonance within the individual (Markman, 2011). The continued inclusion of gender-identity-related diagnoses in the DSM, despite efforts to improve terminology and definitions, bolsters the notion that trans identities are non-normative, aiding in the production and maintenance of discrimination and violence targeting trans people and communities. Further, it communicates the notion that congruence is the norm, suggesting that incongruence is non-normative and problematic (DeCuypere et al., 2009).

These diagnoses have had a profound impact on trans people and on the social work profession. For instance, the World Professional Association for Transgender Health's Standards of Care (WPATH) requires trans people seeking gender-affirming medical care (hormones, chest surgery, genital surgery) to procure a "letter of recommendation" from an "expert" detailing their diagnosis and "readiness" for transition. The same rules do not apply to cisgender individuals who desire hormone therapies or cosmetic surgeries, demonstrating an emphasis on trans-as-pathology "which reifies the idea that the dissonance between the gender performance of an individual and the expectations of society are the result of a psychological problem within the individual rather than a societal problem with defining gender" (Markman, 2011, p. 320). Diagnosing an individual based *solely* on their gender identity or expression directly reinforces the trans-as-pathology narrative and the oppressive, systematic management of trans people and their bodies (Shelton et al., 2019). Even if not responsible for assigning a diagnosis, social workers are often cast in the role of gatekeeper, requiring trans people to prove and defend their gender identities, limiting their bodily autonomy and decision making power.

The Potential of Transfeminism and Social Work

Transfeminism is a useful tool for social workers across varying contexts and in multiple ways. The application of a transfeminist lens has the potential to shift social work education, direct practice, and research such that all women are considered, valued, and positively impacted (Pyne, 2015). Less theory than practice, transfeminism addresses the inherent constraints of binary thinking and biological determinism, while confronting the notion of universal womanhood (Koyama, 2003). When gender is uninterrogated and womanhood considered synonymous with cisgender womanhood, the social work profession may be inadvertently recreating the gendered oppression it is working to confront (Kroehle et al., 2020). Disrupting systems of gender-based oppression requires abandoning the use of "gender" as a stand-in for white cisgender womanhood. Social workers must commit to disrupting centuries old patterns of normative thinking enshrined within social structures. In order to attain gender equity, forms of oppression that police people and their bodies must be confronted—including white supremacy, capitalism, and colonialism (Silva & Ornat, 2016; Simpkins, 2016). The gender binary is a product of these systems of oppression. It is imperative, then, that the social work profession confront the harm of reinforcing systems of binary gender classification.

Similarly, transfeminism encourages a critical examination of trans inclusivity, which is often rooted in transnormativity and capitalist production (Kroehle et al., 2020). A function of transnormativity is the naturalization of trans individuals and their bodies; however, this naturalization occurs within the context of hegemonic gender norms, thereby further excluding those who do not or cannot adhere to these norms (Ruin, 2016). The critical analysis of oppression inherent in transfeminist theory encourages a nuanced examination of trans inclusivity, which has partially emerged out of a need for corporations to do better business, rooted in the neoliberal demands of productivity—both through quelling unrest among employees and through marketing to broader audiences (Kroehle et al., 2020).

Transfeminist history is aligned with the values of social work, as evidenced by its grassroots emergence and the centering of those most marginalized by societal systems and structures. Transfeminism is resistant to the use of social work as a tool for surveillance and assimilation, a frequent function of the profession when tied to the state and charged with operating within the context of neoliberalism, privatization of services, and managerialism (Ferguson, 2009). As such, transfeminism also has the potential to reconnect social work with the values at its core—values that in recent decades have been minimized against the increasing focus on social work as a technical process that can be efficiently streamlined and delivered.

Transfeminism and Structural Social Work

As evidenced by the previous discussion of neoliberalism and privatization, social work practice and education have trended toward the individualization of problems. The result is a focus on eliminating an individual's symptoms rather than alleviating the social problems that frequently underlie an individual's symptoms (George & Marlowe, 2005). Conversely, structural social work connects the personal and political and seeks to identify, examine, and take action toward systems of oppression (George & Marlowe, 2005). This suggests that working to dismantle systems of oppression is both a macro practice *and* a micro practice—one that is necessary to transform that daily lived experience of oppressed populations (Kroehle et al., 2020).

Transfeminism is well aligned with the both/and approach of structural social work. Koyama (2003) writes: "Transfeminism believes in fostering an environment where women's individual choices are honored, while scrutinizing and challenging institutions that limit the range of choices available to them" (p. 247). Transfeminist perspectives move the profession from an assimilationist inclusion framework toward one of equity and justice. The work no longer centers on helping multiply marginalized populations exist within oppressive social structures; rather, the work centers on alleviating the oppressive social structures that result in individual and community suffering. As such, the potential transformative power of structural social work when combined with transfeminism is exactly what the profession needs in order to enact the core values of dignity, self-worth, and social justice.

Embedding Transfeminism in Social Work Education

Scholars in the fields of educational sciences and history have recently explored the pedagogical utility of transfeminism (Carrera-Fernández & DePalma, 2020; Enke, 2016) Given the paucity of social work literature that addresses transfeminism, we draw upon these disciplines for a deeper understanding of the applicability of transfeminism to pedagogical approaches. Scholar Finn Enke highlights the ways in which a transfeminist perspective is affirming of the range of genders represented in the classroom: "Transfeminist perspectives illuminate and affirm the gender diversity that is present in our classrooms and the world while addressing the ways that institutions (including education) continue to devalue women, people of color, people with disabilities, trans people, and poor people" (2016). A pedagogical approach that affirms the identities of the students within the classroom not only is conducive to knowledge development on the part of the students, but also models the type of anti-oppressive practices in which future social workers can engage (Kroehle et al., 2020).

The utility of transfeminist theory extends beyond the content social work educators teach to the processes by which they teach. Freire (2004) teaches us that, as agents of socialization, educational institutions are not value neutral. Rather, when not critically examined, they are often sites of social reproduction. As social work education (by its very nature) leans toward transformation, it is critically important that social work pedagogy engage in a process of critical examination of the oppressive discourses the profession seeks to address. Transfeminist theory makes this type of critical examination possible. Through the intentional reflection of the ways in which the profession of social work writ large and individual educators, students, and practitioners contribute to the production and maintenance of societal hierarchies, power, and oppression, utilizing transfeminism as a pedagogical tool brings us closer to liberation-oriented practice. According to Enke (2016), this is accomplished in both micro and macro ways, including the ways we interrogate (or fail to interrogate) the value we place on histories, production, bodies, property, and nationhood down to the beliefs we hold about "who belongs in our classrooms or what constitutes 'smart' or 'polite' behavior." It is not only the interrogation of values and beliefs, but also the active adoption of and commitment to learning from people across a range of identities, backgrounds, abilities and disabilities, religions, cultures, ages, classes, sexualities, and genders. This is the pedagogical project of transfeminism.

Specifically, Carrera-Fernández and DePalma (2020) suggest that the critical interrogation of normativity provides the foundation from which the infinite possibilities and positions of human diversity may be valued. A transfeminist pedagogy, they suggest, may include the following principles:

1. Understand the range of human diversity—prioritizing human experience and scientific understandings over social constructions
2. Recognize the synergies between cis women's and trans* oppressions, and how these affect all those who transgress normative sex–gender–sexuality.
3. Identify the conscious and unconscious ways in which cisgenderism and sexism, along with obligatory heterosexuality, form part of a hidden school curriculum.
4. Design pedagogies that invite children and young people to critically reflect on oppressive social constructions of sex, gender, and sexuality (p. 758).

Conclusion

This chapter has provided an overview of transfeminism and examined the inherent potential in transfeminist social work education and practice. Though the social work profession has historically participated in the continued oppression of trans people and other marginalized groups through the reinforcement of the gender binary and participation in neoliberalism and managerialism, at its core, the profession of social work is guided by values that are in alignment with transformative and liberatory practices. Transfeminism provides a framework for social work to embrace these values and to work toward justice and equity. All forms of oppression are related and function as reinforcements of each other. To address any one of them as a single issue is an exercise in futility. For instance, to address sexism without also addressing racism, cis/heterosexism, colonialism, capitalism, and so on will do little to unravel the patriarchal system threaded throughout contemporary social structures (Espineira & Bourcier, 2016). Transfeminism is well suited for the project of dismantling these intersecting systems of oppression.

References

American Psychiatric Association. (2013). *Diagnostic and statistical manual of mental disorders* (5th ed.). https://doi.org/10.1176/appi.books.9780890425596

Arfini, E. A. (2020). Transfeminism. *Lambda Nordica, 25*(1), 160–165. https://doi.org/10.34041/ln.v25.631

Austin, A., Craig, S. L., Alessi, E. J., Wagaman, M. A., Paceley, M. S., Dziengel, L., & Balestrery, J. E. (2016). Guidelines for transgender and gender nonconforming (TGNC) affirmative education: Enhancing the climate for TGNC students, staff and faculty in social work education. Alexandria, VA: Council on Social Work Education.

Bowles, D. D., & Hopps, J. G. (2014). The profession's role in meeting its historical mission to serve vulnerable populations. *Advances in Social Work, 15*(1), 1–20.

Carrera-Fernández, M., & DePalma, R. (2020). Feminism will be trans-inclusive or it will not be – Why do two cis-hetero woman educators support transfeminism? *The Sociological Review Monographs, 68*(4), 745–762.

DeCuypere, G., Knudson G., & Bockting, W. (2009). *W.P.A.T.H.: Response of the world professional Association for Transgender Health to the proposed D.S.M. V criteria for gender incongruence*. Retrieved on January 8, 2013 from http://www.wpath.org/documents/WPATH%20Reaction%20to%20the%20proposed%20DSM%20-%20Final.pdf

Dentato, M. P. (2014). *Queer communities (competency and positionality). NASW encyclopedia of social work* (pp. 1–13). Oxford University Press.

Drescher, J. (2009). Queer diagnoses: Parallels and contrasts in the history of homosexuality, gender variance, and the diagnostic and statistical manual. *Archives of Sexual Behavior, 39*, 427–460.

Enke, F. (2016, September 1). *Transfeminist perspectives on history and pedagogy*. Process: A blog for American history. https://www.processhistory.org/transfeminist-perspectives/

Espineira, K., & Bourcier, M. H. S. (2016). Transfeminism: Something else, somewhere else. *Transgender Studies Quarterly, 3*(1–2), 84–94.

Ferguson, I. (2009). Another social work is possible! Reclaiming the radical tradition. In V. Leskošek (Ed.), *Theories and methods of social work: Exploring different perspectives*. Faculty of Social Work, University of Ljubljana.

Freire, P. (2004). *Pedagogy of indignation*. Paradigm Publishers.

George, P., & Marlowe, S. (2005). Structural social work in action: Experiences from rural India. *Journal of Progressive Human Services, 16*(1), 5–24.

Halberstam, J. (2018, January 18). *Toward a trans* feminism*. Boston Review: A Political and Literary Forum. http://bostonreview.net/gender-sexuality/jack-halberstam-towards-trans-feminism

James, S. E., Herman, J. L., Rankin, S., Keisling, M., Mottet, L., & Anafi, M. (2016). The report of the 2015 U.S. transgender survey. Washington, DC: National Center for transgender Equality.

Kcomt, L., Gorey, K., Barrett, B., & McCabe, S. (2020). Healthcare avoidance due to anticipated discrimination among transgender people: A call to create trans-affirmative environments. *SSM – Population Health, 11*, 100608.

Koyama, E. (2003). The transfeminist manifesto. In R. Dicker & A. Piepmeier (Eds.), *Catching a wave: Reclaiming feminism for the twenty-first century* (pp. 244–259). Northeastern University Press.

Kroehle, K., Shelton, J., Clarke, E., & Seelman, K. (2020). Mainstreaming dissidence: Examining binary gender among social work's grand challenges. *Social Work*, swaa 037.

Markman, E. (2011). Gender identity disorder, the gender binary, and transgender oppression: Implications for ethical social work. *Smith College Studies in Social Work, 81*(4), 314–327.

McInroy, L., & Craig, S. L. (2012). Articulating identities: Language and practice with multiethnic sexual minority youth. *Counselling Psychology Quarterly, 25*(2), 137–149.

Pyne, J. (2015). Transfeminist theory and action: Trans women and the contested terrain of women's services. In B. O'Neill, T. Swan, & N. Mule (Eds.), *LGBTQ people and social work: Intersectional perspectives* (pp. 129–149). Canadian Scholars Press.

Ruin. (2016). Discussing transnormativities through transfeminism: Fifth note. *Transgender Studies Quarterly, 3*(1–2), 202–211.

Salamon, G. (2008). Transfeminism and the future of gender. In J. W. Scott (Ed.), *Women's studies on the edge*. Duke University Press.

Serano, J. (2016). *Outspoken: A decade of transgender activism and trans feminism*. Switch Hitter Press.

Sharma, J. (2009). Reflections on the construction of heteronormativity. *Development, 52*(1), 52–55.

Shelton, J., & Dodd, S. J. (2020). Beyond the binary: Addressing cisnormativity in the social work classroom. *Journal of Social Work Education, 56*(1), 179–185.

Shelton, J., Kroehle, K., & Andia, M. (2019). The trans person is not the problem: Brave spaces and structural competence as educative tools for trans justice in social work. *Journal of Sociology and Social Welfare, 46*(4), 97–123.

Silva, J. M., & Ornat, M. J. (2016). Transfeminism and decolonial thought: The contribution of Brazilian travestis. *Transgender Studies Quarterly, 3*(1–2), 220–227.

Simpkins, R. (2016). Trans*feminist intersections. *Transgender Studies Quarterly, 3*(1–2), 228–234.

Social Work Speaks. (2009). *Transgender and gender identity issues*. NASW Press.

Stone, S. (1991). The empire strikes Back: A posttranssexual manifesto. In J. Epstein & K. Straub (Eds.), *Body guards: The cultural politics of gender ambiguity*. Routledge.

Stryker, S., & Bettcher, T. M. (2016). Introduction: Trans/feminisms. *Transgender Studies Quarterly, 3*(1–2), 5–14.

Wagaman, A., Shelton, J., & Carter, R. (2018). Queering the social work classroom: Strategies for increasing inclusion of LGBTQ people and experiences. *Journal of Teaching in Social Work, 38*(2), 1–17.

Weerawardhana, C. (2018). Profoundly decolonizing? Reflections on a transfeminist perspective of international relations. *Meridians: Feminism, Race, Transnationalism, 16*(1), 184–213.

CHAPTER 11

Exploring the Intersection of Queer Disability as Life Story: A Feminist Narrative Approach to Social Work Research and Practice

Jennifer JD Drummond and Shari Brotman

Introduction

This chapter tells the story of Josie, a gender-non-conforming, queer woman living with chronic illness and disability. A feminist narrative methodology elicits Josie's description of her experiences of school, health care and the ways in which she negotiates sexual and gender identity and expression in the context of the many challenges she experiences in living with pain and disability. Through attention to critical moments and

J. JD. Drummond (✉)
Concordia University, Montreal, QC, Canada
e-mail: jennifer.drummond@concordia.ca

S. Brotman
McGill University, Montreal, QC, Canada
e-mail: shari.brotman@mcgill.ca

C. Cocker, T. Hafford-Letchfield (eds.), *Rethinking Feminist Theories for Social Work Practice*,
https://doi.org/10.1007/978-3-030-94241-0_11

turning points in her lived experience of disability, sexuality and gender, this chapter examines how intersecting identities both shape and are shaped by context and experiences of the body as both complex and fluid. Bodily performance and negotiation as tools of resistance to multiple forms of discrimination are also highlighted. Josie's narrative provides a rich account of her experiences embodying multiple marginalized and intersecting social locations, and offers important insights on how to engage in a social work practice informed by a feminist narrative approach, particularly when addressing the intersecting social locations of disability and queer identities. This chapter situates the usefulness of an intersectional lifeline and feminist narrative interviewing as tools for the development of critical, transformative research and practice in social work.

Context

Feminist narrative, as both a research practice and a therapeutic intervention, is a powerful tool for social work. In research, feminist narrative can create conditions for the building of situated knowledge from the perspectives of those who previously had little voice or power in the creation of such knowledge in order to influence change to the social conditions and structures which produce and sustain marginalization and exclusion. In practice, feminist narrative can create and enhance deep alliance-building and shared insight within a therapeutic context in order to support personal and relational growth. Our "feminist theoretical framework" encompasses those feminist theories which contribute to an understanding of the complex ways in which women's experiences are shaped by gendered and intersecting constructs and structures of patriarchy, homophobia and transphobia, and through which women's resistance to domination and oppression are made manifest both in lived experience (Schriempf, 2001) and through alternative forms of praxis (Olesen, 2011).

In this chapter we incorporate an intersectional life course perspective in our analysis (Ferrer et al., 2017) and tools of life story narrative which have been linked to this perspective (Brotman et al., 2020) in order to enhance our capacity to build connections between history and context across time, intersectional identity and interlocking oppressions, and to explicate the relationship between, and encounters with, structures that shape lived experience. Our research story is one of co-construction between ourselves as researchers and, more importantly, in solidarity with Josie, the research participant. On a personal and professional level, we continue to be inspired by both the story shared to us and the potential of feminist narrative to link critical reflection and action, create opportunities for knowledge-sharing and knowledge-making, and facilitate moments of solidarity and support.

The Intersections of Disability, Gender and Sexual Identity in the Literature

Research on the rates of disability and chronic health conditions within the Canadian population has identified that women are overrepresented among the population of people living with disabilities (Burlock, 2017); however, the diversity among these women is still not well understood. In particular, the number of queer women with disabilities has been difficult to estimate. Disability is understood within our work as broad, and includes conditions and illnesses that are chronic, temporary and/or transient. We use the term queer in the current context as an umbrella term that includes a multiplicity of sexual orientations and gender identities, including lesbian, gay, bisexual, transgender, queer and non-binary (LGBTQ+). We do so because this is the way Josie defined herself and so we purposefully engage with the identity label which was meaningful to Josie at the time of these interviews.

Attention to the experiences of sexuality among people with disabilities in the literature has grown over the past two decades and is predicated on an understanding of the positive role that sexuality plays in health outcomes (Organization, 2006). Research on women with disabilities, regardless of their sexual orientation, has pointed to the reality that they often have more difficulty finding romantic, sexual and intimate partners, and participating in these types of relationships (Vaughn et al., 2015). This body of work has been criticized for assuming that people, especially women, with disabilities are asexual (Whitney, 2006), using cis-heteronormative models (Tepper, 2000), and constructing disabled sexuality as inappropriate or non-existent (Crawford & Ostgrove, 2003).

Over the last two decades, the voices of women have emerged to challenge mainstream perspectives used within the disability literature and to explore intersections between gender, sexuality and disability (Walden, 2009; Whitney, 2006; Zitzelsberger, 2005; O'Toole, 2000; Tepper, 2000; Sweeney et al., 2015; Vaughn et al., 2015). Despite this, queer women with disabilities continue to receive minimal attention (Fraley et al., 2007).

Understanding this intersection of identity is essential to providing a more complete picture of the experiences of women with disabilities and to developing responses that support their identities and expressions. People with disabilities and queer people have much in common, including the experience of being subject to sexual oppression and a shared history of injustice (Whitney, 2006). Cultural and social representations of the body and sexuality position both able-bodiedness and cis/heterosexuality as normative, thereby ignoring or devaluing the bodies, sexual and gender identities, and practices of individuals and communities. This positioning reinforces expectations related to women's bodies, sexuality, sexual function and sense of self (Zitzelsberger, 2005; Sweeney et al., 2015; Vaughn et al., 2015). Women with disabilities and queer women have particular challenges regarding expectations of femininity, heterosexuality and normative ideals of the female body that serve to constrain their agency and self-expression (Tighe, 2001; Sweeney et al., 2015).

The literature about queer women with disabilities addresses barriers to sexual relationships, sexual identity and challenges accessing a community (Mona & Gardos, 2000). Queer women with disabilities have been identified as having limited access to relevant sexuality education (Gillespie-Sells et al., 1998) and as being negatively impacted by healthcare providers' assumptions of cis/heterosexuality and limited knowledge of the relationship between minority stress and health (Dibble et al., 2007). The literature clearly demonstrates that the inclusion of the voices of queer women with disabilities is essential to understanding the lived experiences of sexuality and to resisting normative constructions of sexuality and the body that serve to control, marginalize and exclude within multiple sectors such as health and social care services. Themes including empowerment, support within and outside queer communities, and the emotional impact of disability that are explored in the queer disabilities literature serve to render more visible the voices and experiences of queer women with disabilities in unique and context-specific ways (Vaughn et al., 2015). This literature asserts that, in order to illuminate lived experience from situated, subjective positions which pays attention to intersecting identities and interlocking oppressions (Hulko et al., 2020), and which foregrounds queer women with disabilities themselves, an approach sensitive to voice, power-sharing and lived experience across the life course is required.

Doing Feminist Narrative in Social Work Research and Practice

Our feminist narrative approach centred on uncovering the lived experience of queer identity, disability and sexuality (Lyons, 2007). The personal story form (Creswell & Poth, 2018) enabled a full and rich exploration of an individual story, taking into account complex ideas, experiences and critical moments that emerged within that story. The intersectional life story approach was incorporated in order to inform our use of a visual lifeline and enabled an exploration of the intersections between individual lived experience and the relational, social and structural forces which shape that experience across the life course (Brotman et al., 2020). These methods used together are especially relevant for social work researchers, as they include explicit demonstrations of how knowledge is constructed in everyday life through ordinary communicative action and are consistent with social work values of diversity, anti-discrimination and critical consciousness (Riessmann & Quinney, 2005).

Our approach aligns well with feminist narrative as social work practice. Narrative as a therapeutic intervention is based on the premise that meaning emerges from the stories that people share and create about their lived experiences (Combs & Freedman, 2012). Incorporating a feminist lens helps to foreground gender justice in both process and outcomes, exposing multiple and intersecting forms of oppression and dominant discourses that shape women's experience across the life course and centres on the ways in which women both tell their stories and interpret meaning (Banker, 2010). Feminist narrative therapy involves a collaborative process of telling/sharing meaningful moments in order to gain insight and to politicize them (Dumaresque et al., 2018). This process of consciousness-raising facilitates opportunities for clients to account for the often hidden nature of the structural forces which shape lived experience, and to render visible acts of agency in daily life (Brown et al., 2008). As such, feminist narrative can support an empowering re-authoring of one's narrative in the therapeutic encounter, and can lead to social change on a larger scale (Dumaresque et al., 2018).

Feminist narrative (as both research and practice) is inherently reflexive, and transformative in the co-creation of the narrative. This process is called "restorying" (Hollingsworth & Dybdahl, 2007; Creswell & Poth, 2018) and can also include "counter-storytelling" (Hulko et al., 2020) as a means of explicating oppression while simultaneously imagining both personal and structural change. The narrative approach thus involves active collaboration and critical reflexivity (Morley, 2015). In the context of feminist-informed research and practice approaches, the notion of critical reflexivity is used to address the role of the researcher within the research process and requires the researcher to critically reflect upon the nature of research (Elliot, 2005) and its links to social justice and social change efforts (Caine et al., 2018; Morley, 2015). For example, in the context of critical reflexivity, feminist researchers/practitioners have expressed concern about the potential for exploitation of women and other marginalized groups in both academic research and practice interventions, and thus places considerable importance on establishing collaborative, non-exploitative relationships with participants (Sprague, 2005) and clients (Dumaresque et al., 2018). A critical reflexive approach warns against the potential pitfall of simply "pacifying" women through the process of storytelling, but rather affirms women's anger as an important tool of resisting invisibility, exclusion, domination and oppression (Russel & Carey, 2020).

In this study, the authors engaged in a critical self-reflexive process that was designed to explicitly address how social location, personal and professional experience impacted both design and process; the authors set out to share these reflections with the participant to ensure trust and power-sharing over the course of the study (Frisby & Creese, 2011; Sprague, 2005). The authors identify as feminist social workers who are committed to social justice, whether through political activism in the community, professional work or academic scholarship. One of the authors identifies as queer and one as a queer ally. Both have experience with chronic illness and disability and are acutely aware of the impact of these intersecting identities on sexual identity and expression. Issues of discrimination—sexism, homophobia, transphobia and ableism—faced by women with multiple intersecting marginalized social locations are integrally tied to this personal and professional experience. It is for ourselves, our friends and sisters, including the participant of this study, that we became committed to understanding how queer women with disabilities experience and make meaning of sexuality, intimacy and identity.

Research Design in the Current Study

Research that elicits respondent stories about their experiences, understandings of identity and the meaning of their lives brings up particular ethical questions. For example, the data collection process of feminist narrative research involves in-depth interviewing, and can often result in intimate conversations sometimes akin to those that would take place in a therapeutic encounter (Lieblich & Osselson, 1996; Elliot, 2005). As such and given that these studies focus on the retelling of singular accounts, we gave careful attention to anonymity and confidentiality, which were of primary importance, and ethical conduct was assured through adherence to research ethics conditions established by the university ethics review process throughout the duration of the project. We recruited our participant through engaged outreach with local feminist, queer and disability-justice community-based agencies.

Our participant was a 27-year-old queer identified woman at the time of the study who experienced chronic illness and physical disability due to ulcerative colitis, which she described as fluid and shifting between invisible and visible states, depending upon time and circumstances. She also described herself as having a fluid queer identity encompassing both gender non-conformity and sexual minority status. Further description of the participant and her story is included in the Findings section.

A feminist narrative approach requires the interviewer to elicit participant stories, which can be aided by a loosely structured interview guide that encourages a conversational dynamic and space to probe into unanticipated circumstances and responses. A research protocol, an unstructured interview guide and journaling exercises were developed and reviewed with the participant prior to beginning the study and negotiated to ensure relevance. As one example of the negotiation process, the participant rejected the journaling component, which was part of the original study design. Instead, she suggested undertaking more interviews to ensure we had enough space and time to build trust and engage in a meaningful dialogue around the complex themes which emerged in discussion.

Themes were co-structured to lead to open conversation and prioritize the "how," which is important when the research focus is on the ways that people make meaning and participate in creating their lives (Gubrium & Holstein, 1997; Elliot, 2005). The three thematic categories were: (a) how do you describe your identity, (b) how do you express your sexual self; (c) how do others respond to your sexual self? This final theme focused on several key cohorts, including lovers, sexual partners, friends, family, colleagues, queer and communities of people with disabilities, and health and social service professionals. All five interviews lasted approximately two hours each and were audio-recorded and transcribed. A reflective log was kept by the primary researcher/interviewer throughout the

Table 11.1 Step-wise analytic concepts connected

Feminist Narrative Listening Guide (Brown, 1998 as described in Doucet and Mauthner, 2010)	**Connections between concepts** *italics used to emphasize connections	**Intersectional Life Course Life Story (Ferrer, Grenier, Brotman and Koehn, 2017; Brotman, Ferrer and Koehn, 2019)**
Relationally and reflexively constituted narratives (themes, events, time)	↔	*Life events, timing* and structural forces
Tracing narrated subjects (particular narrator or subject and how they speak about themselves and their social world (the "I"))	╳	Globally and *locally linked lives*
Relational narrated subjects' social networks and close and intimate relations	╳	*Identities, categories* and processes of difference (othering)
Structured subjects (structured power relations and dominant ideologies)	↔	*Domination, agency* (resilience) *and resistance*

research process. Analysis was informed by the step-wise concepts of the feminist narrative *Listening Guide* method (Woodiwiss et al., 2017; Doucet & Mauthner, 2008), combined with the intersectional life course/life story method. This paid attention to the integration of personal story across time and place, key events and significant moments across the life course, how meaning is created and shared through storytelling (language, metaphors, etc.) and other representational forms, agency and resistance, as well as relational, structural, social and contextual forces which shape lived experiences (Ferrer et al., 2017; Brotman et al., 2020). These concepts, and their connections, are presented in Table 11.1 (see also Fig. 11.1).

Analysis began after completing three of the five interviews, after which the remaining two interviews focused upon both interviewer and participant sharing/reading of the transcripts and discussing what was deemed as important or required elaboration. This included a focus upon how the participant embodied her identity, maintained agency throughout her struggles and resisted oppression and marginalization.

To clearly understand the relationships between critical moments and events in the participant's life, a lifeline diagram was co-created to represent participant experiences over time. There was a struggle to find a way to represent complex multiple identities of being queer and disabled on one lifeline. Since these two identities were fluid and constantly shifting through time, there were moments of divergence as well as moments where the two intersected; as such, two lifelines were positioned one beside the

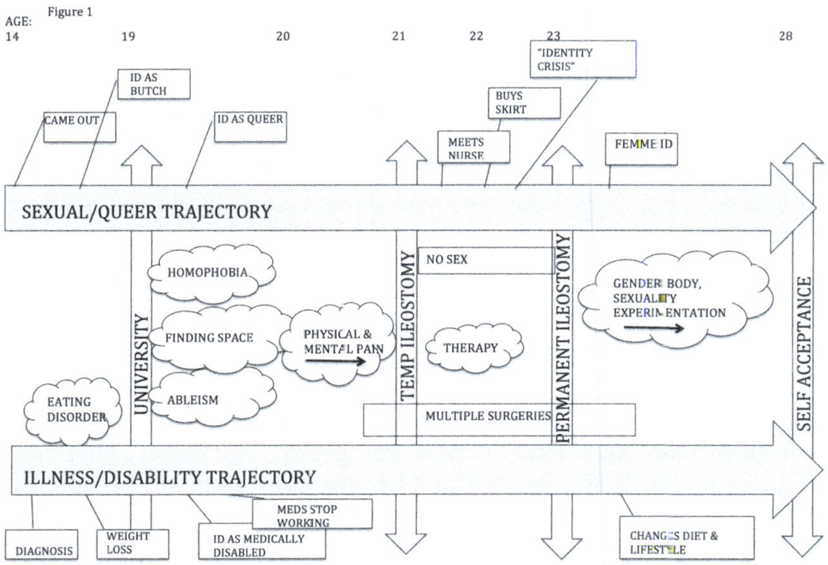

Fig. 11.1 Concepts and connections shaping Josie's experiences

other—one for Josie's queer trajectory and one for her disability trajectory. These two lines are guided through time by the participant's age and include a space for experiences that couldn't be represented as single events on either lifeline. Two rectangles symbolically reference experiences in between both lifelines that span over a specific period of time, while experiences represented in cloud shapes are those that continue throughout time, or weave in and out of Josie's life. The benefit of having a visual representation of this data is that it clarifies the timing of critical moments, and the participant's identities and experiences within a chronological as well as thematic context that acknowledges some experiences as non-linear.

Josie's Story

At the time of the study Josie was a 27-year-old woman who identified as a queer femme and used the words "medically disabled" to describe herself as a person who has struggled with a severe experience of chronic illness. Josie's story began at age 14, when she came out as bisexual. In the first few years after Josie came out, she aligned more closely with a non-binary gender identity and expression, identifying, presenting and calling herself "butch." At 15, Josie experienced symptoms of inflammatory bowel disease, and after testing, was diagnosed with ulcerative colitis. Josie began medication; however, she continued to become progressively ill. She lost weight, then developed an eating disorder, was in pain and became hospitalized a year later. Josie experienced the years that followed as a time of struggle with illness and pain, exploring her understanding of her sexual and gender identity, and negotiating her feelings and ideas about living in and with her body.

At 19, Josie moved to Montreal to start university. Her experiences of homophobia, transphobia and ableism while living in the university residence represented a critical moment in her life and identity. Meeting others who shared some of her experiences of sexuality prompted a shift in her language around the different parts of her identity. She began calling herself queer and medically disabled, and became more aware of the complex ways that her illness and her sexuality would impact her interactions with and in the world. Connecting with supportive queer people provided Josie a sense of community.

One year later, Josie experienced another decline in her health when her medication stopped working, requiring her to undergo surgery. This period of time in Josie's life included extreme physical pain, isolation and uncertainty. Complications following surgery intensified her struggle with her body and, at 23, she decided to have a permanent ileostomy, in the hopes of alleviating pain. The addition of a permanent ileostomy to her body was a huge adjustment for Josie; it impacted her relationship to her body, her sexuality and her gender. Receiving support from family, friends and her queer community, and seeking out knowledge about her body, illness and identity helped Josie maintain strength and agency over the course of her life. In addition, she benefited from adjusting her diet and lifestyle, and experimenting in her sexual life and gender presentation.

Thematic Interpretation of Josie's Story

The impact of illness and disability on the embodiment of identity, as well as on Josie's agency and resistance to social and cultural norms relating to sex, gender and the body, was the main theme that emerged in Josie's narrative. As well, we explore experiences of discrimination across sites. Josie's story represents a unique opportunity to provide insights into how understanding identity as fluid and shifting opens up possibilities for resistance to discrimination among queer women with disabilities. In order to resist conceptions of herself as asexual and to cope with her experiences of illness, Josie developed and maintained strength and agency—which allowed her to resist dominant ideas of sex, gender and the body. A myriad of strategies helped her get to this place of agency, including experimenting with her sexuality and gender presentation, as well as seeking out support from her queer community and family. The factors that Josie saw as limiting her capacity to self-actualize included societal ideals regarding gender, ability, and sexuality and discrimination within school, community and service systems.

Negotiating multiple and intersecting identities often required Josie to resist mainstream notions of the body. These include the female body, the healthy body and the asexual or gender-normative body. She resisted these ideas through her alliance with the queer community, and her appreciation of bodies that she named as "different." She was open to the various ways that identities can be performed, embodied and shifting. Josie's complex understanding of identity as embodied and as fluid is, in itself, a perspective that resists dominant and mainstream understandings of who she is. Josie's perspective challenges the view of identity as fixed, of gender as either male or female, of sexuality as right or wrong and of the body as healthy or not.

Accounting for Experiences of Discrimination in Josie's Story

Josie experienced sexism, ableism, homophobia and transphobia in both overt and covert ways in many of her exchanges with the world around her. For example, Josie had a private room in the university residence, so that she could have a private bathroom to accommodate her illness. She then was singled out as the only student in residence who had a room with a private bathroom. This difference and the attention it generated meant that she became more exposed for her other categories of difference. Her visible queer identity incited homophobic harassment from other women on her floor. She described her experience in residence as "horrible" and said: "Oh my god they made my life a living hell. Oh my god, they called me dyke, on the elevator they would whisper about me but like really loudly."

Josie also described experiences of exclusion and marginalization within the mainstream queer community:

> In the queer community there is sort of this fetishization of mobility, like oh ya I'm gonna go to San Francisco and then I'll go to Vancouver, and then I'll sleep on a couch in Portland, and then I'll go visit my friends .... there is this huge primacy placed on mobility and um people that are in pain, people that have disabilities can't fucking do that...I absolutely cannot sleep on a couch ... if I'm not sleeping my body doesn't work properly ... people don't get it, and you are excluded from community when you can't do those types of things.

Finally, in her interactions with healthcare providers Josie noted an overriding ignorance amongst providers with respect to how her disability may impact her sexuality. In one example, Josie described asking a doctor about how her situation would impact her ability to continue having anal sex:

> I asked the doctor, I came in prepared, because I was thinking about it at home and I was like oh my god, like why hasn't anyone told me this is going to happen? Like if you were having a surgery where they were sewing up your vagina, don't you think they would tell you? But obviously they don't think about it in a sexual way, so I talked to my doctor about it and then I asked him if he had ever had a gay patient, I said well haven't you ever had a ... like he didn't really know what to say.

Resisting Through Sexual Experimentation

Josie's story describes struggling with pain, poor body image and mental health concerns. However, these difficult times are punctuated with positive and pleasurable sexual and intimate experiences. Even in the midst of critical illness, when Josie was hospitalized, she maintained some elements of her sexual life: "It was my idea and she was like 'are you sure?' And I'm like 'ya' and so we had sex in the [hospital] bathroom listening to Ani Difranco on my fisher price cassette player (laughing)."

When Josie received the temporary ileostomy, she was initially concerned about having sex. At this time, an ostomy nurse met with Josie, and one of the points of discussion was sexuality. Josie remembers feeling embarrassed, as well as frustrated, at the nurse's cis-heterosexist assumptions when she described how feminizing her bag with pink lace would make her feel better about herself. Instead, Josie found ways to cover her ileostomy bag to feel more at ease, which included incorporating electrical tape to secure it during sex:

> So I'd have the bag cover and I'd have electrical tape and I'd tape it down, it was sort of sexy … I was in my early twenties, and it really opened my mind up to kinky, I guess I had been sort of vanilla before, well all of a sudden you have to deal with this altered body, you really don't have a choice but to be creative and come up with things that work for you.

Josie described that having a "non-normative" body has been beneficial to her sexual life. "By sort of likening the electrical tape—which is a functional necessity of my disability-to a fun, sexy aspect of my erotic life it makes my ileostomy an addition rather than a hindrance to the whole experience."

Josie structured her sexual life to support her needs. This involved being selective about her partners, ensuring that whom she chooses to engage with sexually will support a positive body image. She explained that because of her particular situation she had to develop excellent communication skills and the ability to talk about sex, her boundaries and her body with partners.

Resisting Through Gender Experimentation

Experimentation also emerged as a theme related to Josie's presentation of her gender as fluid and shifting—moving from a butch to femme identity. Josie needed to find ways to reduce her anxiety about her ileostomy bag's visibility and discomfort caused by pants pressing against the ileostomy, so she decided she would try wearing skirts. Josie's identity as butch became more fluid in response to her need to experiment and find ways to cope with her ileostomy and resulted in new sources of strength and empowerment:

> I love lingerie now, I didn't used to be like this, I used to be really sort of butchy, it's been this slow progression. Like go big or go home. If I'm gonna be wearing skirts I might as well look damn good wearing them.

Over the course of Josie's trajectories of disability and sexuality, there have been many moments of struggle, and more often than not, these difficult experiences were also identified as turning points of growth. Josie developed strength and agency in her life, which she used to cope with pain, to find pleasure and to resist oppression. Allowing herself the freedom to explore and experiment with her gender and sexuality in the ways elaborated above ultimately led her to increased comfort, confidence and happiness.

Discussion

Josie's Story: Lessons for Feminist Narrative-Informed Research and Practice

We chose to focus on Josie's story of agency, resilience and resistance, in order to highlight how the shifting and fluid nature of identity is experienced among queer women with disabilities in their struggles to locate and live their sexual selves. We also addressed Josie's experiences of multiple forms of discrimination, marginalization and ignorance at university, in the healthcare system and within mainstream queer community spaces. Experiences of intersectional discrimination have been noted previously in the literature (Whitney, 2006; O'Toole, 2000). Josie resisted these experiences of discrimination by seeking out alternative communities in which she could get information, feel a sense of belonging and receive support. Foregrounding the relationship between the personal and the structural across time and place was facilitated through the integration of the *intersectional life course perspective* and the *feminist listening guide* as a complex and nuanced analytic framework. The strength of this framework centres on the articulation of a step-wise strategy of analysis along four key concepts, namely, fluidity, time and situated knowledge (context); relationships (to self and others); the interplay of structural forces and lived experience; and resilience and resistance in enacting agency in daily life. Each of these key concepts were used to deepen and enrich our understanding of Josie's life course story in ways which paid attention to her unique subjectivity and positionality. Josie's lived experiences and perspectives on the intersections of queer, disability and gender identities provide a powerful example of agency in confronting interlocking oppressions in daily life.

Josie's story can provide several important lessons for social work.

Social workers, and other helping professionals, have an important role to play in supporting queer women with disabilities' efforts to express their sexuality. First and foremost, addressing discrimination in all its forms and at all levels (i.e. interpersonal, institutional and systemic) is of paramount importance. A lack of attention to and knowledge about queer disabled women's sexuality is a major barrier in creating a space for open discussions. In order to meet the needs of their clients, helping professionals must be aware of their biases and triggers. Additional knowledge about, and comfort in discussing, sexual orientation, gender identity and ways of expressing sexuality among women with disabilities is needed.

It is essential that helping professionals, and the organizations in which they work, incorporate an understanding of the intersectional nature of identity and the impacts of occupying multiple marginalized positions on clients' experiences. Opportunities for education and training on these subjects are needed to increase helping professionals' knowledge about intersecting inequalities and discrimination, and ways in which to challenge oppression at multiple levels. Education and training will also increase helping professionals' comfort in discussing sexuality and identity with clients and will increase their ability to make appropriate referrals to other professionals and/or to possible community connections outside of the healthcare system.

Helping professionals who are well informed about the realities experienced by queer women with disabilities will be in a better position to act as advocates for their clients, to become leaders within their organizations and to provide education and resources when necessary. This could take the form of support groups for queer women with chronic illness or disability, workshops about sexual pleasure, safer sex, building a positive body image and sexual communication skills. Not all clients are as confident, communicative and self-aware as Josie. As helping professionals, we can use Josie's story as one of inspiration, passing on Josie's strategies to our own clients and working alongside them to create lasting change within, and outside of, the healthcare system.

Feminist narrative approaches are well aligned with our commitment to voice, collaboration, empowerment, advocacy and social justice. A feminist narrative approach helps us to use our profession, whether as researchers or clinicians, as an opportunity to amplify voices that are often silenced, and to collaborate with research participants and clients in creating and (re/counter) storying narratives that challenge dominant discourses. Josie's story offers us an example of the power of a politicized narrative of one's experience in the world, and its potential to contribute to personal resistance and resilience in the face of sexism, ableism, cisgenderism and homophobia/transphobia. Engaging in a feminist narrative approach with research participants and clients in social work can make important contributions to a larger conversation about the possibilities created by disrupting dominant narratives.

In concluding, we would like to dedicate this chapter to Josie, who died peacefully surrounded by friends and family in 2018. May her strength, courage and passion live on through the memories held by those whose lives she touched, and in the stories shared through this research. We hope that Josie's story will be an inspiration and call to action for others, as it has been for us.

Acknowledgement This chapter has been adapted and expanded from the original publication Drummond, JD. and Brotman, S. (2014). Intersecting and embodied identities: A queer woman's experience of disability and sexuality. Sexuality and Disability, 32(4), 533–549. Reprinted with permission from Springer Nature Publishing, licence number 4721340225208, received 3 December 2019.

References

Banker, J. R. (2010). Dating is hard work: A narrative approach to understanding sexual and romantic relationships in young adulthood. *Contemporary Family Therapy: An International Journal, 32*, 173–191.

Brotman, S., Ferrer, I., & Koehn, S. (2020). Situating the life story narratives of aging immigrants within a structural context: The intersectional lifecourse perspective as research praxis. *Qualitative Research, 20*, 465–484.

Brown, C., Weber, S., & Ali, S. (2008). Women's body talk: A feminist narrative approach. *Journal of Systemic Therapies, 27*, 92–104.

Burlock, A. (2017). *Women with disabilities. Women in Canada: A gender based statistical report.* Statistics Canada.

Caine, V., Steeves, P., & Clandinin, D. J. (2018). Social justice practice: A narrative inquiry perspective. *Education, citizenship and social justice, 13*, 133–143.

Combs, G., & Freedman, J. (2012). Narrative, postructuralism, and social justice: Current practices in narrative therapy. *The Counseling Psychologist, 40*, 1033–1060.

Crawford, D., & Ostgrove, J. (2003). Representations of disability and the interpersonal relationships of women with disabilities. In M. E. Banks & E. Kaschak (Eds.), *Women with visible and invisible disabilities: Multiple intersections, multiple issues, multiple therapies.* Haworth Press.

Creswell, J., & Poth, C. (2018). *Qualitative inquiry & research design: Choosing among five approaches.* California, Sage Publications.

Dibble, S., Eliason, M., & Christiansen, M. (2007). Chronic illness care for lesbian, gay & bisexual individuals. *Nursing Clinics of North America, 42*, 655–674.

Doucet, A., & Mauthner, N. (2008). What can be known and how? *Narrated subjects and the Listening Guide Qualitative Research, 8*, 399–409.

Drummond, J. D., & Brotman, S. (2014). Intersecting and embodied identities: A queer woman's experience of disability and sexuality. *Sexuality and Disability, 32*(4), 533–549.

Dumaresque, R., Thornton, T., Glaser, D., & Lawrence, A. (2018). Politicized narrative therapy: A reckoning and a call to action. *Canadian Social Work Review, 35*, 109–129.

Elliot, J. (2005). *Using narrative in social research: Qualitative and quantitative approaches.* Sage.

Ferrer, I., Grenier, A., Brotman, S., & Koehn, S. (2017). Understanding the experiences of racialized older people through an intersectional life course perspective. *Journal of Aging Studies, 41*, 10–17.

Fraley, S. S., Mona, L. R., & Theodore, P. (2007). The sexual lives of lesbian, gay, and bisexual people with disabilities: Psychological perspectives. *Sexuality Research and Social Policy Journal of NSRC, 4*, 15–26.

Frisby, W., & Creese, G. (2011). Unpacking relationships in feminist community research: Crosscutting themes. In G. Creese & W. Frisby (Eds.), *Feminist community research: Case studies and methodologies*. UBC Press.

Gillespie-Sells, K., Hill, M., & Robbins, B. (1998). *She dances to different drums: Research into disabled women's sexuality*. London King's Fund

Gubrium, J., & Holstein, J. (1997). *The new language of qualitative method*. Oxford University Press.

Hollingsworth, S., & Dybdahl, M. (2007). Talking to learn: The critical role of conversation in narrative inquiry. In D. J. Clandinin (Ed.), *Handbook of narrative inquiry: Mapping a methodology*. Sage.

Hulko, W., Brotman, S., Stern, L., & Ferrer, I. (2020). *Gerontological social work in action: Anti-oppressive practice with older adults, their families and communities*. Routledge Press.

Lieblich, A., & Osselson, R. (Eds.). (1996). *The narrative study of lives*. Sage.

Lyons, N. (2007). Narrative inquiry, what possible future influence on policy or practice? In J. Clandinin (Ed.), *Handbook of narrative inquiry, mapping a methodology*. Sage.

Mona, L. R., & Gardos, P. S. (2000). Disabled sexual partners. In L. T. Szuchman & F. Muscaralla (Eds.), *Psychologist perspectives on human sexuality*. Wiley.

Morley, C. (2015). Critical reflexivity and social work practice. In J. D. Wright (Ed.), *International Encyclopedia of the Social and Behavioral Sciences* (2nd ed.). Elsevier.

O'Toole, C. J. (2000). The view from below: Developing a knowledge base about an unknown population. *Sexuality & Disability, 18*, 207–224.

Olesen, V. (2011). Feminist qualitative research in the millennium's first decade: Developments, challenges, prospects. In N. Denzin & Y. S. Lincoln (Eds.), *The Sage handbook of qualitative research* (4th ed.). Sage.

Organization, W. H. 2006. *Defining sexual health: Report of a technical consultation on sexual health Geneva: World Health Organization*. https://www.who.int/reproductivehealth/publications/sexual_health/defining_sh/en/. Accessed 9 Aug 2021.

Riessmann, C. K., & Quinney, L. (2005). Narrative in social work: A critical review. *Qualitative Social Work, 4*, 391–412.

Russel, S., & Carey, M. 2020. *Feminism, therapy and narrative ideas: Exploring some not-so-commonly asked questions*. http://www.narrativeapproaches.com/resources/narrative-therapy-archive/329-2/. Accessed 20 Feb 2020.

Schriempf, A. (2001). (Re)fusing the amputated body: An interactionist bridge for feminism and disability. *Hypatia, 16*, 53–79.

Sprague, J. (2005). *Feminist methodologies for critical researchers.* Altimira Press.

Sweeney, K., Horne, S., & Ketz, K. (2015). Sexual orientation, body image, and age as predictors of sexual self-schema for women with physical disabilities. *Sexuality & Disability, 33*, 313–326.

Tepper, M. S. (2000). Sexuality and disability: The missing discourse of pleasure. *Sexuality & Disability, 18*, 283–293.

Tighe, C. A. (2001). 'Working at disability': A qualitative study of the meaning of health and disability for women with physical impairments. *Disability & Society, 16*, 511–529.

Vaughn, M., Silver, K., Murphy, S., Ashbaugh, R., & Hoffman, A. (2015). Women with disabilities discuss sexuality in San Francisco focus groups. *Sexuality & Disability, 33*, 19–46.

Walden, E. (2009). An exploration of the experiences of lesbians with chronic illness. *Journal of Homosexuality, 56*, 548–574.

Whitney, C. (2006). Intersections in identity- identity development among queer women with disabilities. *Sexuality & Disability., 24*, 39–52.

Woodiwiss, J., Smith, K., & Lockwood, K. (Eds.). (2017). *Feminist narrative research: Opportunities and challenges.* Palgrave Macmillan.

Zitzelsberger, H. (2005). (In)visibility: Accounts of embodiment of women with physical disabilities and differences. *Disability & Society, 20*, 389–403.

CHAPTER 12

Invisible Women: Critical Perspectives on Social Work and Gender in Later Life

Trish Hafford-Letchfield

Introduction

This chapter examines the literature on gender and ageing through the lens of critical feminist gerontology. This draws on multidimensional concepts comprising power, economic, social, and symbolic relations operating simultaneously at intrapersonal, interpersonal, institutional, and societal levels (see Connell, 2012). Applying an intersectional lens and feminist perspective to social work practice with women in later life requires us to move beyond gender as one way of explaining differences (Hafford-Letchfield, 2020) towards ensuring advocacy and empowerment to address inequalities and shape supportive social work interactions. Relational theory (Connell, 2012) can also provide insights into the complexity of gender relations in later life, with a clear focus on the uniqueness of a person's circumstances within the context of ageism. The impact of the COVID-19 pandemic has brought societal ageism into sharp relief.

T. Hafford-Letchfield (✉)
School of Social Work and Social Policy, University of Strathclyde, Glasgow, UK
e-mail: Trish.hafford-letchfield@strath.ac.uk

C. Cocker, T. Hafford-Letchfield (eds.), *Rethinking Feminist Theories for Social Work Practice*,
https://doi.org/10.1007/978-3-030-94241-0_12

Swift and Chasteen (2021) highlight how discourse surrounding the COVID-19 pandemic has strengthened the homogeneous view of older people as vulnerable, the social stigma from being an older person, and the exacerbation of both hostile and benevolent expressions of ageism, which must be counteracted.

The wider social work profession and its policy makers (Richards et al., 2014) have not yet grasped the challenge and importance of gerontological social work and its evidence base. This chapter will review some key areas of structural gender inequalities in later life and draw on some of the empirical evidence. I will outline the contribution made by critical gerontology and feminist gerontology and its relevance for social work and social care. This chapter aims to facilitate a better and more nuanced understanding of the specific challenges faced by women in later life to help social workers improve support through purposive gender responsive and age sensitive practice (Crockett et al., 2018).

Key Issues for Gender in Later Life

The Global report on ageism (World Health Organisation, 2021) shows that ageism is prevalent, ubiquitous, and insidious because it goes largely unrecognised and unchallenged. Ageism has serious and far-reaching consequences for people's health, well-being, and human rights and is costly to society. Examples are the association with poorer physical and mental health, increased social isolation and loneliness, greater financial insecurity, decreased quality of life, and premature death. Ageism refers to the stereotypes (how we think), prejudice (how we feel), and discrimination (how we act) directed towards people based on their chronological age. This can be institutional, interpersonal, or self-directed. Ageism at an institutional level refers to the laws, rules, social norms, policies, and practices of institutions that unfairly restrict opportunities and systematically disadvantage individuals based on their age alone. Interpersonal ageism arises in interactions between two or more individuals, while self-directed ageism occurs when ageism is internalised and self-directed. Over the past few decades, many studies have shown that people's representations of old age can even turn into self-fulfilling prophecies, affecting their health, well-being, and even their longevity via physiological, psychological, and behavioural mechanisms (Kornadt et al., 2019). Barrett et al.'s (2020) 'calculated ageism' refers to expressions or sentiments that justify prejudicial beliefs or unfair treatment with positive claims. This has never become as clear as it has during the COVID-19 pandemic where public discourse suggested older generations should sacrifice their lives in order to reduce the economic cost for future generations. The expression that older lives are expendable, set against the benefit of reduced economic burden on younger generations, is a stark expression of ageist and a damaging characterisation of how a society feels about its older generations (Aronson, 2020; Swift & Chasteen, 2021).

Within ageism, gender is a significant factor, as women outnumber men in older age groups in all EU member states. Resulting from a longer life expectancy, there are more women than men in the EU, with 105 women per 100 men (5% more) in 2019 (Eurostat, 2018). For example, of over-80-year-olds, women make up 65.0% of the population and of over-85-year-olds, the proportion of women is 69.0% (Eurostat, 2019b). These statistics are interesting given that research on the combined impact of sexism and ageism (known as gendered ageism) in later life has concluded that older women—relative to older and younger men—bear the brunt of multiple forms of discrimination.

Critical Gerontology and Feminism

When Butler introduced the concept of ageism in 1969, the primary aim was to highlight forms of exposure to marginalisation and discrimination by older people. Butler was concerned with the wider discourse of the time that focused on revealing categorical inequalities as experienced by different social groups based on age, gender, and race/ethnicity and those responses that were rooted in social movements. The term 'gendered ageism' was subsequently introduced by Itzin and Phillipson (1993, 1995), in their study of age barriers at work, where they focused particularly on gender in both the private and public sector. Gendered ageism describes a double jeopardy, where two interacting power systems lead to an increased vulnerability and emphasises the dominance of patriarchal norms combined with a preoccupation with youth that results in a faster deterioration of older women's status compared to that of men. Theoretical work emphasising gender as a relational process has also led to theoretical problematisation of this concept to understand the processes of multiple marginalisations based on age and to see how these affect both men and women of various ages (Krekula et al., 2018).

If we look at the development of critical gerontology this has included the contributions of feminism, but ageing has not been central to feminist theory (Neysmith and Reitsma-Street, 2009). Ray (2008) credited the philosopher, writer, and activist Simone de Beauvoir with inspiring critical gerontologists. de Beauvoir's 'The Coming of Age' (1972) challenged readers to stop denying old age and to combine micro- and macro-analyses in fully recognising that ageism must be overcome at both the individual and the societal levels. Other activists who had an impact on popular and academic audiences include Maggie Kuhn (1977), Barbara Macdonald and Cynthia Rich (1983), Doris Lessing (1983), May Sarton (1984), Betty Friedan (1993), and Eleanor Stoller and Rose Gibson (1994). These contributors have all provided an excellent foundation for feminist gerontology. According to Ray (2008) feminist gerontology has become an area of inquiry where gender studies become central to both the understanding of ageing and the construction of gerontology as a field (Ray, 2008).

Social Work with People in Later Life

The status of social work with people in later life reflects the status of the service-user group itself and the relative prestige given to this area of practice (see BASW, 2018). Within social care, older people are often conceptualised as a problem of demography (Hafford-Letchfield, 2013) and practice has become focused on the challenges associated with supporting those living with disabilities and long-term conditions. Further, globalisation and structural influences continue to disadvantage older people as the ideologies of economic rationalism have radically shaped the organisational and professional landscape of social work practice. Social work however has unique insights into the genuine connection of gender to current circumstances and illuminates a more nuanced understanding of the factors that underpin the connections between lifestyle and identities obscuring the trajectories of those using social care but more than likely structured by previous opportunities in later life. We need to move away from the preoccupation of delivering services against increasingly detailed government guidance and rhetoric to helping people to adjust to personal and social circumstances in a reductionist approach thus reducing complex socio-economic, pathological, or individual and frankly ageist failings (Hafford-Letchfield, 2016).

Social workers might use both gerontology and feminism which share common goals in relation to the development of social consciousness about inequities, utilisation of theories and methods that accurately depict life experiences, and promotion of change in conditions that negatively affect older women. As described in virtually all areas of this book, feminism recognises the intrinsic value of women, their human rights, right to equal treatment, and individuality. However, research and practice evidence in feminist social work still focuses on issues concerned with other age cohorts and likewise, much of the gerontological research and practice evidence has tended to focus on older people without making distinctions by gender (Tesch-Römer & Von Kondratowitz, 2006; Ray, 2008). Given that gender equality issues and empowerment of women are clearly signalled in the social change agenda of the United Nations Millennium Development Goals, wider interest in feminist gerontology in social work is gaining momentum. This may also be partially related to the ageing of feminist social workers themselves as they confront the combination of the myths, stereotypes, and discrimination based on age coupled with the still present myths, stereotypes, and discrimination based on gender (Craciun & Flick, 2016). Further, views on ageing emerge in a certain social context (Tesch-Römer & Von Kondratowitz, 2006) and influenced by gender constructs, stereotypes, and roles existing in that context. Exploring how gender shapes views on ageing may help better understand why some men and women age in a positive way while others do not. It may also support further examination of the gendered nature of power and domination from a collective rather than individual perspective and embed intersectional issues throughout. Further, feminist gerontology views gender as a relational construct that provides women and men with advantages and disadvantages, which can be cumulative and should be seen in a holistic way, which emphasises strengths, oppression, differences, and abilities (Hooyman et al., 2002). By analysing practice with people in later life in a more inclusive way, social workers can identify where they might advocate for social change to reduce inequities. Hooyman et al. (2002) has usefully explicated nine interconnected elements that must be considered in critically analysing policies from a feminist gerontological framework. These elements are inequality and oppression, diversity and ageing, production and reproduction, separate spheres and familism, political and structural change, power and empowerment, holistic and ecological view, solidarity, and alternative methodology.

Socio-Economic Factors in Later Life

Social work is often concerned with resource allocation. Pensions are an important determinant of their beneficiaries' economic independence and pension policies and systems are far from neutral. There is an estimated pension's gap of 39% against women in the EU (Bettio et al. n.d.). Much is already known about the gender pay gaps (Eurostat, 2019a) and economic inequalities between men and women (Stefko et al., 2020) so this is hardly surprising. It is well known that women participate less in the labour market, they work fewer hours and/or years, and receive a lower wage, which can really reduce their career earnings. These gaps can follow through to later life and often reflect the cumulated disadvantages of a career spent in a gender-biased labour market. The structure of pensions is further influenced by gender blind government policies. Krekula et al. (2018) go as far to assert that the institution of retirement as we know it is a male concept and suggests that within that context it is not surprising that its unquestioned heteropatriarchal norms are overlooked in social policies as a gender-neutral approach. In relation to pensions, firstly, there are very long-term structural changes, such as trends in relation to how women have been engaged in employment, and social norms such as divorce, widowhood, and cohabitation between generations. Secondly, historical pension reforms affect current pensions and not all cohorts will be covered by such transitional arrangements. Some women will have lived and worked under one system and will in many cases receive benefits under another. Being a group stuck in the middle, some women often lose protection by the internal operational logic of the system, whether new or old, and can be vulnerable to new kinds of pension risks (Betten et al., undated). This is illustrated in the UK campaign '*Back to 60*' in which women born in the 1950s had their state pensions severely impacted by pension reform, combined with cumulative impact of other losses such as protection in preceding divorce legislation and employment rights and enduring challenges in financial sustainability (Ageing Better, 2017). The Government Work and Pensions Select Committee (House of Commons, 2016) concluded that this group of women had been severely disadvantaged as a result of having to wait longer before they can claim their state pension, being unable to save for retirement, or having to work longer into later life, which placed them in an irrecoverable position.

Other evidence of gender inequalities in financial insecurity in later life follow family circumstances and working patterns of women who still do the majority of caring for children and other family members. Caring determines part-time work, shorter part-time careers, or women spending most of their time caring for their family. Although based on a small number of studies, there is evidence that people from Black and minority ethnic groups are less likely to receive a private pension and more likely to receive pension credits than white men (Krekula et al., 2018). Krekal et al.'s study found that Bangladeshi and Pakistani people appear to be the most vulnerable to financial insecurity in later life, especially women.

These factors affect choice and future access to care, for example, in relation to the UK long-term care market, which has considerable financial impact on individuals and their families who are required to contribute financially, or in some circumstances, fully fund their own care (Hollinrake & Thomas, 2015). In the UK, the need to improve public confidence in the quality of care in the face of this huge expense is also poised against a background of austerity measures (Tanner et al., 2018). Social workers have a role in triggering financial assessments and attempting to balance needs and choice with resources available and may struggle with balancing person-centred support with fair and reasonable resource allocation when promoting 'choice' within limited options (Higgins & Hafford-Letchfield, 2018). Additionally, women themselves may have priorities that do not necessarily sit easily with other organisational policy and procedural requirements. These can result in conceptions of empowerment, which are simplistic or limited. A range of other policy measures can promote gender equity through improvements in the quality and affordability of care, health care, and housing. This will help to counter-balance the disparities in individuals' economic resources in later life and reduce the care burden placed on family members.

Global Inequalities

Sources of discrimination for women in later life often take on many other forms and can be extreme in some regions of the world. For example, in parts of sub-Saharan Africa, accusations of witchcraft are widespread, with older women being persecuted and accused of causing ill luck, disease, or death (Eboiyehi, 2017). In many parts of the world, older widows are socially ostracised or discriminated against. For instance, widows are denied the right to inherit the property they shared with their husbands (Kimani & Maina, 2010). Older women are disproportionately impacted by poverty, disability, and violence. Over a lifetime, gender-based disparities in employment, health care, education, and more negatively influence the well-being of older women (World Health Organisation, 2015). Van Eeuwijk (2006) identified old-age vulnerability in relation to risk to inadequate care provision in Indonesia and found that unmarried women and poor widows were the most vulnerable due to their weaker social position in their family and kin. Studies reviewed by Virokannas et al. (2020) that addressed widowhood highlighted the significance of women's societal and cultural position in producing or protecting them from vulnerability, which increases the risk of social exclusion. Such vulnerability related to the weaker and less protected economic situation of women in certain life situations, and their care needs. It is noteworthy that the concept of vulnerability is not often explicitly discussed as being gender specific within social work (Fineman, 2008, 2010), particularly in relation to the circumstances that generate vulnerability and the responsibility of the state and its institutions in reducing the risks and consequences of vulnerability. Fineman argues for the recognition of the temporal, situational, relational, and structural nature of vulnerability, and Virokannas et al. (2020) recommends that gendered practices and relationships in society must be problematised so as to address design more relevant interventions (Virokannas et al., 2020).

Health and Well-being

In the UK, there is clear evidence of large inequalities in health and wellbeing in later life, related to socio-economic position, ethnicity, gender, and regions. For example, there are stark differences in the levels of frailty related to wealth and these inequalities have widened. Health is impacted by ethnic origin as in the age group 61–70 yrs, where 34% of White English people report fair or bad health, compared with 63–69% of Indian, Pakistani, and Caribbean people and 86% of Bangladeshi people with a potential disproportionate impact on women in those communities (Nazroo, 2015). The focus of descriptive work on gender inequalities in health has been on the higher levels of morbidity experienced by women, which contrast strikingly with their lower risks of mortality. Nazroo (2015) illustrates this contradiction and shows that most of women's higher life expectancy is a consequence of that spent with some disability. This paradox and its persistence across age cohorts remain unexplained although Nazroo's work postulates that gender is related to key determinants of life expectancy and morbidity, be they considered from within a biological or socio-economic framework. These include important (although changing) gender differences in educational levels, occupational types and levels, pension wealth, domestic roles, and health behaviours (see also Arber, 2006). It is also possible that men are more likely to be exposed to hazards that result in mortality and women to hazards that result in morbidity. However, the roles of these and other potentially important factors in explaining gender inequalities in life expectancy and healthy life expectancy, especially as they unfold in later life and change across age cohorts, are woefully under-researched (Arber et al., 2007) and there is a clear need for more thorough investigations (Nazroo, 2015). As caregivers, women report more fatigue and physical conditions, as well as depression including taking more medications to handle these. It has been shown that son caregivers usually dedicate more to tasks such as making arrangements, while caring daughters with hygiene and daily living activities (Nazroo, 2015). Older women are disproportionately affected by dementia primarily because of ageing demographics: women tend to live longer than men do, and dementia is age-related (Knapp & Prince, 2007). Further, individuals aged over 80 living with dementia are twice as likely to be women as men are, and individuals reaching 100 years who are living with dementia are four times as likely to be women as men (World Health Organisation, 2007). European statistics show that there are large differences between women and men for singles aged 65 and over: the share of elderly women living alone (40.1%) was twice the share for men (19.4%) (Eurostat, 2020).

Certainly, research in specific health areas has revealed disparities for older women. Firstly, ageism contributes to poor sexual and reproductive health and is associated with an increase in rates of sexually transmitted diseases (STDs). Epidemiological research from around the world indicates that rates of STDs are increasing in older people, and ageism may have a part to play in this (Hafford-Letchfield, 2021a, b). Older people more generally may be at a greater risk of STDs due to the lack of information and campaigns targeted at them and because they are also less likely to seek diagnosis and treatment in services that are ageist and inaccessible and exclude them from surveillance data and research studies that can help with prevention (Hafford-Letchfield et al., 2021a, b).

As suggested earlier, health in later life is linked to the living circumstances and actions of an individual during their entire life span and there are a range of factors globally that increase women's vulnerability to poor health in later life. Gender discrimination starts with girls having inequitable access to food, restrictions on education, poor care during childbirth, and post-natal care in some countries. The burden of caregiving responsibilities across different generations, domestic violence which may begin in childhood and continue in subsequent relationships, and living solo as a result of losing a partner or not having a partner or children can all severely impact on income, social isolation, and cultural traditions and attitudes. These in turn limit access to health in later life. For example, older women are much less likely than older men to receive cataract surgery in many countries (World Health Organisation, 2007, p. 5). An intersectionality perspective supports the idea of studying gender in interaction with other variables such as social class or education levels in order to understand these multifaceted health disparities between men and women in their later life (Sen & Iyer, 2012).

For care professionals, being able to understand and recognise gender and age differences, how these intersect with racism and socio-economic factors and cultural contexts are important to ensure that women get access to adequate medical care, recognising the associations with mental health and domestic abuse. Taking a gender inequalities perspective means going beyond physical symptoms to explore the sociocultural as well as the biological factors underlying problems that older women may present with to care services. This must ensure that prevention services are age and gender-sensitive particularly to support older women in the management of long-term challenging health conditions such as heart disease, diabetes, arthritis, and dementias.

Giving and Receiving Care

Whilst women are also more likely than men to live the latter part of their life without a partner, they are more likely than men to be involved in caring for other ill or disabled older family members. In a study of women making this transition from partner to caregiver for husbands with cognitive impairments, DeCaporale-Ryan et al. (2016) found that this can be an overwhelming experience and that it was very important to identify the quality of the pre-carer relationships to promote well-being in caregiving. In the UK, as the State has retreated, women have often filled the gaps in services. There are 1.25 million sandwich carers in the UK of which 68% are women. These are people caring for an older relative as well as bringing up a family and 7% provide over 35 hours per week where the oldest sandwich carers (55–64) provide the longest hours with 29% of them providing over 20 hours of caring a week (Age UK, 2019).

Older lesbians face the triple jeopardy of ageism, sexism, and heterosexism, and their experiences are largely invisible (Traise, 2013). Older lesbians who require help due to illness or disability are less likely to have partners or children than their heterosexual peers, and their informal support networks are more likely to be made up of friends than family (Shiu et al., 2016). However, friend informal caregivers typically provide less personal care and receive less support than traditional family caregivers, leaving older lesbians with unmet needs (Westwood and Carey, 2018). Little is known specifically about the experiences of old lesbians within the care system, but there is considerable evidence that lesbians across the life span may experience discrimination (Westwood, 2016). Butlers (2018) study of how older lesbians form relationships with their home carers found a lot of diversity in how each woman experienced her later years, the nature of her support system, and her relationships with formal home care providers. Carers needed to respect their unique life experiences, their current needs and demonstrate a level of competence, sensitivity, and caring as formal caregivers. Traies (2018) suggests that older lesbians have much in common with other women of their generation, as their lives are also affected by factors such as race, class, poverty, and sexism. For many a life lived without children may play its part (Reilly et al., 2018). Although lesbians are reported to have larger networks of friends from which to draw support than gay men (Brennan-Ing et al., 2014), some lesbians will spend their last years without such support. Similarly in a study of how the life course experiences of bisexual older people in their relationship histories (Jones et al., 2018) older bisexuals were found to be at risk of poorer quality care because providers do not always understand or fully consider their needs or relationships, and support networks are not always immediately recognisable within normative frameworks. There are limited safe spaces to express a bisexual identity or interests, this includes within care settings or when in receipt of care at home. There may also be negative perceptions of the support available from adult social care, as well as concerns in approaching services, possibly due to experiences of discrimination. Further, there is likely to be a growing number of trans women entering later life and in need of support services. Waling et al. (2020) explored older Australian trans women's perceptions of residential care, and the rationale behind their current and future decision-making processes about ageing and future care. They identified a range of concerns expressed by trans women about residential care services, specifically related to issues of discrimination, inclusivity, and loss of autonomy. Other concerns included health risks, such as the gendering of diseases including cervical and prostate cancer, which could result in missed examinations, and the expectation for trans people to have more knowledge than their doctors could about their specific health needs.

In summary, adult social care should encourage and recognise that people will be in different relationships, and it is important that practitioners become knowledgeable about what resources are available in the community such as support groups and advocacy and community organisations, integral to offering older women the necessary services and support that demonstrates cultural competence (Keppel, 2006). This includes addressing loss and bereavement experiences in different relationships in later life as well as recognising strengths and resilience accumulated over the life course (Fenge, 2014).

Despite all of these disadvantages, women in later life have been shown to create powerful networks of friendship, neighbourliness, and community that involve spaces of support and solidarity (Reilly et al., 2018) and networks that provide an invaluable framework of support both in difficult situations and when they face the losses that tend to come with later life. These bonds also facilitate a high level of activity and involvement, both in their own private lives and in the care of the community that they create together with other women (Reilly et al., 2018). This reminds us that this is not about women having an innate capacity for intimacy, but rather gender socialisation in providing care, which can become a positive skill later on.

Gender-Based Violence in Later Life

There are barriers to identifying, recording, and referring older people experiencing domestic abuse and a blurring of the definitions and boundaries between adult safeguarding and intimate partner violence (LGA & ADASS, 2015). This makes it difficult to form a reliable and comprehensive overview of the scale and nature of domestic abuse and sexual violence against older people. Women after the age of 65 are more often found to have a higher risk than men of being victims of domestic violence and abuse (Görgen et al., 2009) but safeguarding may be gender neutral leading to inadequate responses and needs a gender-based analysis to understand how discrimination contributes to abuse (Nerenberg, 2002).

De Donder et al.'s (2016) study based on data from 2880 older women in five European countries (Austria, Belgium, Finland, Lithuania, and

Portugal) found that overall 30.1% older women had at least one experience of abuse in the past year. The findings demonstrate that a single emphasis on personal risk factors (e.g., health, coping) is important but too simple. Abuse was reported as multifaceted and embedded in environmental (e.g., loneliness, household income) as well as macro-cultural contexts (e.g., old age dependency ratio). It was established that married women with poor physical and mental health status who adopt a behavioural disengagement coping strategy, lonely women, and those who manage poorly with their household income had a higher risk of being the most severely abused than of not being abused. These results could support the hypothesis that structural inequalities (such as, e.g., income) increase the risk of abuse among older women (Pain, 1997). Likewise, findings relating to poor mental health and disengagement could be interpreted as experiencing low levels of power. De Donder et al. (2016) suggest that empowering older women to a more active way of coping, could have potential benefits in terms of reduced likelihood of abuse and violence occurring. They established evidence that an in-depth understanding of abuse against older women needs differentiation between the different levels of severity. In that sense, different risk factors may or may not contribute to vulnerability to abuse, when different levels of abuse are taken into account. Hence, research, policies, and intervention strategies need to be further developed and devised in a way that consider these complicated and multiple layers of the phenomenon, in order to find appropriate solutions to this complex social and public health problem.

Sexual violence within the context of domestic abuse does not disappear in later life, but it is equally not well understood. Internationally, there have been few attempts to measure or estimate the prevalence of sexual violence against older women. The voices of older people, for example, are not often present in contemporary movements such as #MeToo (Bows, 2018). There is a lack of research exploring the issue and no easily identifiable body of literature informing it (Bows, 2018) and huge barriers reporting in health and social care. Tamutiene et al. (2013) found that the most common reasons for older women not reporting abuse were that they thought the incidents were too trivial, a belief that little could be done to help them, not wanting to involve anyone or thinking that they would not be believed.

Bows's (2018) systematic review of studies that address sexual violence against older people reports impacts on their physical and mental health. These include women in constant need of painkillers because of their injuries, long-term problems such as genital trauma and incontinence since the assault, and a range of psychological problems including anxiety, depression, and flashbacks. One study (Burgess et al., 2000) reported that over half of their sample (11 out of 20) of women sexually abused in nursing homes died within a year of the assault. Other studies report somatic symptoms (such as body aches and pains) linked to the sexual abuse (Soares et al., 2010), as well as sexually transmitted infections and increased HIV risk (Sormanti & Shibusawa, 2008). Ramsey-Klawsnik (2004), in one of the few studies to look at the impact of sexual abuse on both older men and women, identified that older male victims of sexually abusive partners experienced low self-esteem and hopelessness and faced multiple barriers to leaving a long-term violent or abusive partner. These include, for example, emotional, social, and financial attachment to the abuser and a perceived duty to remain to care for the spouse. Other reported impacts on women (Burgess & Clements, 2006) include numbness, appearing withdrawn, sleeping problems, depression, and other psychological effects such as flashbacks, nightmares, anxiety, and fears about leaving home. In some cases, this led to older women moving house or into residential care (Jeary, 2005). Jeary et al. found that either the sexual abuse tended to occur in the victim-survivor's home or in an institutional setting where the victim-survivor lived and that the perpetrator was typically known to the victim-survivor and was either a spouse or other family member.

These particular studies all highlight that professionals may fail to recognise injuries resulting from domestic abuse involving sexual violence, for instance, attributing bruises and fractures to falls and confusion to age-related conditions (Crockett et al., 2018). The reluctance to disclose abuse may ultimately lead to emotional isolation and powerlessness, which further exacerbates the long-term impact on self-esteem and self-worth as well as physical health problems (Wydall et al., 2018). For women from Black, Asian, and minority ethnic communities, there are cultural misconceptions, which normalise certain behaviours, inhibit recognition and access to services and support (Burman et al., 2004). However, the voices of older women from ethnic minorities remain largely unheard, and this clearly represents an additional deficit in our knowledge and understanding in adult social care.

Barriers to help-seeking in the face of sexual violence can include the fear of losing independence—being placed in an institution (Beaulaurier & Taylor, 2001). Family obligations, a sense of loyalty or having a caring role may also mean that older victim-survivors need or want to remain in contact with their abuser, who may also be a long-time partner, grandchild, adult child, or other relative (Crockett et al., 2018). Few services or structures for domestic abuse support and intervention are designed for older adults or accessible to people with differently abled bodies. As such, older women may not feel able to live communally with families or participate in group activities and chores, typical to living in refuge accommodation set up to safely house individuals experiencing domestic abuse and who wish to leave their abusive partner. These settings are also often inaccessible for older adults with care and support needs (Dunlop et al., 2005).

Responding to sexual violence has several implications for social work practice (Bows 2018). Practitioners should be mindful of women presenting with genital trauma, which research shows is more likely in older groups, and sexually transmitted infections (including HIV). As little is known about coping methods, practitioners should be alert to the possibility that problematic alcohol use may be a response by older people to experiencing sexual violence. Crockett et al. (2018) also set out some practice implications for social workers on how to bolster safety and healing for older survivors of intimate partner and sexual violence. In particular, they recommend connecting older women to peer-support circles that can provide survivors with a 'therapeutic space' in which to share their experiences, seek support and build relationships in an affirmative and empowering way (p. 1010). This, they suggest, may be the first chance for being listened-to, that older victims have had, and believed. Crockett et al. referred to grounding practice in a rights-based, victim-defined advocacy that affirms respect for the values, life experience, and culture of an older survivor, not discounting or minimising the marginalisation and invisibility that the older victim-survivor may feel because of victim blaming, ageism, sexism, homophobia or transphobia and other forms of discrimination. They further highlighted how older victim-survivors may be experiencing harm from an intimate partner, adult child, grandchild or other family member, caregiver, acquaintance or stranger. Older women may have experienced assault or abuse many years earlier, but may be only now seeking help for the trauma and its enduring effects. Support plans need to address the impact of multiple traumatic events including historic abuse and establish strong relationships with domestic abuse and sexual-assault programmes in your area to respond to complex cases of violence and abuse against older adults in the community (Crockett et al., 2018, pp. 1009–1010).

Conclusion

This chapter offered a theoretical contribution to the discussions on the multiple factors contributing to the marginalisation of women in later life. Gendered ageism is based on the interaction of age and gender and its potential to highlight the processes and practices of marginalisation. According to a 2020 meta-analysis regarding ageism, 42 (96%) of 44 studies found evidence that being exposed to or internalising negative representations of old age influences psychiatric conditions in older people. Furthermore, older people might internalise that other people (and not themselves) know what is best for their well-being. Such paternalism could negatively affect their sense of autonomy and control (Chang et al., 2020).

Evaluation of gender differences and their consequences for the design of public policy is important for a variety of reasons, including the ability to identify policy options that contribute to greater gender equity and enhanced economic efficiency (Austen, 2016). In relation to social work, Richards et al.'s (2014) examination of how social work education, research and practice engages with gerontological social work suggested that the foundations for informed, competent and critical social work practice with older people are shaky. They assert that it lacks engagement with the distinctive challenges of old age, such as gender, have failed to build on an expanding gerontological knowledge base. Engagement therefore requires a good understanding of the complexities of the physical and the psychosocial implications of ageing for exclusion (Clarke & Griffin, 2008). It should also consider the benefits of person-centred and enablement support and in challenging injustice and inequality that are compounded not only by the intersectionalities of gender and ageing, but through the barriers within health and social care settings and other institutional discrimination.

Research shows how older women may internalise their own ageism and have low expectations of services and of themselves and seeking ways to promote their inclusion and equal treatment are significant concerns for social workers (Richards et al., 2014). Promoting older people's well-being, demands 'flexible, imaginative and transformational approaches that encourage interconnectedness and acknowledge interdependence between older people themselves as well as those working in the caring professions' (Maidment & Macfarlane, 2011, p. 710). The imagery used in services, may put people off or service literature not be available in the services that older people are using, requiring assertive outreach. The physical layout of services and lack of consideration for sensory impairment can exclude older women with disabilities or limited mobility, who speak different languages. Older women as primary care-givers, their family obligations, a sense of loyalty or duty to provide care means that women may be consistently disadvantaged in many ways impacting on their physical and mental health, economic, social isolation.

Social workers need to maintain an explicit focus on challenging both the structure and its systems around the older person as well as on the individual. According to Striar and Biniyamin (2010), this requires giving attention to long-term strategies associated with lobbying, advocacy, coalition-building, increasing social awareness, and supporting social movements that increase social awareness and political processes They also refer to the different ways in which social workers and service users might hold about their situations or problems which is fundamental to the types of assessment or interventions taken or offered and outcomes expected.

References

Age, U. K. (2019). *Breaking point: The social care burden on women*. Age UK.

Arber, S. (2006). Gender and later life: Change, choice and constraints. In J. Vincent, C. Phillipson, & M. Downs (Eds.), *The futures of old Age* (pp. 54–61). Sage.

Arber, S., Andersson, L., & Hoff, A. (2007). Changing approaches to gender and ageing. *Current Sociology, 55*(3), 147–153.

Aronson, L. (2020). Ageism is making the pandemic worse. *The Atlantic*. https://www.theatlantic.com/culture/archive/2020/03/americas-ageism-crisis-is-helping-the-coronavirus/608905/

Austen, S. (2016). Gender issues in an ageing society. *The Australian Economic Review, 49*(4), 494–502.

Barrett, A. E., Michael, C., & Padavic, I. (2020). Calculated ageism: Generational sacrifice as a response to the COVID-19 pandemic. *The Journals of Gerontology: Series B*. Online first. https://doi.org/10.1093/geronb/gbaa132

BASW. (2018). *Capabilities Statement for social workers in England who work with older people*, London, British Association of Social Workers.

Beaulaurier, R. L., & Taylor, S. H. (2001). Social work practice with people with disabilities in the era of disability rights. *Social Work in Health Care, 32*(4), 67–91.

Bettio, F., Tinios, P., & Betti, G. (n.d.). *The gender gaps in pensions in the EU. ENEGE (experts network in European gender equality)*. European Commission.

Betty, F. (1993). *The Feminine Mystique*. Harmondsworth, Penguin.

Bows, H. (2018). 'Recognising the Sexual Victimisation of Older Women'. In Duggan, M. (Ed.), *Revisiting the 'Ideal Victim'* Concept. Bristol: Policy Press.

Brennan-Ing, M., Seidel, L., Larson, B., & Karpiak, S. E. (2014). Social care networks and older LGBT adults: Challenges for the future. *Journal of Homosexuality, 61*(1), 21–52.

Burgess, A. W., Clements, .T. (2006). Information processing of sexual abuse in elders. *Journal of Forensic Nursing. 2*(3), 113–120.

Burgess, A. W., Dowdell, E. B., & Prentky, R. A. (2000). Sexual abuse of nursing home residents. *Journal ofPsychosocial Nursing & Mental Health Services, 38*, 10–18. https://doi.org/10.1201/9781420042375.ch4.

Burman, E., Smailes, S. L., & Chantler, K. (2004). 'Culture' as a Barrier to Service Provision and Delivery: Domestic Violence Services for Minoritized Women. *Critical Social Policy, 24*(3), 32–357.

Butler, S. (2018). Older lesbians receiving home care: Formal and informal dimensions of caregiving. *Journal of Women & Aging, 30*(2), 91–110.

Centre for Ageing Better. (2017). *Inequalities in later life the issue and the implications for policy and practice.* Centre for Ageing Better.

Chang, E. S., Kannoth, S., Levy, S., Wang, S. Y., Lee, J. E., & Levy, B. R. (2020). Global reach of ageism on older persons' health: A systematic review. *PLoS One, 15*, e0220857.

Clarke, L. H., & Griffin, M. (2008). Visible and invisible ageing: Beauty work as a response to ageism. *Ageing & Society, 28*(5), 653–674. https://doi.org/10.1017/S0144686X07007003

Connell, R. (2012). Gender, health and theory: conceptualizing the issue, in local and world perspective. *Social Science & Medicine, 74*(11), 1675–1683. https://doi.org/10.1016/j.socscimed.2011.06.006. Epub 2011 Jun 29. PMID: 21764489.

Craciun, C., & Flick, U. (2016). Aging in precarious times: Exploring the role of gender in shaping views on aging. *Journal of Women & Aging, 28*(6), 530–539.

Crockett, C., Cooper, B., & Brandl, B. (2018). Intersectional stigma and late-life intimate-partner and sexual violence: How social workers can bolster safety and healing for older survivors. *British Journal of Social Work, 48*, 1000–1013.

Cynthia, R., & Barbara, M. (1983). Audiotape collection, 1973-1990; item description, dates. T-338, reel #. Schlesinger Library, Radcliffe Institute, Harvard University, Cambridge, Mass. https://id.lib.harvard.edu/ead/sch01246/catalog. Accessed February 09, 2022.

De Donder, L., Lang, G., Ferreira-Alves, J., Penhale, B., Tamutiene, I., & Luoma, M. (2016). Risk factors of severity of abuse against older women in the home setting: A multinational European study. *Journal of Women & Aging, 28*(6), 540–554.

DeCaporale, R., Steffen, A. M., Samuel, J., Marwitc, L., & Meuser, T. (2016). Dementia spousal caregivers and past transgressions: Measuring and understanding forgiveness experiences. *Journal of Women and Ageing, 28*(6), 510–520.

Doris Lessing (1983). The Diary of a Good Neighbour. In M. Joseph (Ed.), *The Diaries of Jane Somers*, 13–261. London.

Dunlop, D., Manheim, L. M., Song, J., Lyons, J. S., & Chang, R. W. (2005). Incidence of Disability Among Preretirement Adults: The Impact of Depression. *American Journal of Public Health, 95*, 2003–2008. https://doi.org/10.2105/AJPH.2004.050948

Eboiyehi, F. A. (2017). Convicted without evidence: Elderly women and witchcraft accusations in contemporary Nigeria. *Journal of International Women's Studies, 18*(4), 247–265.

Eurostat. (2018). *Ageing Europe - statistics on health and disability*. https://ec.europa.eu/eurostat/statisticsexplained/ (accessed 9/2/2022).

Eurostat. (2019a). *Gender pay statistics*. From https://ec.europa.eu/eurostat/statistics-explained/

Eurostat. (2019b). *Ageing Europe: Looking at the lives of older people in the EU*. European Union https://ec.europa.eu/eurostat/documents/3217494/10166544/KS-02-19%E2%80%91681-EN-N.pdf/c701972f-6b4eb432-57d2-91898ca94893. Accessed 9/2/2022.

Eurostat. (2020). *The life of women and men in Europe: A statistical portrait*. 2020 Edition available from https://ec.europa.eu/eurostat/cache/infographs/womenmen/

Fenge, L. E. (2014). Developing Understanding of Same-Sex Partner Bereavement for Older Lesbian and Gay People: Implications for Social Work Practice, *Journal of Gerontological Social Work, 57*(2–4), 288–304.

Fineman, M. A. (2008). The vulnerable subject: Anchoring equality in the human condition. *Yale Journal of Law and Feminism, 20*(1), 1–24.

Fineman, M. A. (2010). The vulnerable subject and the responsive state. *Emory Law Journal, 60*(2), 251–275.

Görgen, T., Herbst, S., Kortlenga, S., Nagele, B., & Rabold, S. (2009). *Crime experiences and experiences of violence in the lives of older people. Summary of major events of a study on risk of older and dependent people*. Berlin, Germany. Federal Ministry of Federal Affairs. Senior Citizens Women and Youth.

Hafford-Letchfield, T. (2013). 'Social Work, Class and Later Life'. In M. Formosa & P. Higgs (Eds.), *Social Class in Later Life: Power, Identity and Lifestyle*. Bristol: Policy Press.

Hafford-Letchfield, T. (2016). Learning in Later Life: Challenges for social work and social care. London: Routledge.

Hafford-Letchfield, T. (2020). Gender and Relationship Status in Later Life. In: Leal Filho W., Azul A.M., Brandli L., Lange Salvia A., Wall T. (eds) *Gender Equality. Encyclopedia of the UN Sustainable Development Goals*. Springer, Cham. https://doi.org/10.1007/978-3-319-70060-1_104-1

Hafford-Letchfield, T. (2021a). *Sex, intimacy and relationships in later life*. Research in Practice for Adults.

Hafford-Letchfield, T. (2021b). Heterosexual sex, love and intimacy in later life; what have older women got to say? In T. Hafford-Letchfield, P. Simpson, & P. Reynolds (Eds.), *Sex and diversity in later life*. Policy Press.

Higgs, A., & Hafford-Letchfield, T. (2018). 'At what cost? The impact of UK long-term care funding policies on social work practice with older people': A literature review. *Ethics and Social Welfare*. https://doi.org/10.1080/17496535.2018.1505931

Hollinrake, S., & Thomas, W. (2015). Caring relationships and efficient social care provision: Can an ethic of care provide a better Foundation for Responding to care needs in later life? *International Journal of Sociology and Social Policy, 35*(5/6), 419–436.

Hooyman, N., Browne, C. V., Ray, R., & Richardson, V. (2002). Feminist Gerontology and the Life Course, *Gerontology & Geriatrics Education, 22*(4), 3–26.

House of Commons Work and Pensions and Business, Innovation. (2016). *First Report of the Work and Pensions Committee and Fourth Report of the Business, Innovation and Skills Committee of Session 2016–17*. BHS report (parliament.uk) (Accessed 14/2/2022).

Itzin, C., & Phillipson, C. (1993). *Age barriers at work*. London: METRA.

Itzin, C., & Phillipson, C. (1995). Gendered ageism: A double jeopardy for women in organisations. In C. Itzin & C. Phillipson (Eds.), *Gender, culture and organisational change. Putting theory into practice* (pp. 84–94). London: Routledge.

Jeary, K. (2005). Sexual abuse and sexual offending against elderly people: A focus on perpetrators and victims. *The Journal of Forensic Psychiatry & Psychology, 16*(2), 328–343.

Jones, R. L., Almack, K., & Scicluna, R. (2018). 'Older bisexual people: Implications for social work from the 'Looking Both Ways' study', *Journal of Gerontological Social Work , 61*(3), 334–347.

Keppel, B. (2006). Affirmative Psychotherapy with Older Bisexual Women and Men. *Journal of Bisexuality, 6*(1–2), 85–104.

Kimani, E. N., & Maina, L. W. (2010). Older women's rights to property and inheritance in Kenya: Culture, policy, and disenfranchisement. *Journal of Ethnic and Cultural Diversity in Social Work, 19*(4), 256–271.

Kornadt, A. E., Kessler, E.-M., Wurm, S., Bowen, C. E., Gabrian, M., & Klusmann, V. (2019). Views on ageing: A lifespan perspective. *European Journal of Ageing*. https://doi.org/10.1007/s10433-019-00535-9

Knapp, M., Prince., Albanese, L., & Banerjee, S. (2007). *Dementia UK: A Report into the Prevalence and Cost of Dementia*. The Personal Social Services Research Unit (PSSRU), London School of Economics and the Institute of Psychiatry at King's College London, for the Alzheimer's Society.

Krekula, C., Nikander, P., & Wilińska, M. (2018). Multiple Marginalizations based on age: Gendered ageism and beyond. In L. Ayalon & C. Tesch-Römer (Eds.), *Contemporary perspectives on ageism*. Springer.

Kuhn, M. (1977). *Maggie Kuhn on aging: A dialogue*. Philadelphia, PA: Westminster Press.

LGA & ADASS. (2015). Adult safeguarding and domestic abuse: A guide to support practitioners and managers (2nd Ed.). London, London Government Association.

Maidment, J., & Macfarlane, S. (2011). Crafting Communities: Promoting Inclusion, Empowerment, and Learning between Older Women. *Australian Social Work, 64(3)*. https://doi.org/10.1080/0312407X.2010.520087

Nazroo, J. Y. (2015). *Addressing inequalities in healthy life expectancy. Future of an ageing population: Evidence review foresight*. Government Office for Science.

Nerenberg, L. (2002). *A Feminist Perspective on Gender and Elder Abuse: A Review of the Literature*. City College of San Francisco. USA

Neysmith, S., & Reitsma-Street, M. (2009). The provisioning responsibilities of older women. *Journal of Aging Studies, 23*(4), 236–244. https://doi.org/10.1016/j.jaging.2008.03.001

Neysmith, S., Reitsma-Street, M., Baker Collins, S., Porter, E., Cerny, J., Tam, S (2012). *Beyond Caring Labour to Provisioning Work*. Toronto, University of Toronto Press.

Pain R. H. (1997). Social Geographies of Women's Fear of Crime. *Transactions of the Institute of British Geographers, 22*(2), 231–244.

Ramsey-Klawsnik, H. (2004). Elder Sexual Abuse Within the Family. *Journal of Elder Abuse & Neglect, 15*(1), 43–58.

Reilly, E., Hafford-Letchfield, T., & Lambert, N. (2020). Women ageing solo in Ireland: An exploratory study of women's perspectives on relationship status and future care needs. *Qualitative Social Work. 19*(1), 75–92. https://doi.org/10.1177/1473325018796138

Richards, S., Sullivan, M. P., Tanner, D., Beech, C., Milne, A., Ray, M., Phillips, J., & Lloyd, L. (2014). Gerontological social work in the UK ready to meet twenty-first-century challenges? *British Journal of Social Work, 44*, 2307–2324.

Ray, M. (2008). *Critical Issues in Social Work with Older People*. Basingstoke: Palgrave, MacMillan.

Sarton, M. (1984). *At Seventy: A Journal by May Sarton*. W. W. Norton & Company. USA.

Sen, G., & Iyer, A. (2012). Who gains, who loses and how: leveraging gender and class intersections to secure health entitlements. *Social Science & Medicine, 74*(11), 1802–1811. https://doi.org/10.1016/j.socscimed.2011.05.035

Shiu, C., Muraco, A., & Fredriksen-Goldsen, K. (2016). Invisible Care: Friend and Partner Care Among Older Lesbian, Gay, Bisexual, and Transgender (LGBT) Adults. *Journal of the Society for Social Work and Research, 7*(3), 527–546.

Soares, J. F., Barros, H., Torres-Gonzales, F., Ioannidi-Kapolou, E., Lamura, G., Lindert, J., de Dios Luna, J., Macassa, G., Melchiorre, M.G. & Stank, M. (2010). *Abuse and Health among elderly in Europe*. https://www.divaportal.org/smash/get/diva2:377016/FULLTEXT01.pdf. Accessed 09/2/2022.

Sormanti, M., & Shibusawa, T. (2008). Intimate Partner Violence among Midlife and Older Women: A Descriptive Analysis of Women Seeking Medical Services, *Health & Social Work, 33*(1), 33–41.

Stefko, R., Gavurova, B., Ivankova, V., & Rigelsky, M. (2020). Gender inequalities in health and their effect on the economic prosperity represented by the GDP of selected developed countries—Empirical study. *International Journal of Environmental and Public Health Research.* ijerph-17-03555.pdf.

Stoller, E., & Gibson, R. (1994). *Worlds of difference: Inequality in the ageing experience*, 3rd edition. Thousand Oaks, California: Pine Forge Press.

Strier, R., & Binyamin, S. (2010). Developing anti-oppressive services for the poor. A theoretical and organisationalrationale. *British Journal of Social Work, 40*(6), 1908–1926.

Swift, H. J., & Chasteen, A. L. (2021). Ageism in the time of COVID-19. *Group Processes & Intergroup Relations, 24*(2), 246–252.

Tanner, D., Ward, L., & Ray, M. (2018). 'Paying our own way': Application of the capability approach to explore older People's experiences of self-funding social care. *Critical Social Policy, 38*(2), 262–282.

Tesch-Römer, C., & von Kondratowitz, H.-J. (2006). Comparative ageing research: A flourishing field in need of theoretical cultivation. *European Journal of Ageing, 3*(3), 155–167.

Traise, J. (2018). *The lives of older Lesbians.* Basingstoke: Palgrave McMillan.

Van Eeuwijk, P. (2006). Old-age vulnerability, ill-health and care support in urban areas of Indonesia. *Ageing & Society, 26*(1), 61–80.

Virokannas, E., Liuski, S., & Kuronen, M. (2020). The contested concept of vulnerability – A literature review. *European Journal of Social Work, 23*(2), 327–339.

Waling, A., Lyons, A., Alba, B., Minichiello, V., Barrett, C., Hughes, M., Fredriksen-Goldsen, K., & Edmonds, S. (2020). Trans Women's perceptions of residential aged care in Australia. *British Journal of Social Work, 50*, 1304–1323.

Westwood, S. (2016). Dementia, women and sexuality: How the intersection of ageing, gender and sexuality magnify dementia concerns among lesbian and bisexual women. *Dementia, 15*(6), 1494–1514.

Westwood, S.L., & Carey, N. (2018). Ageing with physical disabilities and/or long-term health conditions. In S. Westwood (ed.) *Ageing, Diversity and Equality: Social justice perspectives.* London: Routledge.

World Health Organisation (2007). *Women, ageing and health: A framework for action.* WHO.

World Health Organisation (2015). *World report on ageing and health.* Geneva, WHO.

World Health Organisation. (2021). *Global report on ageism.* WHO.

Wydall, S., Clarke, A., Williams, J., & Zerk, R. (2018). Domestic Abuse and Elder Abuse in Wales: A Tale of Two Initiatives. *The British Journal of Social Work, 48*(4), 962–981.

PART III

Gender in Social Work Practice

In this final section comprising seven chapters, we turn to the importance of feminist theories for working with gender in social work practice. These chapters address the way in which social work engages with gender, sexuality and sexual identities across a range of practice themes. Social workers' commitments to social justice may seem undiminished, but it has attracted debate about how the meaning of feminism and the explicit incorporation of feminist perspectives into social work practice have decreased (Lazzari et al., 2009). This extends to the inclusion of gender in social work policy development (Orme, 2001).

The International Association of Schools of Social Work (IASSW)/International Federation of Social Work mandate for social work lies in its intervention at the points where people interact with their environment and is consistent with the social development paradigm. Social workers utilise a range of skills, techniques, strategies, principles and activities at various system levels to support people or actively creating change. From an emancipatory perspective, the IASSW/IFSW supports social work strategies at increasing people's hope, self-esteem and creative potential to confront and challenge oppressive power dynamics and structural sources of injustices. The holistic focus of social work is universal, but the priorities of social work practice will vary from one country to the next and from time to time depending on historical, cultural, political and socio-economic conditions.

We draw extensively on Stephen Hicks (2015) analysis of the conceptualisation and usage of 'gender' within social work theory, research and practice and how gender is actually theorised for practice. Hicks contends that social work tends to treat gender as a rather static characteristic which must open up to debates that consider materialist, interactionist and discursive accounts, before finally considering what social work practice might learn from these. One example is Hicks' consideration of the notion of egalitarian power relations that do not engage with the power dynamics that always exist between social workers and service users and how important it is to make space outside of those power relations for reflexivity about power in all practices. This theme is taken up in Chap. 13 by Roxana Diamond, Priscilla Dunk-West and Sarah Wendt who consciously use praxis and reflexivity in relation to the issues surfaced in sex-worker-led research. They lead research guided by sex work feminism, a methodology that centres live experience and a 'nothing about us without us' approach. They debate how sex work is positioned and stigmatised in within academia and often presented as inherently exploitative, because it reflects gendered inequalities embedded into social relations. They suggest however that these kinds of anti-sex work dichotomies advanced by 'gender critical' feminists is problematic, because their approaches are simplistic, without nuance and do not centre lived experience or uphold the voices of sex workers. Diamond, Dunk-West and Wendt engage in critically authentic dialogue about how they work together to support sex-worker-led research, focusing on who is asking the research questions and which questions get asked in sex work. These intersecting issues within student and supervisor relationships reflect the nature of knowledge construction in the academy and how these inform feminist social work practice. Such a feminist alliance to support Diamond and sex worker research and its 'whorearchy' is crucial to ensure active and ongoing conversations about power, knowledge and activism in social work practice.

The notion of woman-centred practice (Dominelli, 2002), which seeks equality based on empowerment, listening to the stories and validating the experiences of women, is addressed in Chap. 14 by Linda Bell who explores the many contextual changes that inform women-centred practice. Bell laments the loss of preventative work, the pressure on time because of service bureaucracies affecting practice and how these influence the way in which social workers practice. Bell tries to square these challenges with her experience in researching work with mothers who had experiences of involvement with statutory child protection, of domestic and interpersonal violence and their stories of how they had felt let down or even abandoned by statutory social workers. This was a collective project including practitioners and researchers some of whom worked directly with the women in their practice. The Hummingbirds project, a preventative initiative subsequently described by Bell, articulates how a broadly feminist and gender-specific approach is used to work intensively with these mothers. Bell epitomises the necessity of 'time' or feminist social work practice and unpicks the potentially simplistic assertion about having enough 'time to care'. She reflects on how this influences gendered expectations of good enough parenting in society and the institutional positioning of men in this agenda.

In Chap. 15 by Christine Cocker, we turn to the influence of queer theories for social work practice critiques of feminist social work for their heteronormativity, particularly in relation to the lives of lesbians which have been missing from some feminist accounts (Hicks, 2011, 2015; Brown & Cocker, 2011; Traies, 2016). Cocker explores the discourses around lesbian parenting in particular, to consider whether it remains 'progressive' as an alternative family form, whether it is now so commonplace that it is normalised or whether it occupies another different space altogether. She reflects on how social work practice with lesbian parents has changed and discusses how the learning applies to working with queer families. Lesbian feminists have challenged the heteronormative nuclear family status quo by creating their own families of choice, including having children through self and donor insemination, as single mothers, in couples or collective households. Lesbian feminists have had an important role in influencing the public debates about lesbian parenting. Cocker's main discussion centres on the insider/outsider binary that hitherto preoccupies how lesbian parents are positioned and their increasing move towards occupying the 'in-between' spaces that does not position them as radicals but as ordinary and concerned with everyday actions and interactions that being a parent encompasses (Hicks, 2011). How far social work has improved on its practices and the practice assessments and models it uses to flexibly address people's individuality in a reflexive manner still reflects a heteronormative and cisgendered frame. Cocker offers the SPRIINT model and calls for the rejection of binaries, which promote stereotypes about heteronormative parenting, moving towards valuing the strengths in the diversities of families and family forms.

Jane Dodsworth in Chap. 16 returns to the theme on polarising of feminist theorists' positions about sex work as a form of violent exploitation or as a form of labour and what can be learned to inform and challenge current social work practice from people with lived experience. Her chapter uses a feminist filter to share women's stories of their experiences, how they made sense of them and what they can teach us about future social work practice with people who are at risk of or have experienced child sexual exploitation. Dodsworth advocates for a strength-based, systemic perspective, in order to recognise the structural elements contributing to child sexual exploitation and the accumulation of risk factors throughout the life course. This indicates the need for increased professional awareness of wider factors impacting on 'choices' made and the need for ongoing, trusting, non-judgemental holistic, relationship-based support interventions that promote strengths and increase resilience support.

Social work is a female majority, male-dominated profession (McPhail, 2004). There have been calls within the profession to attract more men into it and to improve the profession's practice with men. This reflects the point made earlier in Bell's chapter (also see Cavanagh & Cree, 1996; Pease, 2006) such as by working with fathers and acknowledging the complexity of men's position within social work in relation to notions of 'masculinity' and the category of 'men' (Hicks, 2015). Scourfield (2001) refers to the professional discourse on fathers in child protection work, which paints them variously as useless, irrelevant, absent and a threat. Jason Schaub in Chap. 17 scrutinises the experiences of social work student men and mobilises feminist theories to illuminate the topic of men's position in social work. He argues that a pro-feminist standpoint is essential to critique the privileged societal power of men and how men's power perpetuates gender inequality with a commitment to destabilise inequality and that men are well placed to critique their positions of power and problematising the continuation of gender inequality. Schaub refers to the position of men in social work as a numerical minority in the social work profession without all the challenges of a minority group, because of their privilege. In his study of undergraduate social work student men, he reflects on touchpoints of discomfort and difficulty in relation to physical contact, discussions on domestic abuse and generally confronting stereotypes about their motivation and potential for caring. Analysis of their experiences poses questions as to how social workers navigate gender boundaries. Scrutinising men's experiences can provide knowledge about the gender boundaries present in social work and how social workers navigate these and develop co-operative working relationships for bringing a gender consciousness into practice.

In Chap. 18, Rebecca Infanti-Milne, Richard Mc Kenny and Lee Walton, coming from a systemic family therapy position, take us into these challenges further and offer their self-reflexivity as therapists when working with women and men in domestic violence and abuse. They draw on a feminist family therapy tradition developed in the USA and UK in which they purposely create safety for women in couple work contexts. This pays close attention to *voice*, to the unvoiced experiences-in-context, of violence, control, threat and fear, which women, mothers of children with social workers, dissociate from in order to survive and cope day to day. Engaging men in social work, particularly after reports of domestic violence and abuse, can give rise to considerable professional anxiety given the risks for those involved and the challenges in managing difficult feelings that such work can evoke, including about professionals' own experiences of violence. Their argument is that when social workers do not engage with men, they inevitably transfer responsibility for safety to women, and their assessments of risk are incomplete. Infanti-Milne et al. advocate for robust training and support for practitioners to support the task of including men in their assessment and therapy work to 'help men voice and live out an ethics of protest, remorse, and generous love'.

The fact that a smaller number of men in the profession hold more institutional power in the social work profession, even though it has a majority female workforce, raises questions about how power works within social work institutions (Hicks, 2015). In our final chapter in this section on gender and social work practice, Adi Cooper and Lynn Romeo take us to the issue of leadership of social work and articulate the significant role of feminist leadership. They explore how it expresses core feminist and social work principles in the process of influencing and managing change. Taking up the debates about workforce inequalities, they explore some of the practical barriers for women moving into senior roles and highlight fundamental challenges about the nature and place of 'feminist leadership'. This draws on feminist theories and the principles of how to 'lead' as an activist and how to enable collective decision-making as opposed to hierarchical decision-making across organisations. Cooper and Romeo assert that challenging rather than reinforcing unequal power structures and relationships to reflect on the criticality of the core feminist concept that 'the personal is political' in relation to the 'use of self' in leadership is critical. Feminist leadership is demonstrated through inclusive and participatory styles, which privileges collaboration, shared decision-making and continuous reflexivity to consider why and what is being actioned and whether the goals of inclusion and social justice remain at the heart of practice.

This chapter concludes this book on feminist theories for social work practice. It is by no means comprehensive but provides a window for further inquiry into research, theory, policies and practices, which can be harnessed for challenging oppression and discrimination within social work. If gender in social work practice is to be seen in its complexity, then this concept needs to be interrogated and expanded. Hicks (2015) raises several issues about what we mean by feminism(s) in current social work contexts and the relevance of *gender* and particularly 'gender specificity' in light of queer and trans scholarship and its intersectionalities with gender, sexualities and intimacies in social work (see also Dunk-West &

Hafford-Letchfield, 2011, 2018). Talking about gender involves powerful linguistic claims in respect of feminist social work practice as the chapters within this section illustrate. According to Hicks (2015), it also has to be conceptualised within materialist or structuralist accounts that focus on institutions, such as the family, ageing and the workplace in order to examine how gender inequality and its intersections are produced and reproduced within such settings and how gender is performed.

Praxis is the dynamic, reciprocal interplay of action, reflection and theory construction grounded in the experiences of women. If feminist principles truly guide the work of social work and its institutions, then praxis and reflexivity would be inclusive and the themes in these chapters all throw light on what inclusivity might mean in the diverse contexts discussed.

References

Baum, N. (2016). The unheard gender: The neglect of men as social work clients. *British Journal of Social Work, 46*(5), 1463–1471.

Brown, H. C., & Cocker, C. (2011). *Social work with lesbians and gay men.* Sage.

Cavanagh, K., & Cree, V. E. (1996). *Working with men, feminism and social work.* Routledge.

Dominelli, L. (2002). *Feminist social work theory and practice.* Palgrave.

Dunk-West, P., & Hafford-Letchfield, T. (Eds.). (2011). *Sexuality and sexual identities in social work, research and reflections from women in the field.* Ashgate.

Dunk-West, P., & Hafford-Letchfield, T. (Eds.). (2018). *Sexuality, sexual and gender identities research in social work and social care: A lifecourse epistemology.* Routledge.

Hicks, S. (2011). *Lesbian, gay and queer parenting: Families, intimacies, genealogies.* Palgrave.

Hicks, S. (2015). Social work and gender: An argument for practical accounts. *Qualitative Social Work, 14*(4), 471–487.

Lazzari, M., Colarossi, L., & Collins, K. (2009, November). Feminists in social work: Where have all the leaders gone? *Affilia, 24*(4), 348–359.

IASSW AIETS. Global definition of social work. https://www.iassw-aiets.org/global-definition-of-social-work-review-of-the-global-definition/

McPhail, B. A. (2004). Setting the record straight: Social work is not a female-dominated profession. *Social Work, 49*(2), 323–326.

Orme, J. (2001). *Gender and community care: Social work and social care perspectives.* Palgrave.

Pease, B. (2016). *Engaging men in feminist social work: Theory, politics and practice.* Routledge.

Scourfield, J. (2001). Constructing women in child protection work. *Child & Family Social Work, 61*, 77–87.

Traies, J. (2016). *The lives of older lesbians: Sexuality, identity & the life course.* Palgrave.

CHAPTER 13

Using Sex Worker Feminisms in Practice to Promote a Peer-Based Methodology; Exploring Personal and Professional Identities in a Research Alliance Centring Sex Worker Lived Experience

Roxana Diamond, Priscilla Dunk-West, and Sarah Wendt

In this chapter we explore peer-based research and the construction of knowledge by and about sex workers. We do this by drawing upon a study into the everyday experiences of sex workers in South Australia. This chapter commences by exploring sex work opposition in the scholarly community. We examine the traditions in which sex work research has been grounded (medico-scientific) and highlight the importance of the asking

R. Diamond (✉) • S. Wendt
Flinders University, Adelaide, SA, Australia
e-mail: roxana.diamond@flinders.edu.au; sarah.wendt@flinders.edu.au

P. Dunk-West
Charles Darwin University, Casuarina, NT, Australia
e-mail: priscilla.dunk-west@cdu.edu.au

C. Cocker, T. Hafford-Letchfield (eds.), *Rethinking Feminist Theories for Social Work Practice*,
https://doi.org/10.1007/978-3-030-94241-0_13

of ethical and peer-based research questions within a feminist and identity-conscious approach. We then move to autoethnographic accounts to highlight the importance of insider research from Roxana's sex worker feminist perspective as the lead author and upon whose research this chapter is based. Priscilla's account follows in which she considers identity in the academy which is then followed by Sarah's reflexive engagement with sex work feminisms as presented to her by Roxana in the development of the research project. We conclude that feminist scholarship should and can support sex workers to be the centre of research about sex workers' lives.

Introduction

Sex work feminism, as the title suggests, is an approach to research that centres lived experience and a 'nothing about us, without us' methodology. Sex work is an umbrella term and can be defined as the exchange of sexual services for money or other rewards (Scarlet Alliance, 2014). The definition of sex work is purposefully broad, ensuring that people who may not identify as sex workers are still captured within the definition. Sex work is a gender-neutral term and the sex work community includes trans and gender diverse workers and workers of diverse sexual orientations (Scarlet Alliance, 2014). The use of the term sex worker reiterates that sex work is work and that sex workers are deserving of the same rights and protections as any other profession (Scarlet Alliance, 2014). Sex work can be seen as queer work, as sex work exists outside of accepted understandings of heteronormativity (Zahra Stardust, 2015).

Dr. Elena Jeffreys, sex worker and activist, stated that sex workers are:

> at the cutting edge of feminist debate … sex work is feminist—being sexually active, putting a value on your sexual interactions, negotiating boundaries and making informed choices about your body. Sex workers aren't subscribing to prescriptive morals about chastity, nuclear family and monogamy—this is the feminist project, is it not? (Holden, 2011, p. 48).

Anti-sex work *feminists'* strong opposition to sex work was epitomised during the so-called sex wars of the 1980s and 1990s (Phipps, 2017). Opposition to sex work within a radical feminist tradition continues to be a perspective that is employed by some academics globally. The argument is that in a patriarchal society with unequal relations between men and women, women cannot make a purely free 'choice' to enter sex work and are therefore being exploited (Gerassi, 2015). We note the ways in which anti-sex work scholarly traditions create dichotomies such as women having a 'choice' or 'no choice'. Similarly, the idea of gender as binary, an idea advanced by the so-called gender critical feminists, is problematic because of the non-acceptance of trans and gender non-binary experiences and identities (Zanghellini, 2020). There are not only epistemological issues to address when approaching sex work research, but historical and methodological considerations assist in framing sex work research which we will now discuss.

Historically, empirical work about sex work was undertaken by non-sex workers, many of whom focused on regulatory health issues such as sexually transmissible infections and drug use (Leaker & Dunk-West, 2011). The consideration of who is asking the research questions and which questions get asked in sex work research are two important and intersecting issues (see Letherby, 2003). There has been increased awareness about the importance of research coming from within sex worker communities rather than the continuation of the academic gaze 'on' sex workers by non-sex working academics. As Desyllas notes:

> Throughout history, in society, and within academia, the aspirations of sex workers are often ignored and their needs have been consistently defined and represented by non-sex workers (Desyllas, 2013, p.772).

Conceptualising Everyday Sex Work

This chapter is based on the first authors' position as a sex worker leading PhD research on sex workers, asking: ***what are the everyday experiences of sex workers in South Australia?*** and the contributions from her supervisors who are not sex workers. The asking of particular research questions is both a deeply ethical and political undertaking and as we shall see, this particular research question is ethically grounded and deeply political. Specifically, this research question:

- is designed and 'owned' by a sex worker and researcher;
- takes the position that sex work is work; and
- rejects pathologising or exoticising sex work through the orientation towards the mundane or day-to-day.

Given the stigma attached to sex work, literature has markedly erased the 'in-between' and 'mundane' aspects of sex work. The 'opponents' of sex work continue to conflate the work with violence and trafficking, denying sex workers autonomy or decision-making. Sex workers are depicted by organised opponents to present sex as a threat to society because they disrupt traditional values about relationships, families and intimacy (Hubbard & Sanders, 2003; Sanders, 2004). Sex workers are also seen as people who need to be saved or are not capable of decision-making (Farley, 2019), particularly if they are women or migrant workers (Kempadoo et al., 2012). Kate Holden, sex worker and author, sums up this crowded landscape in which sex workers are placed:

> lamented by some feminists, lauded by others, lectured by religious groups, legislated by governments; monitored by health services, spurned by mortgage brokers, envied or condemned by friends, invited to write memoirs by publishers, assisted by outreach services; must live under one name and work by another (Holden, 2011, p. 46).

The research project emerged through the first author's (Roxana) experience of being a sex worker. Peer-based research is crucial to ethical empirical studies with any marginalised community. Insiders have access to the communities they research because they are often trusted by the community whereas outsiders are distrusted. 'When research is undertaken by a community member, the tensions that exist between the researcher and participants are alleviated by the united aims of both the researcher and the participants' (Kim & Jeffreys, 2013, p. 39).

Through the framing of the research and the information given to participants, Roxana 'came out' as a sex worker. 'Coming out' often refers to the experiences of the lesbian, gay, bisexual, trans, intersex and queer communities (LGBTIQ) and is simplified as a linear process of sexual development, beginning with repression and or concealment and eventuating with the disclosure of one's identity (Rosenberg, 2018). However, it is also used with the sex work community when sex workers 'out' themselves. Rosenberg (2018) explored the concept of sexual self-discovery and found that it is a lifelong process and sexuality is fluid, dynamic and situational rather than static. Furthermore, she found that sexual discovery and acceptance was an internal journey, where for many of the participants self-acceptance was more important than public disclosure, and thus the author termed this process 'coming in'. Rosenberg (2018) challenges coming out's focus on social visibility and disclosure as the only viable means for queer people to achieve sexual self-acceptance. The notion of 'coming in' is a useful term to describe the internal work that goes on for sex workers when navigating the stigma situated around sex work and internalised 'whorephobia', in conjunction with coming out. Staying 'closeted' has more negative associations connected to it, whereas the concept of coming in values the private, intimate journey of self-acceptance and the acceptance of the realities and difficulties of the external environment.

For LGBTQ people of colour, the experiences of coming out are challenged by social and institutional barriers unlike those that affect the white LGBTQ community (Ghabrial, 2016; Simien, 2007). An intersectional perspective (Crenshaw, 1991) rejects the separability of identity categories and it encompasses perspectives that maintain that identity categories such as gender, age, race, ethnicity, class, and sexuality are mutually constituted and cannot be added together (Simien, 2007; Sullivan & Day, 2019). Sullivan and Day (2019) interviewed two Indigenous transmasculine sex workers, JJ and Jeremy, and found that their lived experience offered insights into how their work was entangled with sex and gender, Indigeneity and identity. Their 'narratives rework considerations of Indigenous transmasculinities engaged in sex work, where they reveal that bodies, gender, emotion, self-image, and race intersect Indigenous transmasculine lives in complex ways that exceed heteronormative, homonormative, and transnormative discourse' (Sullivan & Day, 2019, p. 6). Furthermore, LGBTQ people of colour, similar to sex workers of colour, are regularly exposed to unique and contextual forms of prejudice and stigma based on their sexual identities, where these communities experience stress and an increased likelihood of mental and physical health problems, increased risk of violence, police targeting and persecution (Ghabrial, 2016).

Many sex workers are often able to conceal their identities, and Quinn and Chaudoir (2009) explore concealable identities and propose that depending on how much devaluation and prejudice a person may expect to experience if they are outed, the greater the anticipated stigma and distress. Also, if one's stigmatised identity is closely related to one's sense of self and who they are, even greater the distress (Quinn & Chaudoir, 2009). Lastly, the level of internal distress around one's stigmatised identity is closely related to how culturally devalued the stigmatised identity is, and the greater the cultural stigma, the greater the internalised stigma and distress (Quinn & Chaudoir, 2009). For those with concealable stigmas, such as sex workers, the decision to conceal or reveal one's sex worker status becomes a crucial strategy for managing the impact of stigma on one's life (Koken, 2012). Sex workers are often fearful of coming out or being 'outed' and because of this many sex workers keep their sex worker identities hidden (Bellhouse et al., 2015; Bradley, 2007; Levey & Pinsky, 2014). Being 'out' poses risk and being closeted can create emotional and psychological stress. In Hong Kong, Wong et al. (2011) explored how female sex workers experience stigma and found through conducting semi-structured interviews with 49 female sex workers that sex workers were 'forced' to keep their identities hidden and due to this tended to find themselves withdrawn from support and social networks. Keeping their identities hidden meant further marginalisation and feelings of loneliness (Wong et al., 2011). When sex workers were 'out' they experienced degrading or insulting language and the imposition of labels such as cheap, bad, greedy or shameless (Wong et al., 2011).

However, many sex workers see value in the work they do and Wong et al. (2011) found that some participants felt that sex work was similar to social work. This may have acted as a mitigating factor to relieve some of the internalised stigma and to justify their choice to mitigate experience of stigma from friends and family. Sex workers often are unable to out themselves, or only disclose their status to a select few trusted people in their lives, but in turn risk real and perceived consequences for doing so (Rayson & Alba, 2019; Wolf, 2019). Furthermore, sex work communities and peer organisations have been shown to help alleviate and resist social and institutional stigma (Argento et al., 2011).

A feminist research approach acknowledges the subjective influences of the researcher in their work (Letherby, 2003). In this chapter sex work feminism practiced by the first author includes the intersectional consideration of identities of the supervisory team and the role of the academy in constructing particular knowledges. We now shift to the first person for the second half of this chapter. The following autoethnographic reflexive accounts are intended to provide a feminist approach to understanding sex work research, student and supervisor relationships and knowledge construction in the academy. It is a conversation that represents multiple identities and relationships coming together to research and know the everyday lives of sex workers. We commence with Roxana's account which explores insider, peer-based research and approaching sex work research. Priscilla's autoethnographic account speaks to understanding her identity within the university setting. Sarah's account explores how feminist thinking frames her approach to supervising sex work research.

Roxana

What drew me into sex work research? This is a question I often ask myself and think about. I started my social work degree in 2012, the same year I started sex work. I still work today almost a decade later. My university education and my sex work are closely linked as they both were important places of learning for me. My education as a sex work researcher has come from both my sex work and community, and my formal research education. To this day they continue to weave into each other, however I learn most from other sex workers, past and present, and I build upon already existing knowledge.

Very soon into my degree I became critical of social work as an institution, when I began to understand its problematic legacy and in turn how we as colonisers and outsiders persistently forced and continue to force ourselves onto marginalised communities. I continue to navigate this and despite being a sex worker, queer and a survivor of domestic and sexual violence, I am a white, cisgendered women and come from a financially privileged background. These intersections impact upon how I experience criminalisation and stigma (or don't in some cases) and also mean that I can out myself within an academic setting.

I started my sex work career in a massage parlour, or 'rub and tug' as we called it. This was a rather easy transition for me as I was already working as a 'regular' masseuse, with formal qualifications, where the work was hard and the pay was even worse. Unintentionally, my parents who were both natural health practitioners instilled into me a particular practice and way of conceptualising health. Though I reject many of their beliefs, including an anti-vaccination stance, nevertheless their 'holistic' philosophy eventuated into accepting a beautifully perverted form of body work. One that inevitably resulted in naked body slides and pouring copious amounts of oil on strangers for money. For me working as a sex worker was not too different, at least from my perspective, from some of the teachings that I had grown up with. The final move that pushed me into sex work, as is the case for many sex workers, was financial and happened due to a culmination of things; an ever-present curiosity about sex, a relationship break-up (and needing to pay rent on my own) and starting university; my own 'holy trinity'.

I started out my degree with a saviour complex, as unfortunately many social work students do. However, over time and specifically due to my sex work experience, I started to understand the politics around identity and the importance of appropriate and meaningful consultation and representation. Towards the end of the degree, I became more aware that sex work and sex work activism was an important tool for me to challenge injustice. I wanted to learn as much as I could about sex work, but my social work degree failed to acknowledge criminalised populations, like sex work, in much depth. When I was offered a placement in an honours programme, I knew I had to conduct sex work research.

I truly believe in the value of sex work to the community, but also, it offers its practitioners financial freedom and flexibility like no other work. Though, I felt an injustice as my work was not only not recognised by the law but also was actively criminalised and my moral character questioned. The discussions I had with clients, the tears they shared with me, their vulnerability, to me, showed how dynamic the field of sex work can be, where it can be both horny and healing all at once. Or none of these things, as was sometimes the case. During honours, I started to understand that not only is sex work stigmatised, I already knew this personally, but from an academic perspective little was written about sex work by sex workers themselves, and in fact huge bodies of work focused on problematising sex work. In 2015, I also started getting involved with SIN, our local peer, government funded sex work organisation, and sat on their committee working towards incorporation for the organisation. Since I have been a permanent fixture on the SIN board as now we are an independent organisation. During my honours programme, SIN offered me a desk, laptop and fed into the research process from the start. Very quickly I discovered the world of sex work activism and have been inspired by whores ever since.

Sex work politics and ethics became important to me, and this fed into my passion for community work, sex work feminism and law reform. Encapsulating sex work feminism to me means also understanding that 'sex work is work, and all work sucks' (Origin unknown). Academia is merely a space for me to formalise what the community already knows to be true. Now as a PhD student, my passion for the sex work community has continued to grow. My main aims for research, along with other sex workers, are for sex work to be recognised as real work and to achieve the full decriminalisation of sex work, with no sex worker left behind, including migrant sex workers (DeCat, 2019).

Outsider research is a pervasive problem for the sex work community. People are fascinated by sex workers and want to 'save prostitutes', so it is unsurprising that more and more people are drawn to sex work research. The stigma situated around sex work is often strategically used to justify outsiders asserting their 'expertise' on 'prostitution' and often these critiques of sex work are disguised as 'rescue' and appeal to those with conservative agendas. Sex worker Zahra Stardust (2017) points out how unjust it is that due to stigma sex workers are taken away from important community work when having to respond to sensationalist arguments created by detractors. Instead, this energy could be directed at peer education, community building and creating safer spaces for sex workers. Researchers and people in positions of power who advocate for laws and policies that sex workers condemn use their privilege and power to destabilise community action and further stigmatise sex workers.

Within the sex work community a 'whorearchy' exists where privileged sex workers can be more visible within the community (me) without the associated risks of state-based violence that sex workers of colour, Aboriginal and Torres Strait Islander peoples, trans and gender diverse workers, street-based workers and migrant sex workers experience (Brooks, 1999). State-based violence is directly linked to the stigma sex workers experience and is compounded by intersecting factors such as the criminalisation of drug use, colonisation, racism and migration policy. Additionally, poverty or homelessness results in the over policing of visible forms of sex work, targeting those who do not have the privilege to conduct work behind closed doors. Outsiders do not understand the complexity of the 'whorearchy' and perpetuate the dichotomy of sex work. The more privileged voices become heard and used to gain insight into the 'glitz and glamour' of sex work as they are more palatable and 'exciting'. Whereas marginalised voices become fodder for outsiders to co-opt sex workers' lived experiences to sensationalise 'prostitution' or advocate for law reforms that sex workers do not support. This itself creates a division where privileged voices are seen to advance advocacy whereas marginalised voices are harmful to it. However, both narratives are often used to deny our realities and the legitimacy of sex work. The media and those who digest it prefer easy messaging over more complex and at times conflicting depictions of sex work.

Peer work and research is important because sex workers are better placed to speak on insider issues. Sex workers can get to the core of 'issues' pertaining to sex work because of their lived experience rather than focusing on dominant constructions of and stigmatised assumptions about sex work. Furthermore, when sex workers have worked directly with sex worker organisations and within peer-based activist spaces they are better placed to understand the discussions that are being had at the community level.

Sex work research led by peer researchers mean that they have a personal experience working within the sex industry and because of this lived experience deeply understand sex work stigma and sex work as work. This understanding is important as outsiders who speak on sex work often, even unintentionally, further harm sex workers through ignorance, especially when a paternalistic, saviour complex approach is the rationale for wanting to research sex work. Despite this, sex workers are often explored by outsiders, those with no sex work experience, meaning they are not connected to sex work peer spaces and sex worker organisations. Sex work being explored by outsiders is commonplace, however marginalised communities are critical of this approach as it often results in participants being viewed as objects to be studied, rather than people with real lives (Kim & Jeffreys, 2013). In an academic space research has the power to influence policy outcomes for sex workers and if appropriate consultation does not occur, which is often the case, this well intended pursuit quickly turns sour (Jeffreys, 2010).

Researching sex work is difficult for both peer researchers and outsiders because of challenges that arise from differing spaces and competing motivational factors (Dewey et al., 2019). For example, how academia treats sex work and sex workers differs depending on discipline and political leaning and varies significantly to how sex workers and sex work organisations speak on sex work. The challenges specific to sex work research often centres around who can claim to be an expert on sex work, who has authority to speak on sex work, for example if lived experience of sex work ought to outweigh 'empirically' derived knowledge (Dewey et al., 2019). According to Australian sex worker organisations, insider knowledge should be privileged and sex workers seen as the experts (Jeffreys, 2010; Kim & Jeffreys, 2013; Scarlet Alliance, 2014). Researchers cannot be accountable to community if they are not connected to that community. Ethics specific to sex work have been outlined by the sex work community and are important because they have a meaning outside of academia. Being an insider to sex work also does not mean I am an 'expert' on all things of sex work, even when situated within an academic space as a PhD candidate exploring sex work. Universities are quick to position me as a 'bourgeoning expert' on sex work research.

There are two important papers written by sex workers on sex work research and ethics within the Australian setting. Jeffreys (2010) discusses how sex workers are critical of outsiders and explored reflexive practices developed by sex workers for conducting ethical research with sex workers. Positive research experiences for sex workers are unfortunately rare and this has motivated sex workers to lead projects and take research into their own hands, for example my honours research (Baratosy & Wendt, 2017). Jeffrey's work is valuable because it points to the importance of participatory action research with sex workers and acknowledging the importance of peer research. Kim and Jeffreys (2013, p. 62) remind us that 'researching marginalised groups is challenging and fraught with ethical issues and these issues can be exacerbated when the researcher is an outsider'. When researching migrant sex workers, 'outsiders have the potential for their personal beliefs and moral views around migration, sex work, race, gender and sexuality to influence research methodology, analysis, interpretation and outcomes' (Kim & Jeffreys, 2013, p. 62). Sex work research ethics should not only be directed by University Ethics committees because they sit outside of the sex work community and are not an authority on sex work ethics. Alongside this, researchers must consider and utilise the guidelines sex workers have developed, work with sex workers and seek guidance from sex worker organisations.

Working within my supervision team has been incredibly rewarding, because I have worked alongside supervisors who have supported me throughout this process. While enacting my peer-based research the personal is very much political, and I have benefited from having those I work with understand that.

Priscilla

Who am I? How does one answer such a question? When I think about identity, I think about who I am and the roles I play and notice that thinking about identity takes up a lot of space in my professional life. In fact, I am so interested in notions of the self that I research identity, theorise social work identity (Dunk-West, 2018) and focus on sexual identity in my scholarship. I find explorations of identity rich and complex. Is identity a list of 'things'? If I were to list my intersectional identity categories, I am a white cisgender woman, queer, invisibly disabled and middle-aged. I come from a working-class background in Adelaide, South Australia. These are some of the aspects to my identity but does it tell you who I am or what I bring to supervision with PhD students? Important to my identity across personal and professional contexts—or 'how' I am—is the idea of allyship. For example, although I am part of queer communities, there are many gender non-binary and trans people who are subjected to violence, discrimination and abuse not only within broader society, but within their own communities.

Allyship, for me, means standing in solidarity with people who are not from the same social or cultural groups as one's self. For me it means listening to people, hearing what they want and need and stepping up and/or speaking out against discrimination.

The term 'allyship' is not without its critics and requires reflexive engagement with not only one's actions, behaviours, values, and so on, but also requires an examination into the question: who am I? What do I know, and how have I come to know? In this sense (and here I am drawing from Foucauldian traditions), knowledges are constructions from a range of factors such as biographical events and experiences but, importantly, require some unravelling. This is captured in a recent article about social movements, allyship and settler-colonist practices and language (Kluttz et al., 2020) and the authors argue that non-Indigenous people should not self-identify as an ally to Indigenous people. Though this paper explores Canadian relations, there are some similarities to the Australian context. The researchers capture the ways in which their assumptions were challenged, provoking 'unlearning':

> However, as we interviewed activists who spent time at Standing Rock, listened to Indigenous scholars on the subject, and read scholarship from (mainly) Indigenous educational researchers, we changed our minds. We came to question our assumptions, became 'un-settled' in our thoughts and identities, experienced emotionally laden disorienting dilemmas, and began to rethink the premises of allyship, both theoretically and in practical terms. We began a transformative process of un-learning and learning, resulting in the 'unsettling' of allyship (Kluttz et al., 2020).

In this sense, allyship, to me, remains an important concept relating to how I am in the world, rather than a label I attach to myself. Notwithstanding such critiques of allyship, questions about who I am cannot occur without acknowledging that I grew up on unceded Kaurna land. The history of settler-colonisers should therefore always be a reflexive point when I think about my personal and professional identity.

In my professional life, I am a continuing academic in Australia and part of my role is to supervise research higher degree students. In this chapter we have considered the role of academia in knowledge building and considered who asks research questions. I remain acutely aware of the power involved in the asking and framing of research questions. Being part of the supervisory team for this research into everyday sex worker lives in South Australia has not only been rewarding but has provided an opportunity to enable me to reflect upon my supervisory style. I am acutely aware that because I am employed in academia, I have access to systems of knowledge and privilege. For me, practising allyship means drawing from my previous work with sex workers and being open that I am a non-sex worker in this research space. Working with sex work researchers means that I can offer insights into the workings of academia, hear about and act on (when appropriate for me to do so) legislative reform activities and encourage scholarship in an area that is underrepresented in non-pathologising scholarship. Employing feminist approaches in the student and supervisory relationships, for me, means applying an intersectional lens, understanding power and privilege and recognising how and for whom knowledge is produced and situated. It means understanding who I am, for whom the academy speaks and what can be done to disrupt and resist oppressive forces.

Sarah

As a commencing social worker and academic I was influenced by radical, structural ideas of gender that emerged from feminism. The arguments that sex work is inherently exploitative because it reflects gendered inequalities embedded into social relations was familiar to me. Yet Roxy and I decided to work together during her honours research and now her PhD. Our early conversations acknowledged our different experiences and influences. Roxy wanted to contribute to knowledge about sex work as an insider and I needed to suspend any views I had on sex work as I was coming from a position of not knowing.

I remember questioning whether I was the 'right' person to supervise Roxy's research. However, this very question raises the privilege and status I held in the academy to make this decision. Why would I say no? To hide from my own discomfort. Or out of respect for insider research? These questions also come from a deep questioning of myself as a researcher, feminist and social worker against a backdrop of evolving and often contested feminist theorising in the academe. In the end, it was Roxy and I that decided despite this complex terrain, it is our commitment to feminism and our professions that enabled us to partner to research sex work. The key strength of feminist theorising is its contested nature and history, being open to reflection, and embracing contestation. If this were not present in feminism, there would be no possibility for change and development of new ideas. Roxy convinced me, we needed some new ideas about sex work and sex work research. However, it was Roxy that would lead this research, and I would partner with her to navigate and understand this contested terrain, with a joint aspiration to build on existing knowledge with new insight and contribution. The diversity of feminist thought is exciting!

It was the everyday lives of sex workers that Roxy argued needed to be explored as a way to cut through the long-entrenched debates about sex work. I was not a sex worker but I was committed, through feminist post-structural ideas, to explore difference and specificity, local contexts and local narratives of the everyday. It is this focus that allows us to move beyond binary opposites and taken-for-granted propositions of gender. It is the local stories that can disrupt the dominant discourses is what feminist post-structuralism taught me. It was these ideas that I worked with in my own community—a rural community—when I researched domestic violence. I was an insider (a farmer's daughter) and was tired of the dominant discourses that 'rural' women were traditional and conservative. It was the local stories from women, that I was able to facilitate, that disrupted such constructions. Women told stories of a hard work ethic, self-reliance and pride that challenged ideas of powerlessness and passivity. It was these stories that showed insight into how women understood and survived domestic violence (Wendt & Zannettino, 2015). Feminist post-structuralism enabled me to examine how we are all involved in 'doing gender' and how we all create and recreate our identities as gendered and sexual beings (Bradley, 2013). In sum, even though I was not an insider to Roxy's community, it was the importance of insider research that I valued with her. I was committed with Roxy that feminism is politically for women and should make sense of women's situations and lives in all its ambiguity, contestations and richness and I valued in the insider perspective.

Even though we shared the commitment of contributing to feminist thought and valued insider perspectives, I recognise this does not dissolve power that resides in the student supervisor relationship. But like Priscilla, I am committed to allyship. I come to this partnership with years of experience navigating the standards of the academy such as complying with the ethical standards of human research ethics committees when researching highly sensitive topics and hard to reach populations. I have interviewed hundreds of women who have experienced violence and abuse, men who have used violence, and practitioners working in high crisis and risk. Face to face interactions with participants of my own research has enabled much writing and advocacy. My motivations with Roxy and Priscilla are to give voice to personal, experiential and emotional aspects of existence (Ramazanoglu with Holland, 2002, p. 155). Roxy's research comes from experiences of sex workers in the everyday, without this knowledge, we cannot generate knowledge and move the debates on sex work. I have been challenged and it is important for social work and feminism to be challenged in how we think about sex work.

Conclusions: Towards a New Feminist Alliance to Research

This chapter points to the centrality of identities and usefulness of feminist praxis to supporting peer-based research. The autoethnographic accounts provide examinations into the intersectional identities involved in producing knowledge and supporting research in academia. Roxana is out as a sex worker peer and researcher, supported by her supervisors to continue to come out. We see this as a major achievement in and of itself, especially given that sex work at the time of writing is still criminalised in the jurisdiction where this research is occurring. Bridge-building between academia and sex worker researchers can happen in a single research project. In order for this to occur, however, reflexive and ongoing engagement with the tension points between academia and sex work needs to occur. This can entail the supervisory team understanding the role that historical research undertaken by non-sex workers has had in the production of stigma as well as a commitment to encourage and facilitate sex worker voices within the academy. Such a feminist alliance to sex work research ought to also involve active and ongoing conversations about power, knowledge and activism.

We argue that the recognition of access to power and privilege ought to inform collaborations in which a peer-based methodology is central. Through the literature, discussion and reflection we have examined the ways in which knowledge is produced in academic settings and argue that feminist scholarship should and can support sex workers to lead research about sex workers' lives. Strategies used by the supervisors and explained in the autoethnographies include recognising 'non-sex worker' as an identity, listening to sex workers and being prepared to step aside and let sex workers step forward; and sex worker feminism in practice.

Acknowledgement Sex work feminism guided this chapter from its conception to its conclusion and meant that feedback and input was sought out from the sex work community. This exemplifies 'nothing about us without us' and I encourage those undertaking sex work research to follow this tradition and to actively centre sex workers in their approach to research.

Those who contributed were:

Dr. Anonymous: sex work researcher: read the chapter and suggested edits.
Kat Morrison, SIN General Manager: read the chapter and suggested edits.
Dr. Elena Jeffreys: read the chapter, suggested edits and pulled the chapter together in the conclusion.
Jules Kim, CEO of Scarlet Alliance: read the chapter and suggested edits.

References

Argento, E., Reza-Paul, S., Lorway, R., Jain, J., Bhagya, M., Fathima, M., & O'Neil, J. (2011). Confronting structural violence in sex work: Lessons from a community-led HIV prevention project in Mysore, India. *AIDS Care, 23*(1), 69–74. https://doi.org/10.1080/09540121.2010.498868

Baratosy, R., & Wendt, S. (2017, May). "Outdated laws, outspoken whores": Exploring sex work in a criminalised setting. In *Women's studies international forum* (Vol. 62, pp. 34–42). Pergamon.

Bellhouse, C., Crebbin, S., Fairley, C. K., & Bilardi, J. E. (2015). The Impact of Sex Work on Women's Personal Romantic Relationships and the Mental Separation of Their Work and Personal Lives: A Mixed-Methods Study. *PLoS One, 10*(10), e0141575. https://doi.org/10.1371/journal.pone.0141575

Bradley, H. (2013). *Gender* (2nd ed.). Polity.

Bradley, M. S. (2007). Girlfriends, wives, and strippers: managing stigma in exotic dancer romantic relationships. *Deviant Behavior, 28*(4), 379–406. https://doi.org/10.1080/01639620701233308

Brooks, S. (1999). Sex work and feminism: building alliances through a dialogue between Siobhan Brooks and Professor Angela Davis. (Symposium: Economic Justice for Sex Workers) (Interview). *Hastings Women's Law Journal, 10*(1), 187.

Crenshaw, K. (1991). Mapping the margins: Intersectionality, identity politics, and violence against women of color. *Stanford Law Review, 43*(6), 1241–1299. Retrieved from www.jstor.org/stable/1229039. https://doi.org/10.2307/1229039

DeCat, N. Z. (2019, March 28). The racism of decriminalisation. *Tits and Sass.* http://titsandsass.com/the-racism-of-decriminalization/

Desyllas, M. C. (2013). Representations of sex workers' needs and aspirations: A case for arts-based research. *Sexualities, 16*(7), 772–787. Retrieved from https://journals.sagepub.com/doi/abs/10.1177/1363460713497214. https://doi.org/10.1177/1363460713497214

Dewey, S., Crowhurst, I., & Izugbara, C. O. (2019). *The Routledge international handbook of sex industry research.* Routledge International Handbooks. Web.

Dunk-West, P. (2018). *How to be a social worker: A critical guide for students.* Palgrave Macmillan.

Farley, M. (2019). Fact-Free Rationalizations Used to Promote Legal Pimping. *Arch Sex Behav, 48*(7), 1901–1902. https://doi.org/10.1007/s10508-019-01542-8

Gerassi, L. (2015). A heated debate: Theoretical perspectives of sexual exploitation and sex work. *Journal of Sociology and Social Welfare, 42*(4), 79–100. Retrieved from https://pubmed.ncbi.nlm.nih.gov/26834302. https://www.ncbi.nlm.nih.gov/pmc/articles/PMC4730391/

Ghabrial, M. A. (2016). "Trying to Figure Out Where We Belong": Narratives of Racialized Sexual Minorities on Community, Identity, Discrimination, and Health. *Sexuality Research and Social Policy, 14*(1), 42–55. https://doi.org/10.1007/s13178-016-0229-x

Holden, K. (2011). Sex work and feminism. *Meanjin, 70*(1), 46–54.

Hubbard, P., & Sanders, T. (2003). Making space for sex work: Female street prostitution and the production of urban space. *International Journal of Urban and Regional Research, 27*(1), 75. https://doi.org/10.1111/1468-2427.00432

Jeffreys, E. (2010). Sex worker-driven research: Best practice ethics. Challenging Politics: Critical Voices.

Kempadoo, K., Sanghera, J., & Pattanaik, B. (2012). Trafficking and prostitution reconsidered: New perspectives on migration, sex work, and human rights (2nd ed.). Boulder, Colo: Paradigm Publishers.

Kim, J., & Jeffreys, E. (2013). Migrant sex workers and trafficking - insider research for and by migrant sex workers. *ALAR: Action Learning and Action Research Journal, 19*(1), 62–96.

Kluttz, J., Walker, J., & Walter, P. (2020). Unsettling allyship, unlearning and learning towards decolonising solidarity. Studies in the Education of Adults, 52(1), 49–66. Retrieved from doi:https://doi.org/10.1080/02660830.2019.1654591.

Koken, J. A. (2012). Independent female escort's strategies for coping with sex work related stigma. *Sexuality & Culture, 16*(3), 209–229.

Leaker, M., & Dunk-West, P. (2011). Socio-cultural risk? Reporting on a qualitative study with female street-based sex workers. *Sociological Research Online, 16*(4), 9. Retrieved from http://www.socresonline.org.uk/16/4/9.html.

Levey, T. G., & Pinsky, D. (2014). A Constellation of Stigmas: Intersectional Stigma Management and the Professional Dominatrix. *Deviant Behavior, 36*(5), 347–367. https://doi.org/10.1080/01639625.2014.935658

Letherby, G. (2003). *Feminist research in theory and practice*. Open University Press.

Phipps, A. (2017). Sex wars revisited: A rhetorical economy of sex industry opposition. *Journal of International Women's Studies, 18*(4), 306.

Quinn, D. M., & Chaudoir, S. R. (2009). Living with a concealable stigmatized identity: the impact of anticipated stigma, centrality, salience, and cultural stigma on psychological distress and health. *J Pers Soc Psychol, 97*(4), 634–651. https://doi.org/10.1037/a0015815

Ramazanoglu, C., & Holland, J. (2002). Feminist methodology: Challenges and choices. Sage.

Rayson, J., & Alba, B. (2019). Experiences of stigma and discrimination as predictors of mental health help-seeking among sex workers. *Sexual and Relationship Therapy, 34*(3), 277–289. https://doi.org/10.1080/14681994.2019.1628488

Rosenberg, S. (2018). Coming In: Queer Narratives of Sexual Self-Discovery. *J Homosex, 65*(13), 1788–1816. https://doi.org/10.1080/00918369.2017.1390811

Sanders, T. (2004). A continuum of risk? The management of health, physical and emotional risks by female sex workers. *Sociol Health Illn, 26*(5), 557–574. https://doi.org/10.1111/j.0141-9889.2004.00405.x

Scarlet Alliance. (2014). *Principles for model sex work legislation*. Retrieved from http://www.scarletalliance.org.au/library/principles_2014

Simien, E. M. (2007). Doing Intersectionality Research: From Conceptual Issues to Practical Examples. *Politics & Gender, 3*(2), 264–271. https://doi.org/10.1017/S1743923x07000086

Stardust, Z. (2015). *Critical femininities, fluid sexualities and queer temporalities: erotic performers on objectification, femmephobia and oppression. In Queer sex work* (pp. 91–102). Routledge.

Stardust, Z. Z. (Producer). (2017). The stigma of sex work comes with a high cost. The Conversation. Retrieved from http://theconversation.com/the-stigma-of-sex-work-comes-with-a-high-cost-79657

Sullivan, C. T., & Day, M. (2019). Indigenous transmasculine Australians & sex work. *Emotion Space and Society, 32*. https://doi.org/10059110.1016/j.emospa.2019.100591

Wendt, S., & Zannettino, L. (2015). *Domestic Violence in Diverse Context; re-examining gender*. London: Routledge.

Wolf, A. (2019). Stigma in the sex trades. *Sexual and Relationship Therapy, 34*(3), 290–308. https://doi.org/10.1080/14681994.2019.1573979

Wong, W. C., Holroyd, E., & Bingham, A. (2011). Stigma and sex work from the perspective of female sex workers in Hong Kong. *Sociol Health Illn, 33*(1), 50–65. https://doi.org/10.1111/j.1467-9566.2010.01276.x

Zanghellini, A. (2020). Philosophical problems with the gender-critical feminist argument against trans inclusion. *SAGE Open, 10*(2), 2158244020927029. Retrieved from https://journals.sagepub.com/doi/abs/10.1177/2158244020927029. https://doi.org/10.1177/2158244020927029

CHAPTER 14

Does Feminist Social Work Practice Need Time? Gender, Parenting and Changing Times for Social Work

Linda Bell

Introduction

Social workers' commitments to social justice may seem undiminished, yet arguably there is less emphasis on explicit practice around gender and empowerment. *'We were very much proactive about leading agendas around feminist practice'*. This quote from an interview with one of the English social workers I interviewed in 2018 (Bell, 2020) suggests that earlier in her professional career, there was more inclination, as well as more time and other resources, to enable social workers to take up a proactively feminist agenda.

This raises several issues. Firstly, about what we mean by 'feminism(s)' in current social work contexts. How relevant is *gender* and particularly 'gender specificity' to feminism now? Can being 'gender-specific' sometimes be seen as discriminatory by social workers today? And what further

L. Bell (✉)
Middlesex University, London, UK
e-mail: l.bell@mdx.ac.uk

C. Cocker, T. Hafford-Letchfield (eds.), *Rethinking Feminist Theories for Social Work Practice*,
https://doi.org/10.1007/978-3-030-94241-0_14

issues would be raised by focusing more clearly on intersectionality/ies (see e.g. Bilge, 2013). By 'gender specificity', I follow definitions from other research, for example, Piller, Gibley and Peled's (2019) broad description of 'gender-specific intervention' (GSI), where professionals are working with people categorised by their gender or a gender-based role ('mother' and 'father').

Secondly, examining changed ways of working throws the spotlight onto resources, particularly time. Whether in practical terms, conceptual aspects or in terms of policy expectations for social work intervention, time may have implications, especially for gender-specific forms of social work.

In this chapter I will explore these issues using my own published collaborative research with social workers and their colleagues; this can be defined as 'gender-specific' by providing support for women whose children have been taken into public care. In the context of other published examples, we will examine how changed ways of working could contribute towards altering social workers' sensitivities towards using feminist approaches in practice.

'Feminism(s)' in Current Social Work Contexts: How Relevant Are 'Gender' and 'Gender Specificity' to Practice Now?

Cree (2018: 4) suggests that 'Social work has [always] had an uneasy relationship with feminism' citing, for example, the work of Maynard (1985), whose analysis of case files exemplified why some earlier feminists (starting in the 1960s–1970s) had become critical of social workers for neglecting women's rights (see also Phillips & Cree, 2014). Marietta Barretti's reviews of women and feminism in social work journals (Barretti, 2001, 2011) covering the periods 1988–1997 and 1998–2007, based on content analyses, reveal a decrease during the second time period of references in these journals to either women or feminism; she also notes that key referents to these terms related mainly to mothers/care giving, health (including mental health) and violence.

Moves towards pluralism within post-structuralist and post-modernist turns have indicated greater focus within feminism(s) on intersectionality/ies. How 'rights' could interact with these multiple social divisions and differences, both theoretically and in practice, is a matter of continuing debate (see e.g. Mehrotra, 2010; Soldatic & Meekosha, 2012). Cree (2018) argues that feminists in the Global North (compared to those in the South) seem to have lost their way in 'a deluge of identity politics' (p. 6), despite continuing to challenge essentialising discourses. She considers that debates within feminism around issues such as intersectionality threaten to fracture what 'feminism' itself means. Others, such as Bilge (2013, p. 420), also challenge certain notions of intersectionality where, in her view, these may 'shut down' hearing about specific oppressions, particularly racism:

> careful intersectional thinking must always account for different meanings, purposes, and audiences. Intersectionality does not create a shopping list of categories that can be deployed to shut down discussion of specific oppressions.

These ideas may also underscore what Harris is discussing when he suggests that social work is necessarily a 'contingent activity' (Harris, 2008, p. 663) 'conditioned by and dependent upon the context from which it emerges and in which it engages'. Earlier social work commentators charted policy changes brought about, for example, during the development of the New Labour political project in the UK in the late 1990s. Featherstone (2006) suggested that, during that government administration, 'gender' as an issue came to be seen as less crucial and, in her view, gender inequalities may thus have been overlooked in contemporary policy developments:

> ......the uninterrogated and widespread use of the term *parent* is problematic because it can obscure the kinds of strategies needed to ensure that the differing and often complexly changing needs and desires of men and women are addressed.... (Featherstone, 2006, p. 300)

She argues that aiming policy developments at working with 'parents' rather than with 'mothers' and 'fathers' ignored gender-differentiated investments in fathering and mothering; structural barriers to gender equality in childcare also went unchallenged. Significantly, she argues that these developments related to *power* and particularly to the exercise of power through use of language in both policy and practice.

A key context for social work with families that emerged over recent decades is the rise of what has been termed 'parenting culture' (Lee et al.,

2014). This is exemplified by the rise of the concept of 'parenting' used within professional practices including social work, as noted by Featherstone. 'Parenting culture' has many other implications, notably development of what some argue are unhelpful, deterministic approaches to what is seen as appropriate 'parenting' (Lee et al., 2014, pp. 9–10; Lee & McVarish, 2014). In the context of 'parenting culture', Lee et al. (2014) and Lee & McVarish (2014) point to links between children's emotional development, brain structure, cognitive development and parental nurturing practices which, they suggest, involve 'the promotion of a biologized and *gender-neutral* version of parent-child attachment' (*my italics*) (Lee & McVarish, 2014, p. 11). Others have also suggested inherent difficulties in relying on the kinds of logic privileging brain science when developing early years policy and, by extension, difficulties this may lead to in social work practice with families (see e.g. Edwards et al., 2015; Wastell & White, 2012). These developments are also interesting given what is described as the 'female dominated' nature of social work; current numbers of registered practitioners (Social Work England, 2020, p. 23) are estimated to have around 82% identifying as female (see Bell, 2020, p. 95). Implications for social workers who do not identify as female have been discussed (Christie, 2006) although 'female domination' of the profession, in 'power' terms, is also contested (see McPhail, 2004).

In this broad context, where do social work practitioners stand on adopting practices based on ideas about *ignoring perceived differences* based around gender (which others and I would define as 'gender neutrality')? Paradoxically, if we consider social work with families, particularly work with parents, what is striking is how often, when social workers work with 'mothers' they may see this as a proxy for 'parents'. This complaint is borne out by much of the literature about social work with fathers who, it is said, are often neglected in their parental role(s) by social workers and other practitioners (see e.g. Bell et al., 2020; Brandon et al., 2017; Nygren et al., 2019; Philip et al., 2019; Scourfield, 2014). Should social workers who consider themselves to be feminists then simply be more attentive to *fathers'* needs? Would this help to address the deep structural issues inherent in patriarchal societies, as also noted by Featherstone (2006, 2010) and others? We need to interrogate attempts to address these structural issues by asking how far is 'family' and 'childcare' still perceived by practitioners as 'women's responsibility'? Philip, Clifton and Brandon (2019) discuss how what they term a broader, overall lack of 'gender sensitivity' in social work practice may be at the root of some difficulties when working with families. They suggest:

> the value of holding in mind that fathers and mothers encounter different expectations, sanctions, opportunities, and constraints around their parenting is important to emphasize, despite social work's long-standing concern with equality. ... Our point here is that the established approach of gender "neutrality" may have the unintended consequence of ignoring important differences in men and women's experiences of parenting. (p 2305)

Although there is not the space in this chapter to critically interrogate all these assumptions around gendered experiences and parenting, does this argument imply, therefore, that practitioners should be providing separate, 'gender-specific' intervention for mothers and for fathers when working with parents? Others, writing in different international contexts, have in contrast questioned the value of concentrating on 'gender-specific' intervention, as well as suggesting the need for further research. A qualitative study relating to adolescent girls who are perceived to be 'at risk' (Piller et al., 2019) involved 15 Israeli professionals who had all worked with 'gender-specific' interventions (GSIs). The researchers found that, when interviewed, these professionals made assumptions about the benefits of GSIs for adolescent girls who they perceived to be 'at risk' (the authors abbreviate them as ARAGS), but that there was little research evidence to support the professionals' assumptions. They assumed, for example, that exclusion of males led to greater safety for the girls. The authors argue instead that female solidarity (seen as positive by these professionals) could be leading to neglect of intersectional characteristics, and to other unintended consequences for the intervention itself (Piller et al., 2019: p. 74):

> the failure of GSI to acknowledge gendered heterogeneity may damage the intervention

Issues with this study include giving very little information about the definition of being 'at risk' and what the GSIs being discussed actually involved. In order to examine such claims about the merits or disadvantages of such GSIs, I therefore present below details of our own 'gender-specific' study 'Mothers Apart' and discuss the relevance or otherwise to feminist ways of working.

Another strand to this issue is explored by Burnette and Renner (2017) who discuss the significance of patriarchal colonialism in the lives of indigenous American women who were experiencing violence and sexual abuse. In those contexts, the necessity of listening to the voices of women themselves is emphasised, something that my co-researchers and I also considered significant for our own project (see Bell et al., 2020; Lewis-Brooke et al., 2017).

Many researchers continue to suggest that women's (and men's) rights are what should be a key underpinning issue for social workers, and especially for those aiming to work from a feminist perspective (Ife, 2012; Turner & Maschi, 2015). But as noted above, links between 'rights' and 'intersectionality' within feminism may not be easy to disentangle. Potential for conflict between the rights of children and rights of parents is also expressed in much of the research literature relating to parenting (see e.g. Broadhurst & Mason, 2013; Lewis-Brooke et al., 2017; May-Chahal et al., 2006; Poso et al., 2014). Areas of social work which may bring these potential conflicts to the fore through the development of overlapping, yet potentially incompatible priorities, are child protection and work with families experiencing interpersonal/domestic violence (see Burnette & Renner, 2017; Witt & Diaz, 2019). For example, Witt and Diaz suggest that social workers:

> are often unable to empower mothers whilst, simultaneously, trying to safeguard children, due to procedural diktats and the lack of adequate resources to perform this task. (p. 216)

Robbins and Cook (2018), whose research involved focus groups with women experiencing interpersonal/domestic violence discuss how, in these situations, the system is set up to regard such violence and coercive control as 'hurdles that mothers must overcome' (p. 1666); gaining trust in such situations is thus a key issue for social workers, but one that they may struggle to achieve. Instead, Robbins and Cook view these experiences as a trauma through which women should be supported. Their overall conclusions are that this support may be more likely to succeed if it is coming from the voluntary/ not for profit sector, because women may find it easier to develop trusting relationships with staff within that context. This issue of *trust* with users being easier to maintain by staff working outside statutory (state) services was also significant for our own 'Mothers Apart' project. In our example, these staff formed a not for profit initiative that, despite being closely linked to the (state) local authority, worked separately in providing support to mothers and their families whose children had been taken permanently into public care.

Time to Care?

Social workers have reportedly gone from being 'in the community' to 'sitting behind a computer in the office' (Bell, 2020, p. 132; see also Yuill and Mueller-Hirth (2019)) and from developing long-term preventive work processes to providing short-term crisis interventions, especially when working with families. Could all these contextual changes have contributed towards altering social workers' sensitivities towards using feminist approaches in practice?

There is much discussion about the necessity for social workers and others having enough 'time to care', in the face of resource limitations, but from a theoretical perspective, we first need to unpick this potentially simplistic assertion. Sociologists, philosophers and other theorists have frequently asserted that 'time' does not only consist of 'clock' time, linked to Newtonian and Kantian concepts (in which time is seen as a uniform and universal continuum, which can be counted or measured). Kant furthermore saw time and space as the two basic preconditions for experience (see discussion in Boscolo and Bertrando (1993), p. 32 onwards); more recent philosophers, such as Elias, have also explored various aspects of the philosophy and nature of time, adding to our understanding of these issues (for background see Baiasu, Bird and Moore (eds) (2012); de Carvalho, 2018). Some theorists have explicitly linked feminist approaches to these ontological and epistemological aspects of time (e.g. see Adam (1989, 2002) and Davies (1990, 1994)), including the idea that in 'care' work, different notions of time, frequently linked to gender, are at play. There are tensions between these differing concepts of time, designated by researchers in various comparative ways such as 'clock time' or 'process time'. According to Davies, 'process time' is more flexible than clock time and is where tasks can take 'the time they need to take' (Davies, 1994). It relies on caring being an 'unlimited' aspect of relationships, outside the scope of time seen as commodified or a measurable resource. Both Davies and Adam discuss the hidden, gendered nature of these different temporalities, but they also acknowledge the dominating nature of 'clock' or linear forms of time, especially in industrialised societies.

Fahlgren (2009) discusses many of these issues in relation to social work, drawing upon the theoretical developments put forward by Davies and Adam, amongst others. Using a Swedish social work example relating to her own experience, Fahlgren explores, through discourse analysis, a situation in which a child was removed from his parents into foster care. In this example, many issues are raised for social workers, particularly through aspects of what we might identify as their need to professionally exercise both 'care' and 'control'. She suggests that there are a number of contradictory yet simultaneously operative discourses at play within social work, which can lead to different and incompatible meanings (and thus practice(s)). One of Fahlgren's key arguments in this paper is that since social work is often concerned with change (e.g. encouraging people to change their behaviour) then it is mainly following a discourse of 'linear' time (broadly relating to 'clock' time in Davies' formulation) anchored in the concepts of 'past', 'present' and 'future'. Fahlgren suggests (p. 212):

> Once social work interventions have been put into action, one expects improvements/normalization to result. If this fails, there is a risk of having to use more forceful intervention, as in the case of the boy who was dragged from his mother's arms [her social work example].

She describes the predictions that would result in that case, arising from the social work focus on linear time (p.212):

> First, there is fear of downhill change or deterioration—the risks for his future development. Second, there is a desired upward change, or development—he should be taken into care and placed in a foster home.

Fahlgren expands on how mothers and fathers may be regarded by social workers in the kind of situation she describes and identifies a 'normality discourse' which

> refers to the way in which the exercise of authority and social control aim to create, maintain and recreate the socially and culturally defined 'normal' order. The 'normality discourse' includes the constitution of a *gender order*, in which the mother is also upheld as the obvious mothering figure while the father is left more or less out of the narrative.

In another example drawing on these ideas about differing temporalities, Yuill and Mueller-Hirth (2019) present a small-scale, qualitative interview-based study with British social workers. They explore what they designate as 'paperwork time' (linked to a linear trajectory) alongside what they identify as 'compassionate time'. 'Compassionate time' is more developmental and cyclical than 'paperwork time', requiring 'slower engagement', particularly while working with service users. Yuill and Mueller-Hirth's message is that whilst their interviewees said they wished to work more often in 'compassionate time', this temporality was largely 'fleeting and contingent' (p. 1548), and so this aim was rarely achieved. This also applied to having sufficient time for professional development (on this issue see also Bell et al., 2017, for examples from Denmark, England and Norway). Although Yuill and Mueller-Hirth concur with Fahlgren about the significance of linear temporality in social work, to which 'paperwork time' is linked, they propose that this temporality also

> reflects the structural changes that social work has undergone in the neoliberal era and its culture of speed, creating a workplace dense with multiple demands, particular temporal expectations and short time frames. (p. 1548)

These comments chime with the views of social workers I interviewed in 2018, (Bell, 2020), where a social worker told me there was no longer the time or space to proactively lead or develop feminist practice. I will now discuss an example relating to social work, which brings together many of these ideas about gender and time already discussed. This project aimed to take enough time to both support and listen to the 'voices' of mothers who were or had been social work service users. In this sense the project was taking a feminist stance, as well as being a 'gender-specific' intervention through its focus on mothers (even though the mothers involved did ask for the work to be extended to fathers, as 'they are parents too'). This focus on 'voice' and on subjective experience (as 'mother') is the nearest I can come to defining this overall project as directly 'feminist', although I acknowledge that it is also embedded within the gendered expectations of 'good enough' parenting within this society.

'Mothers Apart': A Partnership Action Project

'Mothers Apart' was a partnership-based action project bringing together staff in a London borough with university researchers (Bell et al., 2020; Lewis-Brooke et al., 2017; Bell, 2017, (chapter 3 'Project D')). This project aimed to both provide and research improved support for mothers and their families where children had been taken permanently into public care. The 'Mothers Apart' project was innovative in the sense that it involved professional intervention with these families (the 'Hummingbirds' pilot) whilst there was also a significant independent research element involving myself and another researcher who were not social workers. This partnership project was challenging since it involved the development of effective collaborative working between staff from different types of institutions (UK public sector/local authority and a university), where there were differing priorities and expectations. Nevertheless the partnership provided members of the team with useful experiences and the opportunity to collaborate on the provision of appropriate funding for the overall project (see Bell et al., 2018).

The 'Hummingbirds' pilot service was set up and operated outside this particular London borough's mainstream Children's Social Care (CSC) service; this was partly because mothers who had 'lost' children in this way were often distrustful of professionals, and those we interviewed also said that they would not work directly within the CSC (for a similar example, see Robbins & Cook, 2018). Lewis-Brooke et al. (2017) have described how the support offered to local mothers through the 'Hummingbirds' pilot initiative operated to provide group work and other individualised forms of support. The authors wrote about how this initiative was begun collaboratively and followed research-based interviews with ten local mothers. The interviews were originally developed by the project team because local authority members wanted to place mothers' own voices at the centre of this developing action-project. Bell et al. (2020) elaborates on the methodology we used to elicit these mothers' 'voices' and the implications of working with what they were able to tell us in order to underpin and develop the pilot intervention.

A key aim of the pilot itself was 'feminist' to the extent that:

> Our intention was that the service ['Hummingbirds'] sent messages that built self-esteem in the women from the relationships built with workers, and continued recognition that women had agency and were likely to help each other as much as the workers did. (Lewis-Brooke et al., 2017, p. 10)

The development of the 'Hummingbirds' pilot was time consuming in its process (see also Davies (1990, 1994)) and its trajectory also fits well into a model of professionals working with 'compassionate time' (Yuill & Mueller-Hirth, 2019). In contrast, mothers' experiences of involvement with child protection workers in the CSC had often been fraught with difficulties and clashes over procedures, timescales and 'paperwork', with mothers feeling let down or even abandoned by statutory social work staff (Bell et al., 2020). In many ways the 'Mothers Apart' project epitomises why feminist social work practice (and research) needs 'time'. Both social work interventions and action projects of this kind, even though they may be time-limited overall, are also necessarily longitudinal and intensive for individuals, particularly when aiming to take a broadly 'feminist' approach. In its first year of operation the 'Hummingbirds' initiative was able to work intensively with 11 women (Lewis-Brooke et al., 2017, p. 12). This was a voluntary process for participants (since 'Hummingbirds' was not explicitly part of the local authority's CSC) and in order to operate successfully the initiative involved various kinds of individual support, which came from social workers and also from other professionals concerned with education or employment services. Where appropriate for individual women, they were able to join in with group meetings that were based on building *a supportive network with other women* who had 'been through' the same traumatic experiences, with these participant groups facilitated by professional workers. The project team noticed that:

> women continued to firmly identify themselves as 'mothers' throughout these painful experiences. (Bell, 2020, p. 116)

To this extent, 'Hummingbirds' was a 'gender-specific' intervention (Piller et al., 2019), and it was valued as such by women participants. However, the professional team members had also learnt from similar projects involving parents of all genders in other countries (see, in particular, Norwegian experience, reported in Seim & Slettebø, 2011; Slettebø, 2013).

The research element of the 'Mothers Apart' project, which was crucial to exploring the kinds of approach that could be most useful to these women, connected with their own perceived identities as 'mothers' and was explicitly feminist, drawing on methodological practice focused on women's 'voices' (as explained in Bell et al., 2020; see also Mauthner and Doucet (1998), whose work had in turn drawn from the work of Carol Gilligan (1993). Despite following clear ethical guidelines and attempting to build rapport with our interviewees, there were occasions when the project researchers found themselves in awkward situations or felt they were colluding with systems that were not what they would have chosen (see Philip & Bell, 2017). For example, some women interviewed by the researchers did question why fathers were not also included in the 'Hummingbirds' support that was being offered by practitioners. Limited funding and other resources available within the local authority meant that extending the initiative for fathers was never achieved, despite support for the idea from professionals.

Conclusions

In this chapter, I have explored how relevant gender and particularly 'gender specificity' can be to feminist social work practice. I have concentrated on examples relating to parenting and domestic /interpersonal violence and I concur with Robbins and Cook (2018) that social workers' attempts to deal simultaneously with both child protection and women's empowerment can be fraught with difficulty. Nevertheless, gender-specific interventions (such as 'Hummingbirds', developed as part of the 'Mothers Apart' project) can, I believe, serve a useful purpose in empowering women as mothers, despite critiques of this kind of approach (see Piller et al., 2019).

I suggest that time factors, including underpinning ontological and epistemological aspects of time, have implications for these gender-specific forms of social work. Furthermore, although GSIs such as 'Hummingbirds', seen in a 'feminist' context, may allow mothers whose children have been taken into care their 'own space' and extended time to share coming to terms with their experiences, they do have limitations. There are wider issues concerning the position of men in families and child care which need unpicking, especially with regard to their relations with professionals. This is why denying support to fathers (albeit through lack of time, appropriate staffing and funding to enable provision of support) and taking a 'gender neutral' approach can risk reinforcing gender stereotypes, allowing 'mothers' to simply become conflated with 'parents'. Despite tacit agreement by professionals and by mothers that fathers may also need support, acknowledgement that these needs might be different from mothers' needs is important; but these developments are clearly still a work in progress (see also Brandon et al., 2017; Philip et al., 2019).

Ideas for Social Work Practice

- Explore ideas about how to provide empowering initiatives which take account of all genders and intersectionalities, whilst also providing specific services where appropriate; become more inclusive and also sensitive to differing needs of those identifying as 'mothers' and 'fathers'.

- Co-productive initiatives that involve service users will enhance the above aims.
- Recognise the potentially disempowering aspects of short-termism and 'linear' approaches in social work and instead develop ways of enabling social workers and others to develop their capacity for what is termed 'compassionate time'.
- Value partnerships between statutory and voluntary/not for profit sector initiatives; greater flexibility may be available through such 'partnership' arrangements (providing that secure funding is available) which could enable professionals to successfully develop their inclusiveness, feminist practice and use of 'compassionate time'.

Acknowledgements I acknowledge the partnership working of all my colleagues on the 'Mothers Apart' project: Sarah Lewis-Brooke, Dr. Rachel Herring, Sioban O'Farrell-Pearce, Theresa So (all from Middlesex University); Lynne Lehane, Nikki Bradley, Francoise Cosgrove, Jo Prosser, Karen Quinn, Brian Sharpe (all of London Borough of Tower Hamlets) and Pat Oparah (Royal London Hospital, Whitechapel). The project team would particularly like to thank 'Hummingbirds' staff and all the mothers who were interviewed and/or who were 'Hummingbirds' participants.

References

Adam, B. (1989). Feminist social theory needs time; reflections on the relationship between feminist thought, social theory and time as an important parameter in social analysis. *The Sociological Review, 37*(3), 458–473.

Adam, B. (2002). The gendered time politics of globalization: Of shadowlands and elusive justice. *Feminist Review, 70*(1), 3–29. https://doi.org/10.1057/palgrave.fr.9400001

Baiasu, R., Bird, G., & Moore, A. W. (Eds.). (2012). *Contemporary Kantian metaphysics: New essays on space and time*. Palgrave Macmillan

Barretti, M. A. (2001). Social work, women, and feminism: A review of social work journals, 1988–1997. *Affilia, Journal of Women and Social Work, 16*(3), 266–294. https://doi.org/10.1177/08861090101600302

Barretti, M. A. (2011). Women, feminism, and social work journals 10 years later: 1998–2007. *Affilia, Journal of Women and Social Work, 26*(3), 264–277. https://doi.org/10.1177/0886109911417688

Bell, L. (2017). *Research methods for social workers*. Palgrave Macmillan.

Bell, L. (2020). *Exploring social work: An anthropological perspective*. Policy Press.

Bell, L., Herring, R., & Annand, F. (2020). Fathers and substance misuse: A literature review. *Drugs and Alcohol Today, 20*(4), 353–369. https://doi.org/10.1108/DAT-06-2020-0037

Bell, L, Lewis-Brooke, S, Herring, R, Lehane, L, & O'Farrell-Pearce, S. (2018). *Reflecting on 'Mothers Apart': What have we learnt from this partnership project since 2014?* Paper presented at the 8th European Conference for Social Work Research, Edinburgh, April 2018.

Bell, L., Lewis-Brooke, S., Herring, R., Lehane, L., & O'Farrell-Pearce, S. (2020). Mothers' voices: Hearing and assessing the contributions of 'birth mothers' to the development of social work interventions and family support. *British Journal of Social Work.* https://doi.org/10.1093/bjsw/bcaa138

Bell, L., Nissen, M., & Vindegg, J. (2017). The construction of professional identity in social work: Experience, analytical reflection and time. In B. Blom, L. Evertsson, & M. Perlinski (Eds.), *Social and caring professions in European welfare states: Policies, services and professional practices* (pp. 37–51). Policy Press.

Bilge, S. (2013). Intersectionality undone; saving intersectionality from feminist intersectionality studies. *Du Bois Review, 10*(2), 405–424 https://doi.org/10.10170/S1742058X13000283

Boscolo, L., & Bertrando, P. (1993). *The times of time: A new perspective in systemic therapy and consultation.* Norton.

Brandon, M, Philip, G., & Clifton, J. (2017). *Counting fathers in: Understanding men's experiences of the child protection system.* Report. University of East Anglia/Nuffield Foundation.

Broadhurst, K., & Mason, C. (2013). Maternal outcasts: Raising the profile of women who are vulnerable to successive, compulsory removals of their children – A plea for preventative action. *Journal of Social Welfare and Family Law, 35*(3), 291–304. https://doi.org/10.1080/09649069.2013.805061

Burnette, C., & Renner, L. (2017). A pattern of cumulative disadvantage: Risk factors for violence across indigenous women's lives. *British Journal of Social Work, 47*(4), 1166–1185. https://doi.org/10.1093/bjsw/bcw075

Christie, A. (2006). Negotiating the uncomfortable intersections between gender and professional identities in social work. *Critical Social Policy, 26*(2), 390–411.

Cree, V. (2018). Feminism and social work: Where next for an engaged theory and practice? *Aotearoa New Zealand Social Work, 30*(3), 4–7.

Davies, K. (1990). *Women and time. The weaving of the strands of everyday life.* Avesbury.

Davies, K. (1994). The tensions between process time and clock time in care work: The example of day nurseries. *Time & Society, 3*(3), 277–304.

De Carvalho, E. R. (2018). Norbert Elias and the philosophical controversy surrounding the nature of time. *Time and Society, 27*(2), 155–175.

Edwards, R., Gillies, V., & Horsley, N. (2015). Brain science and early years policy: Hopeful ethos or 'cruel optimism'? *Critical Social Policy, 35*(2), 167–187. https://doi.org/10.1177/0261018315574020

Fahlgren, S. (2009). Discourse analysis of a childcare drama: Or the interfaces between paradoxical discourses of time in the context of social work. *Time & Society, 18*(2–3), 208–230.

Featherstone, B. (2006). Why gender matters in child welfare and protection. *Critical Social Policy, 26*(2), 294–314.

Featherstone, B. (2010). Writing fathers in but mothers out!!! *Critical Social Policy, 30* (2001, 2011) (2), 208–224.

Gilligan, C. (1993). *In a different voice: Psychological theory and women's development*. Harvard University Press.

Harris, J. (2008). State social work: Constructing the present from moments in the past. *British Journal of Social Work, 38*(4), 662–679.

Ife, J. (2012). *Human rights and social work: Towards rights-based practice*. Cambridge University Press.

Lee, E., Bristow, J., Faircloth, C., & Macvarish, J. (Eds.). (2014). *Parenting culture studies*. Palgrave Macmillan.

Lee, E., & McVarish, J. (2014, March).*The uses and abuses of biology: Neuroscience, parenting and family policy in Britain*. A key findings report. https://doi.org/10.13140/2.1.3750.5926.

Lewis-Brooke, S., Bell, L., Herring, R., Lehane, L., O'Farrell-Pearce, S., Quinn, K., & So, T. (2017). Mothers Apart: An action research project based on partnership between a Local Authority and a University in London, England. *Revista de Asistentă Socială, XVI*(3), 5–15. www.swrev.ro ISSN 1583-0608

May-Chahal, C., Bertotti, T., Di Blasio, P., Cerezo, M. A., Gerard, M., Grevot, A., Lamers, F., McGrath, K., Thorpe, D. H., Thyen, U., & Al-Hamad, A. (2006). Child maltreatment in the family: A European perspective. *European Journal of Social Work, 9*(1), 3–20.

Maynard, M. (1985). The response of social workers to domestic violence. In J. Pahl (Ed.), *Private violence and public policy: The needs of battered women and the response of the public services*. Routledge/Kegan Paul.

McPhail, B. (2004). Setting the record straight: Social work is not a female- dominated profession: Commentary. *Social Work, 49*(2), 323–326.

Mehrotra, G. (2010). Toward a continuum of intersectionality theorizing for feminist social work scholarship. *Affilia: Journal of Women and Social Work, 25*(4), 417–430.

Mauthner, N., and Doucet, A. (1998). Reflections on a voice-centred relational method: Analysing maternal and domestic voices. In J. Ribbens & R. Edwards (Eds.), Feminist dilemmas in qualitative research: Public knowledge and private lives (pp. 119–146). Sage Publications, Inc https://doi.org/10.4135/9781849209137.n8

Nygren, K., Walsh, J., Ellingsen, I., & Christie, A. (2019). What about the fathers? The presence and absence of the father in social work practice in England, Ireland, Norway, and Sweden – A comparative study. *Child & Family Social Work, 24*(1), 148–155. https://doi.org/10.1111/cfs.12592

Philip, G., & Bell, L. (2017). Thinking critically about rapport and collusion in feminist research: Relationships, contexts and ethical practice. *Introduction to Co-Edited Special Issue, Women s Studies International Forum, 61*. https://doi.org/10.1016/j.wsif.2017.01.002

Philip, G., Clifton, J., & Brandon, M. (2019). The trouble with fathers: The impact of time and gendered-thinking on working relationships between fathers and social workers in child protection practice in England. *Journal of Family Issues, 40*(16), 2288–2309.

Phillips, R., & Cree, V. E. (2014). What does the "fourth wave" mean for teaching feminism in 21st century social work? *Social Work Education: The International Journal, 33*(7), 930–943. https://doi.org/10.1080/02615479.2014.885007

Piller, S., Gibley, J., & Peled, E. (2019). The value and rationale of gender-specific intervention with at-risk adolescent girls. *Child and Family Social Work, 24*(1), 69–76. https://doi.org/10.1111/cfs.12582

Poso, T., Skivenes, M., & Hestbaek, A. (2014). Child protection systems within the Danish, Finnish and Norwegian welfare states – Time for a child centric approach? *European Journal of Social Work, 17*(4), 475–490.

Robbins, R., & Cook, K. (2018). 'Don't even get us started on social workers': Domestic violence, social work and trust—An anecdote from research. *British Journal of Social Work, 48*(6), 1664–1681. https://doi.org/10.1093/bjsw/bcx125

Scourfield, J. (2014). Improving work with fathers to prevent child maltreatment. *Child Abuse & Neglect, 38*(6), 974–981.

Seim, S., & Slettebø, T. (2011). Collective participation in child protection services: Partnership or tokenism? *European Journal of Social Work, 14*(4), 497–512.

Slettebø, T. (2013). Partnership with parents of children in care: A study of collective user participation in child protection services. *British Journal of Social Work, 43*(3), 579–595. https://doi.org/10.1093/bjsw/bcr188

Social Work England. (2020). Annual Report and Accounts HC 539 Crown Copyright July 2020.

Soldatic, K., & Meekosha, H. (2012). Moving the boundaries of feminist social work education with disabled people in the neoliberal era. *Social Work Education, 31*(2), 246–252. https://doi.org/10.1080/02615479.2012.644975

Turner, S., & Maschi, T. (2015). Feminist and empowerment theory and social work practice. *Journal of Social Work Practice, 29*(2), 151–162. https://doi.org/10.1080/02650533.2014.941282

Wastell, D., & White, S. (2012). Blinded by neuroscience: Social policy, the family and the infant brain. *Families Relationships and Societies, 1*(3), 397–414. https://doi.org/10.1332/204674312X656301

Witt, L., & Diaz, C. (2019). Social workers attitudes towards female victims of domestic violence: A study in one English local authority. *Child and Family Social Work, 24*(2), 209–217.

Yuill, C., & Mueller-Hirth, N. (2019). Paperwork, compassion and temporal conflicts in British social work. *Time and Society, 28*(4), 1532–1551.

CHAPTER 15

Lesbian Parenting: Rebellious or Conformist?

Christine Cocker

Introduction

Changes to legislation in many westernised nations in the world has made it easier for lesbians, gay men, bisexual, queer and trans/non-binary (LGBT+) persons living in these countries to become parents, whether this be via birth, surrogacy, fostering, adoption, step parenting, co-parenting or any other means. However, discourses about families continue within a heteronormative and cisgendered frame, which affects how LGBT+ people create 'family' and family structures are then understood. Chambers (2007) states that the concept of heteronormativity, '*reveals institutional, cultural and legal norms that reify and entrench that normativity of heterosexuality. In other words, 'heteronormativity' tells us that heterosexual desire and identity are not merely assumed, they are expected*' (Chambers, 2007, pp. 664–65). The impact of this is that, even with the significant social change that has occurred within many western nations to validate LGBT+ relationships, the complex family forms and structures that queer families create in their lives are not valued in their own right.

C. Cocker (✉)
School of Social Work, University of East Anglia, Norwich, UK
e-mail: Christine.cocker@uea.ac.uk

C. Cocker, T. Hafford-Letchfield (eds.), *Rethinking Feminist Theories for Social Work Practice*,
https://doi.org/10.1007/978-3-030-94241-0_15

Despite changes in the legal position for lesbian parents in the UK over the last 50 years, homophobia, transphobia and heterosexism still exist. The research community itself is not immune from these biases. The literature that has developed since the 1970s illustrates what Clarke (2008) refers to as reflecting 'a liberal equality perspective' (p118), which is largely reductionist and seeks to identify differences between various types of families and the resulting impact on the children growing up within these families. As far back as 1987, the research findings about lesbian mothers being 'just like other mothers' were critiqued from a feminist viewpoint as 'mak[ing] us once more invisible…'(Pollack, 1987, p. 316). Subsequent feminist writing has added much to this earlier viewpoint about lesbian parenting being 'as good as' heterosexual parenting, in response to heteronormative critiques of alternative families.

This chapter will explore the discourses around lesbian parenting in particular, to consider whether it remains 'progressive' as an alternative family form, whether it is now so commonplace that it is normalised or whether it occupies another different space altogether. I provide a brief overview of the history of lesbian parenting in the UK over 50 years before considering how various discourses over this time attached meaning to lesbians having and raising children. I will reflect on how social work practice with lesbian parents has changed and discuss how the learning applies to working with queer families.

Rebellious or Conformist?

In the early 1970s, the seven core demands of the women's liberation movement included: 'The right to a self-defined sexuality. An end to discrimination against lesbians' (Sisterhood and After Research Team, 2013). In the 1970s and early 1980s, many children of lesbians, who were conceived in heterosexual relationships, were routinely removed from their mother's care by the Courts in divorce proceedings, and placed with their fathers. These actions were due to concerns about the effect of a mother's lesbianism on her children. However, in divorces involving heterosexual parents, children almost always remained with their mothers (Hanscombe & Forster, 1982). Judges relied on the psychoanalytic research of the day, which presented a pathological model of homosexuality, despite it having been removed from the Diagnostic and Statistical Manual of Mental Disorders (DSM)-II in 1973 as a 'sexual deviation' (Clarke, 2008). Lesbian motherhood was seen as '*a contradiction in terms, a deviation from the path of natural heterosexuality. Lesbian mothers were widely assumed to transmit their deviance to their children*' (2008, p. 121). In this hostile environment, understandably feminists both organised to defend lesbian mothers' rights to be able to parent their children (Rights of Women Lesbian Custody Group, 1986) as well as challenge these hetero-sexist misconceptions.

During the 1980s lesbian feminists challenged the heteronormative nuclear family status quo by creating their own families of choice, including having children through self and donor insemination, as single mothers, in couples or collective households (Saffron, 1986). The feminist foci on sexual choice, women's health and reproductive rights provided a context for lesbians to think what had been previously unimaginable and find ways to have children outside of heterosexual relationships. In the 1980s, a small number of social science researchers began investigating the outcomes for children who were raised by lesbian parents. In the UK, the work of Susan Golombok (Golombok et al., 1983) showed that the children of lesbians were just as likely as children growing up in heterosexual families to have good mental health, to have positive relationships with peers and to have good relationships with both male and female adults. Golombok's longitudinal work compared the psychological, sexual, educational and social outcomes of children raised by lesbians alongside children raised by single heterosexual mothers (Golombok, 2000; Golombok et al., 2003; Golombok et al., 1983; Golombok & Tasker, 1996). The history of lesbian parenting in the United States is similar, with the work of US researcher Charlotte Patterson (Patterson, 1992, 2005, 2017) echoing Golombok's findings. Golombok and Patterson both found through their separate longitudinal research beginning in the 1980s that the outcomes for children raised in lesbian families were broadly similar to children raised in single (heterosexual) carer families. These findings contributed to changing legal practices regarding the removal of children from their lesbian mothers. However, even in the 1980s, the idea of comparing a lesbian parent to a single female heterosexual parent as the socially acceptable 'norm' was viewed as deeply problematic (Pollack, 1987) due to the reductionist way in which parenting activities were understood.

The social and political landscape of the 1980s was particularly challenging as the Government, with Margaret Thatcher as Prime Minster between 1979 and 1990, brought in a conservative approach to social policy and a backlash against radical politics and left wing Labour. This saw many Labour local authorities challenged and undermined (Cooper,

1994). The conservative view of family was very traditional. Thatcher famously said, '*there is no such thing as society. There are individual men and women, and there are families. And no government can do anything except through people, and people must look to themselves first*' (Thatcher, 1987).

At the same time as lesbians were losing their children in the Courts, lesbians were also creating families of choice. Additionally in the 1980s the AIDS epidemic hit, which had a catastrophic impact on sexual politics. The public homophobic backlash to AIDS as a 'gay disease' provoked a political response from the Conservative Thatcher Government. Continuing to promote traditional nuclear family norms, the Government labelled and demoted lesbian and gay families as 'pretended family relationships' under Section 28 of the Local Government Act 1988. Local authorities, particularly schools, were not able to undertake any activities that might be seen as promoting homosexuality as acceptable (Cooper & Herman, 1995). Section 28 had an impact on lesbian and gay families through the late 1980s and 1990s and has left a lasting shadow on public discourses, reinforcing perceptions that they are 'not good enough'. Clarke (2008) suggests that at this time there was a fracturing of LGBT+ politics, with a focus on AIDS/HIV activism, and consequently many issues affecting lesbians were side-lined by mainstream LGBT+ communities. Within lesbian feminist circles, lesbians having and raising children had occupied a contested space, with strong views expressed about whether all parenting was conventional and heteronormative, whether lesbians could challenge these normative understandings of parenting and family, particularly when raising boys (Clarke, 2005).

Despite the heteronormative hegemony of Section 28 in the UK, lesbians continued to create alternative families, with the aid of self-help manuals published on both sides of the Atlantic (e.g. Pies, 1985; Saffron, 1986, 1994), and lesbian and gay parenting handbooks (e.g. Martin, 1993; Pollack & Vaughn, 1987). Lesbians choosing to create families were making personal decisions that had political implications. Whilst the Golombok studies began to change discourses around lesbian parenting within the general population, particularly in cases where private family life was being assessed by public institutions, such as Courts and Local Authorities, these continued to offer an assimilationist account of lesbian parenting (Cocker & Hafford-Letchfield, 2021; Clarke, 2008; Gabb, 2005), which has powerfully occupied public discourses. Clarke (2008) identifies two images of lesbian mothers within the literature: 'the "just-as-good-as lesbian mother" and the "bad lesbian mother"'(p. 122). In his research investigating social workers assessments of lesbian prospective adopters, Hicks (2000) created three typologies of 'good' and 'bad' lesbians: the 'man-hater'; the 'militant'; and the 'pedestal of virtue'. He argued that the lesbians that were most likely to be approved as foster carers or adopters were:

> *able to deal with anti-lesbian prejudices of birth families, some children/young people and panel; were not threatening to gender norms; were integrated with heterosexuals; … likely to know men, have male role models, present positive images of men; be integrated into the wider (heterosexual) community and family; … where child care abilities were emphasised as paramount, over issues of sexuality; were non-militant, non-radical, not political, not too 'feminist'.* (Hicks, 2000, p. 164)

This publicly acceptable face of lesbian parenting hardly portrays it as a politically rebellious act. Gabb (2005) suggests that another major flaw in the literature exploring lesbian parenting up to that time is its lack of reference to the complexities in the way lesbians and their families use time, places and spaces that are predominantly straight/heterosexual. She comments that little is known about the ways in which lesbian parent families inhabit and use heteronormative space (p.424), whether this be schools, GP surgeries, holiday destinations or other community spaces. Issues of intersectionality also apply, with much of the early lesbian parenting literature dominated by white middle-class perspectives (Gabb, 2004). This dismisses and silences the diversity of experiences, privileging a particular type of lesbian mother, most often the birth mother, leaving little room for exploration of the different ways in which lesbian families create kinship across class and ethnicity in particular.

Conformist or Rebellious?

Since 2000, the numbers of lesbians, gay men and trans/non-binary people having families have markedly increased. This is often referred to as the 'lesbian baby boom' or the 'gayby boom' era, and the literature moves away from a sole focus on lesbian motherhood to include LGBT+ persons becoming parents via fostering and adoption as well as via birth and surrogacy. The enormous loss of life within the gay community because of HIV/AIDS in the 1980s and 1990s may have motivated LGBT+ people to find ways to have their own families. Clarke (2008), referencing Green and Bozett (1991) and Kahn (1995), refers to this as many 'coming under the spell of pronatalism' (p. 122). In 2020 one in six (n = 570/3440) adoptions in England were to same sex couples (Department for Education, 2021). This shows not only the significant contribution that same sex adopters are making in the adoption field but also the diversity of ways in which same sex couples (in this instance) are choosing to have families, which is not always via birth.

There have been considerable changes within English law over the first two decades of the twenty-first century, which directly support LGBT+ families. Section 28 remained law until 2003, when it was finally removed from statute books. Five other pieces of legislation also helped to fundamentally change the legal landscape for LGBT+ families. Firstly, the age of consent was equalised at 16 in England in 2001. Secondly, the Adoption and Children Act 2002 enabled lesbians and gay men in partnerships (as distinct from civil partnerships which were introduced two years later) to jointly adopt, and for the partner of the birth parent of a child (or children) to apply to adopt in the same way as a heterosexual step parent would. Thirdly, the Civil Partnership Act 2004 enabled lesbians and gay men to obtain legal recognition of their relationships. Fourthly, the Human Fertilisation and Embryology Act 2008 made it possible for the non-biological same sex parent to be entered on the birth certificate of their child. Finally, the Equality Act (Sexual Orientation) Regulations 2007 prohibited discrimination on the grounds of sexual orientation in providing goods, facilities and services to the public. Adoption agencies in England then had to assess the suitability of LGBT+ applicants who approached their agency. This is not the same thing as saying that all LGBT+ individuals have the right to become adopters, rather they then had the right to an assessment as a couple, not just as a single adopter (H. C. Brown & Cocker, 2011).

These legal changes have transformed opportunities for same sex couples to have children. There is a level of acceptability of lesbian motherhood in the 2020s that did not exist 40 years previously. Lesbians are more aware that they have choices and being a mother is something they can achieve in their lives, should they want to. Because there are more lesbians having and raising children, visibility is greater. There are more lesbian parents at the school gates collecting their children and living their lives in communities all over the country. Giddens (1992) referred to these changing family forms as 'relationship innovators', often within a hostile environment. However, the framing of lesbian motherhood in legislation and much of the existing research used to describe lesbian families continues to adopt a conformist approach where lesbian families are 'just like' or 'just as good as' heterosexual families but with same sex parents. The danger is that the particular qualities of lesbian families continue to be rendered invisible within these discourses (Clarke, 2008; Gabb, 2018), and this does not speak to the realities of their experiences.

Since 2000, the literature has changed direction, from examining the sex and gender behaviours and 'choices' of children raised by lesbians towards an examination of how same sex families were constructed and how they lived their lives. The unique aspects of same sex family structures and forms have been explored by a number of academics who highlight families of choice and patterns of queer kinship through identity, familial and social relationships (Weeks et al., 2001), egalitarian relationships within same sex parenting activities (Mellish et al., 2013) and how lesbian mothers manage their sexual identities (Gabb, 2005). From her own considerable research, Golombok (2000) commented, 'It is what happens within families, not the way families are composed, that seems to matter most' (2000, p. 101). Hicks's (2011) exploration of same sex parenting within everyday contexts has theorised how concepts and social categories are produced and practiced, such as kinship, family, race, gender and sexuality. He argues that same sex parenting is more complex than describing it either as an assimilative position or a radical act; rather it is 'how the extraordinary is lived in ordinary circumstances' (p. 2). Many lesbian parents constantly fight for their space and their families to be recognised for what they are and not how much they resemble heterosexual families (Cocker, 2011; Hicks, 2011, 2014; Hicks & McDermott, 2018).

Heteronormativity also carries with it the constant threat and experience of homophobia for same sex families and transphobia for trans/non-binary families (Cocker et al., 2019; Hafford-Letchfield et al., 2016; Hicks & McDermott, 2018). This is not something that heterosexual families experience and marks out these families as different. Since this is (unfortunately) an altogether common experience for LGBT+ people, they are adept at negotiating those spaces (Cocker & Brown, 2010; Cocker et al., 2019). In addition, how children come to be in lesbian and gay families frequently differs from heterosexual families. There is an assumption that the children of heterosexual parents are birth children. This is not the same for children of lesbian and gay parents, even when they are birth children. A further assumption in most school environments is that all children are being raised within heterosexual families. This means that same sex families are constantly 'coming out' within school and other public service settings. This is a recurring situation lesbian and gay parents face each year with every new class teacher and at the school gates with other parents. They are required to continuously manage the heteronormative interface between private and public spaces within the home and community for their children (Cocker & Brown, 2010). This would similarly apply to trans/non-binary parents and their children (Hafford-Letchfield et al., 2019).

The fostering and adoption literature has provided an opportunity to explore the experiences of lesbians, gay men, trans/non-binary and other queer adopters who have been assessed by social workers as prospective adopters and foster carers (Brown & Rogers, 2020; Cocker, 2011; Hicks, 2005; Mellish et al., 2013; Skeates & Jabri, 1988; Wood, 2016, 2018). Agencies are involved assessing prospective applicants to ensure that they have the necessary qualities, skills and knowledge to care for children who have experienced abuse and neglect in their early years of life. These agencies represent the State entering the homes of LGBT+ applicants, spaces which are usually private and protected from the gaze of public scrutiny (Cocker & Brown, 2010). This English literature shows that whilst lesbians' experiences of the assessment process have improved markedly in line with the legal and social changes that have occurred within wider society, heteronormative and gender normative practices remain in terms of the kinds of families being approved as foster carers and adopters (Brown & Rogers, 2020; Hicks, 2011).

There is a further major problem within the literature as Gabb (2018, p. 1014) reminds us:

> *Whilst time moves forward and many contemporary equality rights may be indeed progressive, we should not lose sight of who gets written in and who gets written out of the story of LGBTQ parenthood that gets told...socio-economic resources and cultural capital remain as salient as ever.*

It is white middle-class perspectives that continue to dominate the literature about lesbian motherhood and broader LGBT+ parenting, certainly within the UK context (Clarke, 2008; Gabb, 2004). Diversity has been 'sorely overlooked' (Clarke, 2008, p. 125), which means there is very little information that describes the experiences of lesbian mothers from ethnic minority backgrounds, lesbians who are working class, who have a disability, a health condition, who are older, or any intersection of these factors. We know very little about these families. Whilst Gabb (2018) identifies some US-based literature that identifies how these factors 'cut through this sexuality-defined cohort' (p. 1003), this must be more of a priority for researchers moving forward. '*Whilst time moves forward and many contemporary equality rights may be indeed progressive, we should not lose sight of who gets written in and who gets written out of the story of LGBTQ parenthood that gets told*' (Gabb, 2018, p. 1014).

Feminist Rejection of Binaries

Since the 1970s lesbian feminists have had an important role in influencing the public debates about lesbian parenting. This has involved writing, protesting and supporting lesbian mothers (Clarke, 2008). Clarke argues that this is a somewhat different position from a lot of feminist research, where 'motherhood' most often only includes white, middle-class, heterosexual women as the dominant group of mothers.

Pollack (1987) offered a searing commentary on the 'lesbian mothers are "as good as" heterosexual mothers' framing of the debate, arguing that ultimately this view further invisibilised lesbian mothers. The weakness in this perspective is that no matter how good lesbian families are, they will always be viewed as second best to heterosexual families. Subsequently many of the same arguments about lesbian parents in the 1980s have been used to describe trans/non-binary parents and the so-called 'risks' they pose to children (Hafford-Letchfield et al., 2021). The discourse of 'a child's best interests' is a powerful one and comes up again and again in public debates about same sex families—see: Gabb (2018) and Hosking and Ripper (2012), as does the need for female and male role models (Cocker, 2011; Gabb, 2018). Have we not learned anything from the debates over the past 50 years?

Ideas about family continue to be contested within LGBT+ communities. Particular criticisms exist around LGBT+ people embracing civil partnerships and same sex marriage, and the rise in the numbers of LGBT+ people becoming parents, including via adoption, precisely because of the assimilation and heteronormative representation. These are arguments lesbian feminists have been making for years. Parenting as a lesbian continues to be seen as a deeply political issue (Clarke, 2005), both within and outside LGBT+ communities. Spade and Willse (2013) argue that marriage does nothing for the status of the majority of same sex parents, as it is a tool of gendered social control, material distribution and protection of material wealth. The danger of marriage equality is that it legitimises certain types of LGBT+ partnerships and family forms and creates a barrier between those that are 'state sanctioned' and those lesbians and gay men who choose to live their lives outside of this legal structure. Their families and roles may not be valued in the same way (Cocker, 2015). This danger applies equally to those families who are seen as 'legitimate' and those who are not.

Much of the discussion within this article has centred on the insider/outsider binary that hitherto preoccupies research undertaken with lesbian parents. Moves towards theorising the 'in-between' spaces occupied by lesbian parents and others who embrace alternative family forms encourage a debate that does not position these parents as radicals or reactionaries, but acknowledge the political act that LGBT+ parenting is, alongside the ordinary and everyday actions and interactions that being a parent encompasses (Hicks, 2011). The literature is moving towards identifying and exploring 'queer parenting' as a concept/approach that challenges gender binaries, in the way lesbians and gay men challenged the sex binaries many years before, and these are important developments in feminism and feminist research, as well as in social work and social work research (Brown & Rogers, 2020; Hafford-Letchfield et al., 2019; 2020; Walls et al., 2018; 2019).

The social work profession needs to improve on its current practices. LGBTQ+ issues have always been located outside of the mainstream in terms of teaching and research (H. C. Brown & Cocker, 2011; Mason et al., 2020). Over the years a number of publications have provided frameworks for social work assessment, particularly in the fostering and adoption field (C. Brown et al., 2018; Hill, 2009; Mallon & Betts, 2005). There are ongoing debates about whether these are necessary or whether existing practice frameworks should be flexible enough to address people's individuality in a reflexive manner (Cocker & Brown, 2010). Any robust assessment should be able to cover areas unique to people's individual experiences even when these might be particular to the lives of LGBT+ individuals, such as coming out, homophobia and transphobia. For fostering and adoption assessments, sexuality is explored in LGBT+ assessments, because it stands outside a heteronormative experience. Sexuality does not feature in social work assessments generally, except when considering sexual harm or abuse (Dunk-West & Hafford-Letchfield, 2011). This ignores sexuality as a fundamental aspect of human identity and life. All adults have a sexuality. The SPRIINT approach assesses sexuality in relationships in its broadest sense within fostering and adoption assessments, not just in LGBT+ relationships (Cocker & Brown, 2010). SPRIINT is an acronym that stands for:

- **S**exual orientation;
- **P**revious sexual relationship histories;
- **R**elationships (current);
- **I**ntimacy (the expression of this with each other);
- **I**ntegration into the community;
- **N**ot so nice bits, which involves the assessor exploring the long-term nature of relationships; coping with difficulties, stress, disagreements, and so on;
- **T**hinking: about the patterns and the gaps within the stories being told.

This model is flexible enough to apply to all applicants, regardless of sexual orientation and gender expression. It enables practitioners to analyse the content of what applicants discuss with them to synthesise the material (Cocker & Brown, 2010). This is an example of how an assessment process can explore areas relevant to parenting in all families.

For social workers working with lesbians who are mothers, grandmothers, foster carers, extended family members and connected persons, being aware of the effects of heteronormativity and cis-normativity on their own thinking and behaviours as well as the effect on the lived experiences of these women is vital. Similarly, this self-awareness would assist when working with LGBT+ people. Counteracting myths and stereotypes about lesbian parents needing 'male role models', not accepting the binaries of 'good' and 'bad' lesbian parents when assessing prospective fostering and adoption applicants, and valuing the strengths in the diversities of families and family forms is essential.

Conclusion

Lesbian parenting occupies a different social and political space now than it did 50 years ago. However, despite lesbian families generally being more accepted in the mainstream, and these family relationships supported through legal frameworks, the journey to this point has been fraught with difficulties, which at its worst saw many lesbians lose custody of their children because of their sexuality. For younger lesbians today, they know that having children is a possibility and choice they can make in their lives. However, their experiences will be different to heterosexual parents creating a family, with their parenting continuing to be judged by heterosexual norms. The experiences of lesbian parenting provide learning for other alternative families. Creating alternatives to heteronormative and cisnormative nuclear family norms remains a rebellious act within a patriarchal heteronormative society. However much reform provides a cloak of acceptability in some places and spaces, the overarching political environment is not always conducive to promoting diversity and equalities, and so these protections are fragile, and are not guaranteed for those who are marginalised and oppressed or lack resources. Most research undertaken to date in these areas privileges one kind of family—the white middle-class lesbian family. It is time for other families to have their stories told and their experiences and struggles recognised. For social workers the key message is to be 'culturally aware' and understand the history and context for lesbian parents and other LGBT+ families. Developing self-awareness, not applying conventional heteronormative models of family and being aware of the different homo/cisphobic experiences that LGBT+ parents have as part of their ordinary and every day family life will assist social workers to not be part of the problem of reinforcing normalisation, which colludes with oppression of LGBT+ families.

References

Brown, C., Andrew, C., & Adams, P. (2018). *Practice note 69. Assessing and supporting transgender foster carers and adopters.* Coram BAAF Adoption and Fostering Academy.

Brown, C., & Rogers, M. (2020). Removing gender barriers: Promoting inclusion for trans and non-binary carers in fostering and adoption. *Child and Family Social Work, 25*(3), 594–601. https://doi.org/10.1111/cfs.12731

Brown, H. C., & Cocker, C. (2011). *Social work with lesbians and gay men.* SAGE.

Chambers, S. A. (2007). "An incalculable effect": Subversions of heternormativity. *Political Studies, 55*, 656–679.

Clarke, V. (2005). Feminist perspectives on lesbian parenting: A review of the literature 1972–2002. *Psychology of Women Section Review, 7*(2), 11–23.

Clarke, V. (2008). From outsiders to motherhood to reinventing the family: Constructions of lesbian parenting in the psychological literature—1886–2006. *Women's Studies International Forum, 31*(2), 118–128. https://doi.org/10.1016/j.wsif.2008.03.004

Cocker, C. (2011). Sexuality before ability? The assessment of lesbians as adopters. In P. Dunk-West & T. Hafford-Letchfield (Eds.), *Sexual identities and sexuality in social work: Research and reflections from women in the field* (pp. 141–162). Ashgate.

Cocker, C. (2015). Social work and adoption: The impact of civil partnerships and same sex marriage. In N. Barker & D. Monk (Eds.), *From civil partnership to same sex marriage: Interdisciplinary reflections* (pp. 97–114). Routledge.

Cocker, C., & Brown, H. C. (2010). Sex, sexuality and relationships: Developing confidence and discernment when assessing lesbian and gay prospective adopters. *Adoption & Fostering, 34*(1), 20–32.

Cocker, C., & Hafford-Letchfield, T. (2021). LGBT parenting. In SJ Dodd (ed) The Handbook of Social Work and Sexualities Routledge, NY.

Cocker, C., Hafford-Letchfield, T., Ryan, P., & Barran, C. (2019). Positioning discourse on homophobia in schools: What have lesbian and gay families got to say? *Qualitative Social Work, 18*(5), 800–817. https://doi.org/10.1177/1473325018767720

Cooper, D. (1994). *Sexing the city: Lesbian and gay politics and the activist state.* Rivers Oram Press.

Cooper, D., & Herman, D. (1995). Getting 'the family right': Legislating heterosexuality in Britain, 1986–91. In D. Herman & C. Stychin (Eds.), *Legal inversions: Lesbians, gay men and the politics of law* (pp. 162–179). Temple University Press.

Department for Education. (2021, 19 February). *Children looked after in England including adoptions—Reporting year 2020.* Retrieved from https://explore-education-statistics.service.gov.uk/find-statistics/children-looked-after-in-england-including-adoptions/2020

Dunk-West, P., & Hafford-Letchfield, T. (2011). *Sexual identities and sexuality in social work: Research and reflections from women in the field.* Ashgate.

Gabb, J. (2004). Critical differentials: Querying the incongruities between research on lesbian parent families. *Sexualities, 7*(2), 171–187.

Gabb, J. (2005). Locating lesbian parent families: Everyday negotiations of lesbian motherhood in Britain. *Gender, Place and Culture, 12*(4), 419–432. https://doi.org/10.1080/09663690500356768

Gabb, J. (2018). Unsettling lesbian motherhood: Critical reflections over a generation (1990–2015). *Sexualities, 21*(7), 1002–1020. https://doi.org/10.1177/1363460717718510

Giddens, A. (1992). *Modernity and self-identity: Self and society in the late modern age.* Polity Press.

Golombok, S. (2000). *Parenting: What really counts?* Routledge.

Golombok, S., Perry, B., Burston, A., Murray, C., Mooney-Somers, J., Stevens, M., & Golding, J. (2003). Children with lesbian parents: A community study. *Developmental Psychology, 39*, 20–33.

Golombok, S., Spencer, A., & Rutter, M. (1983). Children in lesbian and single parent households: Psychosexual and psychiatric appraisal. *Journal of Child Psychology and Psychiatry, 24*, 551–572.

Golombok, S., & Tasker, F. (1996). Do parents influence the sexual orientation of their children? Findings from a longitudinal study of lesbian families. *Developmental Psychology, 32*, 3–11. https://doi.org/10.1037/0012-1649.32.1.3

Green, G., & Bozett, F. (1991). Lesbian mothers and gay fathers. In J. Gonsiorek & J. Weinrich (Eds.), *Homosexuality: Research implications for public policy* (pp. 197–214). Sage.

Hafford-Letchfield, T., Cocker, C., Manning, R., & McCormack, K. (2020). Trans and nonbinary parenting. In S. K. Kattari, N. E. Walls, K. Leo Kattari, & M. K. Kinney (Eds.), *Social work and health care with transgender and nonbinary individuals and communities*. Routledge.

Hafford-Letchfield, T., Cocker, C., Rutter, D., Manning, R., & McCormack, K. (2021). Doing the right thing and getting it right: Professional perspectives in social work on supporting parents from gender diverse communities. *International Journal of Transgender Health, 22*(1–2), 154–166. https://doi.org/10.1080/26895269.2020.1831417

Hafford-Letchfield, T., Cocker, C., Rutter, D., Tinarwo, M., McCormack, K., & Manning, R. (2019). What do we know about transgender parenting?: Findings from a systematic review. *Health and Social Care in the Community, 27*, 1111–1125. https://doi.org/10.1111/hsc.12759

Hafford-Letchfield, T., Cocker, C., Ryan, P., & Melonowska, J. (2016). Rights through alliances: Findings from a European project tackling homophobic and transphobic bullying in schools through the engagement of families and young people. *British Journal of Social Work, 46*(8), 2338–2356.

Hanscombe, G., & Forster, J. (1982). *Rocking the cradle: Lesbian mothers, a challenge in family living*. Sheba Feminist Publishers.

Hicks, S. (2000). Good lesbian, bad lesbian...Regulating heterosexuality in fostering and adoption assessments. *Child and Family Social Work, 5*(2), 157–168.

Hicks, S. (2005). Lesbian and gay foster care and adoption: A brief UK history. *Adoption and Fostering, 29*(3), 42–56.

Hicks, S. (2011). *Lesbian, gay and queer parenting: Families, intimacies, genealogies*. Palgrave.

Hicks, S. (2014). Deconstructing the family. In C. Cocker & T. Hafford-Letchfield (Eds.), *Rethinking anti-discriminatory and anti-oppressive theories for social work practice*. Palgrave.

Hicks, S., & McDermott, J. (2018). *Lesbian and gay fostering and adoption: Extraordinary yet ordinary* (2nd ed.). Jessica Kingsley Publishers.

Hill, N. (2009). *The pink guide to adoption for lesbians and gay men*. British Association for Adoption and Fostering.

Hosking, G., & Ripper, M. (2012). In the best interests of the (silenced) child. *Australian Feminist Studies, 27*(72), 171–188. https://doi.org/10 1080/08164649.2012.648261

Kahn, K. (1995). The politics of family. In K. Kahn (Ed.), *Frontline feminism, 1975–1995: Essays from Sojourner's first 20 years* (pp. 132–137). Aunt Lute Books.

Mallon, G., & Betts, B. (2005). *Recruiting, assessing and supporting lesbian and gay carers and adopters.* British Association for Adoption and Fostering.

Martin, A. (1993). *The lesbian and gay parenting handbook: Creating and raising our families.* Harper Collins.

Mason, K., Cocker, C., & Hafford-Letchfield, T. (2020). Sexuality and religion: From the court of appeal to the social work classroom. *Social Work Education, 1–13.* https://doi.org/10.1080/02615479.2020.1805426

Mellish, L., Jennings, S., Tasker, F., Lamb, M., & Golombok, S. (2013). *Gay, lesbian and heterosexual adoptive families: Family relationships, child adjustment and adopters' experiences.* British Association for Adoption and Fostering.

Patterson, C. (1992). Children of lesbian and gay parents. *Child Development, 63*(5), 1025. https://doi.org/10.2307/1131517

Patterson, C. (2005). *Lesbian and gay parenting.* American Psychological Association.

Patterson, C. (2017). Parents' sexual orientation and children's development. *Child Development Perspectives, 11*(1), 45–49. https://doi.org/10.1111/cdep.12207

Pies, C. (1985). *Considering parenthood: A workbook for lesbians.* Spinsters/Aunt Lute.

Pollack, S. (1987). Lesbian mothers: A lesbian-feminist perspective on research. In S. Pollack & J. Vaughan (Eds.), *Politics of the heart: A lesbian parenting anthology.* Firebrand Books.

Pollack, S., & Vaughn, J. (1987). *Politics of the heart: A lesbian parenting anthology.* Firebrand Books.

Rights of Women Lesbian Custody Group. (1986). *Lesbian mothers' legal handbook.* The Women's Press.

Saffron, L. (1986). *Getting pregnant our own way: Guide to alternative insemination.* Women's Health Information Centre.

Saffron, L. (1994). *Challenging conceptions: Planning a family by self-insemination.* Cassell.

Sisterhood and After Research Team. (2013). *Women's liberation: A national movement.* Retrieved from https://www.bl.uk/sisterhood/articles/womens-liberation-a-national-movement#

Skeates, J., & Jabri, D. (1988). *Fostering and adoption by lesbians and gay men.* London: Strategic Policy Unit. Retrieved from.

Spade, D., & Willse, C. (2013, September 6). Marriage will never set us free. *Beyond Capitalism.*

Thatcher, M. (1987, October 3). Prime Minister Margaret Thatcher, talking to Women's Own magazine. *Women's Own.*

Walls, N. E., Kattari, S. K., & DeChants, J. (2018). Transmasculine spectrum parenting: Beyond a gendered fatherhood. *Social Work Research, 42*(3), 223–236. https://doi.org/10.1093/swr/svy020

Walls, N. E., Kattari, S. K., Speer, S. R., & Kinney, M. K. (2019). Transfeminine spectrum parenting: Evidence from the national transgender discrimination survey. *Social Work Research, 43*(3), 133–144. https://doi.org/10.1093/swr/svz005

Weeks, J., Heaphy, B., & Donovan, C. (2001). *Same sex intimacies: Families of choice and other life experiments.* Routledge.

Wood, K. (2016). 'It's all a bit pantomime': An exploratory study of gay and lesbian adopters and foster-carers in England and Wales. *British Journal of Social Work, 46*(6), 1708–1723. https://doi.org/10.1093/bjsw/bcv115

Wood, K. (2018). Families beyond boundaries: Conceptualising kinship in gay and lesbian adoption and fostering. *Child & Family Social Work, 23*(2), 155–162. https://doi.org/10.1111/cfs.12394

CHAPTER 16

Child Sexual Exploitation, Victim Blaming or Rescuing: Negotiating a Feminist Perspective on the Way Forward

Jane Dodsworth

Introduction

This chapter explores issues of victimhood and agency, consent and constraint for women who experienced sexual exploitation as adolescents in order to consider how the meaning given by them to those experiences shaped the choices they felt able or unable to make. Current discourses on this issue are largely derived from a gendered, patriarchal narrative that seeks to victimise or to blame. There has been a shift in thinking in terms of how child sexual exploitation is perceived, that is, as abuse/exploitation rather than 'child prostitution'/criminality. However, the wider narratives informing perceptions still result in victim blaming by some professionals, the media and the wider public and consequently in self-blaming by those who exploited both children and adults.

J. Dodsworth (✉)
University of East Anglia, Norwich, UK
e-mail: Jane.dodsworth@uea.ac.uk

C. Cocker, T. Hafford-Letchfield (eds.), *Rethinking Feminist Theories for Social Work Practice*,
https://doi.org/10.1007/978-3-030-94241-0_16

By using feminist research, epistemologies (or theories of knowledge) and feminist theory, this chapter will examine the impact of the dominant narratives on the lives of 12 women who experienced sexual exploitation as children and sex work as adults (Dodsworth, 2015). It will seek to determine how the sense they made of environmental, familial and individual factors informed by overarching gendered narratives impacted on them. It is crucial, from a feminist perspective, to listen to the voices of those directly involved; they are the experts in their own lives. The aim in the research outlined was to undertake it 'with' women who had experienced sexual exploitation and sex work not 'on' them. In conclusion, this chapter will consider the messages for policy and for social work practice.

The Feminist Victim/Agent Dichotomy

The neo-abolitionist movement spreading through Europe perpetuates this perception in terms of adult sex workers (Fitzgerald & McGarry, 2018) and feeds into a narrative that disempowers those involved in selling sex from a sense of their right to have agency, make choices and be heard. As Carpenter (2000) observes, sex work seems to engender some of the most difficult issues in feminism. Feminists are at an impasse because their conceptual dualism; victim or agent, puts them in a bind, because they cannot, she suggests, both support and critique sex workers simultaneously (2000).

However, Carpenter argues that these dichotomous conceptualisations ignore the fact that sex workers are not a homogenous group and the options are therefore not mutually exclusive but should incorporate the complexity of the issue and allow for the various voices of sex workers to be heard and validated in informing policy. As Nagle (1997) cited in Weatherall and Priestley (2001: 325) argued, much of feminist thinking and research on sex work is inadequate because it has ignored the personal experiences and opinions of people involved in sex work. The same is true for children and young people at risk of, or experiencing, sexual exploitation, who also have a right to be heard and their need for agency understood in order that service provision is appropriate, effective and found useful by them.

The feminist critique has, as Kissil and Davey (2010) note, created a shift from pathologising sex workers to considering social discourses as constructing the institution of sex work. However, they suggest that in the struggle to protect sex workers as a marginalised and vulnerable group, the sex worker as an individual has been forgotten. This has resonance for our understanding of working with young people at risk of, or experiencing, sexual exploitation. If we do not listen to them we cannot begin to build a relationship of trust and hear what they feel is important to them. As research with this vulnerable group has highlighted, professionals need to engage children in relationships of trust if they are to be successful in enabling children to explore and address risky behaviours (Lefevre et al., 2017).

An Interpretivist Feminist Perspective

Feminist politics on sex work over the last 20 years has, as FitzGerald and McGarry (2018) argue, become increasingly polarised and ideological. The opposing views of abolitionists[1] and human rights lobbyists/sex positivists have led to deep divisions in feminist thinking and a focus on morality and ideology rather than on the wider structural issues impacting upon the lives of those who become involved as children and as adults, or on the meanings given to those factors by individuals. Without an understanding of the ecological, familial and individual factors that determine the pathways taken by individuals who have been sexually exploited, and an acknowledgement of the importance of their perspectives, it is impossible to develop meaningful policy and practice.

The polarising of feminist theorists' positions about whether selling sex should be seen as violent exploitation or as a form of labour is challenging. However, there remains much of value to be learnt from feminist perspectives to inform and challenge current social work practice. This is particularly true in terms of hearing and respecting experts by experience. There are, as Dominelli notes, several principles that feminists share regardless of their overall analysis and calls for action. These include integrating the personal and political dimensions of life; respecting the diversity encompassed by women; seeking more egalitarian forms of social relationships; and transforming the existing social order for it serves badly the needs of men, women and children (Dominelli, 2002: 3) This is a perspective vital to good social work practice.

Gelsthorpe's (1990: 90) analysis of feminist research identified several key themes; a topic relevant or sympathetic to women; an acknowledgment of the 'subjectivity of the researcher and the researched'; the recognition of issues of power and control; a rethinking of the relationship between researcher and researched; and that the subjective experiences of doing the research are recorded. Brooks-Gordon (2006: 245) adds a further theme; a commitment to social action in order to bring about changes that improve the conditions under which women and the marginalised live.

In choosing the most appropriate methodology to undertake feminist research that informs practice, ontological factors (i.e. showing the relations between the concepts and categories in a subject area or domain) and epistemological factors (i.e. investigating the origin, nature, methods and limits of human knowledge) need to be considered. This is because the traditional scientific positivist paradigm and the interpretivist, humanist, phenomenological paradigm have differing perspectives and assumptions about the social world. The positivist paradigm holds that the methods of the natural sciences produce the most reliable and objective knowledge and that these methodological principles should be applied to the study of the social world. The interpretivist paradigm rejects the idea that external truth about the social world can be revealed by these methods and suggests that 'what people do has to be interpreted in the light of the meanings, motives and intentions behind the actions' (O'Connell Davidson & Layder, 1994: 31).

As Seale (1998: 17) notes, interpretivists tend to favour qualitative rather than quantitative methods 'because on the whole researchers find that people's words provide greater access to their subjective meanings than do statistical trends'. My aim in the research discussed in this chapter was to focus on understanding the subjective experience of being involved in sexual exploitation and sex work and to develop theory based on those subjective accounts. Therefore, what fits best is an interpretive epistemology in which what is important is how people understand their worlds and how they create and share meanings about their lives (Rubin & Rubin, 1995: 34).

Many feminist researchers have challenged the 'objective' nature of positivist research which reflects a rather androcentric world view in which the experiences of women are often omitted or distorted. As Ribbens and Edwards (1998: 2) note 'there is a danger that the voices of particular groups, or particular forms of knowledge are drowned out, systematically silenced or misunderstood as research and researchers engage with dominant academic and public concerns and discourses'. A feminist approach, Stanley and Wise (1993) argue, means 'accepting the essential validity of other people's experiences'. Indeed, as Reinharz (1992) suggests, there are multiple realities.

Intrinsic to feminist research is an acknowledgement that it is inevitable that the researcher's own experiences and consciousness will be involved and are an integral part of the research process (Stanley & Wise, 1993: 58).

> *We believe that all research is 'grounded' in consciousness because it isn't possible to do research (or life) in such a way that we separate ourselves from experiencing what we experience as people (and researchers) involved in a situation.* (Stanley & Wise, 1993: 160)

This perspective is one that fits well with social work perspectives, values and skills particularly in terms of the use of self as a tool for gaining and building trust in often fraught circumstances with vulnerable and understandably defensive people. In order to gain access to, and crucially the trust of, those involved in this often covert and risky world it is necessary to be prepared to give something of one's self which in turn impacts on process and outcome for the researched, the researcher and the research forming a part of the reflexive analysis. This is a similar process to that of the successful development of a trusting relationship between service user and social worker.

Research into sexual exploitation and sex work, by its nature, encompasses experiences of both wider structural and individual factors that require an interactional psychosocial level of analysis. It is in this sense a piece of real-world research (Robson, 1993) which aims to understand stories of lived lives and, in turn, to use that understanding to contribute to practical outcomes. Additionally, as Coy (2008: 4) notes,

> *There is a recognized lack of knowledge about the lives of young women in the sex industry that is based on their voices.* (Coy, 2008: 1408)

Child Sexual Exploitation

> *Child sexual exploitation is a form of child sexual abuse. It occurs where an individual or group takes advantage of an imbalance of power to coerce, manipulate or deceive a child or young person under the age of 18 into sexual activity (a) in exchange for something the victim needs or wants, and/or (b) the financial advantage or increased status of the perpetrator or facilitator. The victim may have been sexually exploited even if the sexual activity appears consensual. Child sexual exploitation does not always involve physical contact; it can also occur through the use of technology.* (Department for Education, 2017)

Definitions of child sexual exploitation (CSE) have undergone a paradigm shift over the last two decades as a consequence of feminist researchers and key children's charities arguing that it is abuse not criminality and that children cannot consent to their own abuse (Barnados, 1998). Whilst the issues in terms of victimhood and agency, consent and constraint are more complex than that, it is clear that CSE occurs within a complex array of wider structural, familial and individual risk factors (Hallett, 2016). It is also recognised as adversely affecting children's physical, mental and sexual health, educational achievements, social and economic contributions and later parenting capacity, and increasing their risk of self-harming, drug and alcohol problems and anti-social and criminal behaviours (Department for Education, 2017).

Whilst is clear that a child cannot consent to their own abuse it has been argued that the pendulum may have swung too far in terms of perceptions of victimhood and agency. As Melrose (2013) observes, the language surrounding CSE has resulted in the issue becoming almost totally synonymous with grooming, and the danger is that it is becoming reduced to a problem of children being groomed and exploited by adult men. She argues that the idea that at least some young people may exercise agency (albeit in severely constrained circumstances) and become involved under their own volition has become unimaginable within the parameters of debate established by the discursive formation of the grooming model.

By viewing CSE in this individualistic way young people are seen as victims/objects and the issue is reduced to one of individual morality/immorality diverting attention away from the wider structural issues which constrain some young people's choices Melrose (2013) suggests.

> *In order to provide young people with the most appropriate support, practice responses need to be developed from the concrete conditions in which young people are subject to sexual exploitation, rather than applying abstract 'models' that fail to capture the lived experience of the young people concerned.* (Melrose, 2013: 155)

Additionally, as Pearce (2013: 16) argues, 'instead of being understood as victims of abuse, sexually exploited young people (particularly the 16–18 age group) are invariably perceived to be consenting active agents, albeit constrained, about their relationships. As such they are seen to carry responsibility for what happens to them and consequently the blame for the abuse that follows' (Pearce, 2013: 163). Young people experiencing

CSE are therefore, it appears, either blamed or rescued, with neither option addressing the wider issues. Pearce argues for a 'social' model of consent which enables consent to be contextualised. Thereby, 'condoned consent' in which some professionals fail to recognise CSE can itself be challenged, removing the sense of blame from the victim.

What is evident from much of the research undertaken on CSE is that the issue is complex; young people become sexually exploited for a far-ranging number of, often interconnected, reasons. As Hallett (2016) notes, there is now a significant body of literature outlining risk and vulnerability indicators, recognising that CSE is more complex than individual problems and events and cannot be separated from wider familial, socio-economic factors.

A key focus of my own research was to develop an understanding of how the meaning ascribed to risk and protective factors influenced perceptions of victimhood and agency as some young people did not regard themselves as victims of exploitation. This resonates with Sidebotham's view that the fact that a young person may exercise some agency in engaging in sexual activity does not mean that they may not also be the victim of sexual exploitation. 'Nor that they are necessarily culpable for any harm they may suffer as a result' (2013: 52) A key factor is to hear what they have to say.

The Research

What can listening, through a feminist filter, to women's stories of their experiences teach us about future social work practice with people who are at risk of or have experienced child sexual exploitation? The research outlined in this chapter used a feminist perspective to explore, with women who had had those experiences, the sense they made of their lived lives through childhood, sexual exploitation and for some, sex work as adults.

From this exploration key themes emerged about how the meanings they attached to key experiences, particularly adverse experiences in childhood, influenced how they interpreted the world around them, their sense of self-worth or the lack of it and the pathways they chose or felt forced to take. For some it was a maladaptive downward pathway but for others far less so (Dodsworth, 2014, 2015) For some it was imbued with a sense of being blamed and consequently self-blaming, for others a sense of being, or needing to be, 'rescued' was either embraced or rejected with differing outcomes.

Three major themes, or groupings, emerged from listening to the stories of 24 women who were, or had been, involved in sex work, 12 of whom (the focus of this chapter) were under 18 when they were sexually exploited. The themes are about a sense of choice or lack of choice of victimhood or agency and about the direction taken on individual pathways.

'*No choice but to stay involved*'

For the women in the first *no choice but to stay involved* themed group, the pathway they feel almost predetermined to take started in childhood experiences of neglect, rejection and abuse. Staying involved in sex work, following experiencing CSE seemed, their narratives indicated, to be something they felt they were powerless to change. For these women the meanings they attached to significant relationships and critical events in lives lived largely in poverty and disadvantage fed into a negative sense of self and others (Bowlby, 1969, 1973), already damaged by negative childhood experiences.

All five of the women in this group who had been sexually exploited as children had backgrounds of absent or rigid parenting, or time spent in care. They spoke of longing for affection and approval and sought to find it wherever they could.

> *She has hurt me she has let me down all the time. I am ashamed of the things she does.* ('Nina'—White British—coerced by her mother at the age of 12 into being sexually exploited)

> *One minute my Dad can be fine and we can get on like a house on fire and the next minute he will hate me and call me a scabby little smack head* (Ursula—White British—as a teenager, drifted into being sexually exploited to fund her heroin habit)

> *I think I was looking for a father figure, I was looking for some support off him* ('Dawn'—White British—describing the 45-year-old 'boyfriend' who, with his friends, sexually exploited her from the age of 15)

> *I was tall and skinny and a lot like my Dad and I always felt rejected, I felt like because I was like him, I was a bastard because of the horrible person he was.* ('Paula'—White British—pregnant at 15 by her abuser)

Victims of CSE are too often met with victim blaming responses and judgements that frame the child as promiscuous and badly behaved (Webb & Holmes, 2015: 15). This, in turn, feeds into the self-blaming that these young people already interpret from their early abusive childhood experiences. They have limited experience of secure, loving relationships from which to make self-protecting or positive life choices and this perhaps explains why they continue down such maladaptive pathways and why they see consent where there is coercion and affection where there is abuse.

> *On the street I'm not really no one-* ('Izzie'—White British—abused from an early age by her parents, 'boyfriends' and men wanting to buy sex)

> *Some of them just see me as an object and then others just see me as a victim* ('Dawn'—White British—speaking about the men wanting to buy sex but also probably about her own sense of self)

The women in this group had such a low sense of self-esteem, having internalised negativity from early childhood onwards compounded by societal and professionals' attitudes that they were not agentic about their lives. They feel they have no choices to make nor any right to make them. As Pearce et al. (2002) suggest there is a cyclical process in which low self-esteem may predispose to involvement in CSE and involvement feeds into an already negative self-image.

How can feminist social work practice ensure that the pathways of similar young people do not take the same downward trajectory? As Dominelli (2002: 162) notes, 'feminists' concerns with the rights of children have focused on the significance of adultist power relations in oppressing children. These must be deconstructed if children's voices are to be heard on a par with those of adults. Doing so is part of a broader strategy whereby feminists endorse the rights of all oppressed groups to have their voices heard'. Listening to the voices of those with direct experience is key in finding a way forward that has meaning to those involved.

'*It's what I do, not who I am*'

For the women in the second *It's what I do not who I am* themed group, the pathway they felt they had a choice about and a right to take was one, to an extent, given wider structural factors, determined by them. Some women chose to continue selling sex as adults, seeing it as a job. Others were, at the time of interview, not involved but felt that whether they returned was something about which they had agency and a right to choose.

These women had very similar backgrounds to those in the first group so what enabled their more agentic view of their pathway choices? The differences appeared to be that they retained a wider range of significant relationships and had greater personal strengths and crucially wider sources of resilience that may have led to a greater sense of self-efficacy and self-esteem and self-protection (Rutter, 1985; Sroufe, 1997).

> *My Dad would always make things better. It didn't matter if I couldn't do it I just had to try, that's all he asked of me … he would always make me feel better about everything.*('Odelle'—Mixed Race British—despite experiencing bullying and racism as a child felt that she retained control)

In its current conception Hallett (2017: 26) argues that CSE emphasises that children and young people who are sexually exploited cannot consent. They are sexually exploited because they are groomed, and it is *this* which places them as victims of abuse. In so doing this obscures young people's agency defining them as helpless, but as the history of this issue has shown, necessarily so because it is their agency which has been (and remains) problematic. This means that they are in some way responsible for their own abuse, the ongoing binary of 'sad' or 'bad' and the continuing problem of victim blaming which can be seen in some of the recent wider sexual abuse inquiries like Rotherham (Jay, 2014).

As Lefevre et al. (2017) argue, more needs to be understood about professional approaches to building trusting relationships, particularly when children feel (often with good cause) that professionals are not always able to balance their need for protection and guidance with the right to a voice and to make agentic choices about their own lives (2017: 1).

This group was able, more clearly, to blame those responsible for the abuse rather than internalising a sense of themselves as powerless victims.

> *I was so messed up, so confused … but it's made me the person I am I suppose, all my experiences. When I was in prostitution, I wasn't a victim as such, I was a victim when I was a child. I don't class myself as a victim.* ('Amy'—White British—sexually abused by her father as a small child and sexually exploited by his friends from age 15, continued into sex work as an adult but is now a sex workers' project worker)

> *Looking back now I was a bit of a rebel. Then X says, 'I know where we can get some easy money' and I just thought 'Oh Yeah'. Also I was always a bit of a tomboy and I thought to myself, well that's easy enough to do and that was it* ('Bella'—White British—who began, what she saw, as choosing to sell sex in her early teens. She continues as an adult part-time and runs a specialist support project).

> *I thought I can do that, no problem, the only thing I was worried about was the actual walking the streets. I was always worried about that and I've never done that, I've never needed to, I've always managed without.* (Tina—White British—aged 51 has been selling sex from home since she was 17/18)

As Melrose (2004) notes, the model of drift recognises that young people can, and do, make decisions for themselves and, particularly in circumstances of socio-economic disadvantage, may decide that sex work is a viable survival strategy for them. As Cooney and Rogowski argue, we should shift from 'child saving' to promoting resilience and support. One way of doing this, they suggest is a critical feminist approach to working with families drawing on a strengths-based, systemic perspective, which recognises the structural aspects of families' environments (2016: 1442).

'*It's not for me*'

For the women in the third *It's not for me* themed group life paths were very similar to those of the other two groups except that these women felt that they had no choice but to stop involvement in selling sex Their childhood experiences of rejection appeared to have left them terrified of further rejection if their sex worker identity was exposed. They saw themselves as survivors with self-determination and agency, but they utilised that agency differently. The role of sex workers was incompatible with how they wanted to see themselves, but they did not condemn others for continued involvement.

> *At the end of the day if I didn't do prostitution I wouldn't be where I am today, or work in the line of work I'm doing now. So yes, I've turned prostitution and my drug habit and made something positive out of it.* ('Mary'—White British—who described herself as deciding to sell sex at 16 but decided she had to give it up at 19)

> *It was closing in on me because there was girls that knew that I was big and bad, and they are 'ho's. I'd stand there and look at them like tramps but really, I'm doing it myself so I knew it was closing in for me, my time was coming to an end.* ('Gina'—Black British—who sold sex from age 15 but stopped at 18 in case others found out and condemned her)

> *I came back from X and that's when I stopped workin' because I thought I don't think this is the lifestyle that I basically want. … I mean all that stuff I got involved in it was just to survive because I never had any money to survive. I'm better off without that money because now I can't kill myself or get involved in drugs or nothing like that. So, I prefer to survive and know that I'm a survivor … and that's on Social Security love! At £60 a week! Ha ha.'* ('Fiona'—mixed race British—began selling sex at age 14 and stopped because she wanted a better life for herself at 22)

Interestingly all the women in this group recalled the importance to them of getting support from specialist workers who had had similar experiences and would not judge them and, in turn, taking on that role themselves.

As Warrington and Brodie (2019: 121) argue, the issue of engaging with young people's accounts and perspectives of CSE relates not only to recognition of such abuse but also on how best to respond to it. They suggest that professionals are liked and valued when they show an ability to listen, show understanding and care, are warm and friendly, use humour and are knowledgeable about the issues involved. It is similarly appreciated when they treat young people as individuals, keep them up to date with what is happening and allow them to make choices and so work in what might be termed a participatory way that recognises young people's competence and individual rights(2019: 127).

Indeed, as Cossar et al. (2013) suggest, children are more likely to recognise they were being abused or neglected, and to tell us about it within the context of a close and trusting relationship of duration with a professional, where they feel listened to and believed rather than judged or patronised. Coy (2009: 26) also suggests that including the views of children and young people in decision making is the foundational principle of childcare law and policy and yet research highlights failures of these aspirations to the detriment of young people.

Conclusions

Despite increasing evidence of the need for the sort of practice outlined above (Beckett & Pearce, 2018), as Cooney and Rogowski (2016) argue, current neo-liberal managerialism which has far less emphasis on preventative work dominates practice and is at odds with critical feminist practice which includes building trusting relationships with young people and their families. So, in considering child sexual exploitation and issues of victim blaming and rescuing how can we negotiate a feminist perspective on the way forward?

What is clear is that people involved in CSE and sex work are not a homogenous group so whilst there are commonalities, 'one size' of social work approach will not fit all. We need to develop our skills in listening to what those directly involved are saying. They are experts in their own lives.

One of the difficulties for professionals in understanding the importance of young people needing to make agentic choices has been that too narrow a definition of CSE has developed, seeing it largely as an issue of male grooming. This has resulted, particularly for older young people, in victim blaming and a lack of recognition of the wider structural factors that impact on self-esteem and limit choices (Pearce, 2013, Hallett, 2016).

Reisel notes that 'there has been intense controversy about the appropriate response to child sexual exploitation, with debates in the UK particularly hinging on the meaning of consent and coercion. For professionals with a duty to safeguard young people from child sexual exploitation, a key site of tension is how to avoid limiting young people's agency without placing them at risk' (2017: 1292). She argues that practitioners' difficulties in sifting different aspects of choice resonate with the wider contradictions within liberal discourse between demands for protection and self-determination.

It is a complex dilemma but as Cooney and Rogowski (2016: 142) suggest, we should shift the focus from 'child saving', to promoting families' resilience and support. One way of doing this they argue is by utilising a critical feminist approach drawing on a strengths-based, systemic perspective, in order to recognise the structural elements of families' environments. This has resonance with my research findings that indicate that whatever the age of a person's first involvement in selling or exchanging sex, there are several determining factors in who will be most vulnerable and who most able to 'manage'.

These are the accumulation of risk factors in early childhood; the individual, familial and environmental resources available to them; and the 'meanings' they ascribed to risk and protective experiences throughout their lives. This indicates the need for increased professional awareness of the wider factors impacting on 'choices' made and the need for ongoing, trusting, non-judgemental holistic, relationship-based support interventions that promote strengths and increase resilience support (Dodsworth, 2014, 2015).

Trust is as, Lefevre et al. (2017) argue, instilled where young people feel their rights, views, agency and confidentiality are respected and their participation promoted. They note that 'more needs to be understood about professional approaches to building such trusting relationships, particularly when children feel (often with good cause) that professionals are not always able to balance their need for protection and guidance with their right to a voice and to make agentic choices about their own lives' (2017: 1).

As Turner and Masch (2015: 15) note, incorporating feminist and empowerment approaches in practice will provide social workers with the knowledge, values and skills most likely to promote human rights and social justice. Kissil and Davey (2010: 16) suggest that the feminist debate has been successful in bringing to the fore the importance of social discourse in understanding sex work. They argue that feminists need to create a synthesis in the dialectic of the right to choose and the right to protection, within a new framework that can include both. As this chapter has argued, synthesis in the dialectic must include the voices of those directly involved' because only then will we be able to develop interventions that are appropriate and meaningful for the young people we are working with.

> '*Listening is the key 'cos people make out they listen, but they don't. Most people just feed you a load of shit that they talk, but if I know that you are really listening then I will tell you, you know'* ('Gina')

> *Like with the (specialist project) they're there…them girls know that they have someone, who understands and accepts them for who they are. Stop putting labels on things and just look into that person and don't confront people…. I think it needs more people, not confronting their drugs or alcohol issues or prostitution, or whatever issues but seeing the person inside and actually talk to the person.'*('Amy')

References

Barnardo's. (1998). *Whose daughter next? Children abused through prostitution*. Barnado's.

Beckett, H., & Pearce, J. (2018). *Understanding and responding to CSE*. Routledge.

Bowlby, J. (1969). *Attachment and loss, volume 1: Attachment*. Hogarth Press.

Bowlby, J. (1973). *Attachment and loss, volume 11: Separation, anxiety and anger*. Hogarth Press.

Brooks-Gordon, B. (2006). *The Price of sex prostitution, policy and society*. Willan Publishing.

Carpenter, B. J. (2000). *Rethinking prostitution. Feminism sex and the self*. Peter Lang Publishing.

Cooney, L., & Rogowski, S. (2016). Towards a critical feminist practice with children and families: Child sexual exploitation as an exemplar. *Practice, 29*(2), 137–149.

Cossar, J., Brandon, M., Bailey, S., Belderson, P., Biggart, L., & Sharpe, D. (2013). *'It takes a lot to build trust': Recognition and telling: Developing earlier routes to help children and young people*. Office of the Children's Commissioner.

Coy, M. (2008). Young women, local authority care and selling sex. *British Journal of Social Work, 38*, 1408–1424.

Coy, M. (2009). Moved around like bags of rubbish nobody wants': How multiple placement moves can make young women vulnerable to sexual exploitation. *Child Abuse Review, 18*, 254–266.

Department for Education. (2017). *Child sexual exploitation: Definition and a guide for practitioners, local leaders and decision makers*. DfE. Available at: www.gov.uk/government/publications/child-sexual-exploitation-definition-and-guide-for-practitioners

Dodsworth, J. (2014). Sexual exploitation, selling and swapping sex: Victimhood and agency. *Child Abuse Review, 23*, 185–199.

Dodsworth, J. (2015). *Pathways into sexual exploitation and sex work: The experience of victimhood and agency*. Palgrave Macmillan.

Dominelli, L. (2002). *Feminist social work theory and practice*. Palgrave.

FitzGerald, S. A., & McGarry, K. (2018). *Realising Justice for Sex Workers: An Agenda for Change*. London: Rowman and Littlefield International Ltd. London.

Gelsthorpe, L. R. (1990). *Sexism and the female offender*. Hants Gower.

Hallett, S. (2016). An uncomfortable 'comfortableness': 'Care': Child protection and child sexual exploitation. *British Journal of Social Work, 46*, 2137–2152.

Hallett, S. (2017). *Making sense of child sexual exploitation: Exchange, abuse and young people*. Polity Press.

Jay, A. (2014). *Independent inquiry into child sexual exploitation in Rotherham (1997–2013)*. Metropolitan Borough Council.

Kissil, K., & Davey, M. (2010). The prostitution debate in feminism: Current trends, policy and clinical issues facing an invisible population. *Journal of Feminist Family Therapy, 22*, 1–21.

Lefevre, M., Hickle, K., Luckock, B., & Ruch, G. (2017). Building trust with children and young people at risk of child sexual exploitation: The professional challenge. *British Journal of Social Work, 47*(98), 2456–2473.

Melrose, M. (2004). Young People abused through prostitution: some observations for practice. *Practice, 16*(1), 17–29.

Melrose, M. (2013). Twenty-first century party people: Young people and sexual exploitation in the new millennium. *Child Abuse Review, 22*, 155–168.

Nagle, A. (1997). *Whores and other feminists.* Routledge.

O'Connell Davidson, J., & Layder, D. (1994). *Methods sex and madness.* Routledge.

Pearce, J. (2013). What's going on ? To safeguard children & young people from sexual exploitation: A review of local safeguarding children boards 'work to protect children from sexual exploitation. *Child Abuse Review, 23*, 214–226.

Pearce, J., Galvin, C., & Williams, M. (2002). *It's someone taking a part of you: A study of young women and sexual exploitation.* National Children's Bureau.

Reinharz, S. (1992). *Feminist methods in social research.* Oxford University Press.

Reisal, A. (2017). Practitioners' perceptions and decision-making regarding child sexual exploitation-a qualitative vignette study. *Child and Family Social Work, 22*, 1292–1301.

Ribbens, J., & Edwards, R. (1998). *Feminist dilemmas in qualitative research.* Sage.

Robson, C. (1993). *Real world research.* Blackwell.

Rubin, H. J., & Rubin, I. S. (1995). *Qualitative interviewing-the art of hearing data.* Sage Publications.

Rutter, M. (1985). Resilience in the face of adversity protective factors and resistance to psychiatric disorder. *British Journal of Psychiatry, 147*, 598–611.

Seale, C. (Ed.). (1998). *Researching society and culture.* Sage Publications.

Sidebotham, P. (2013). Culpability, vulnerability, agency and potential: Exploring our attitudes to victims and perpetrators of abuse. *Child Abuse Review, 22*(3), 151–154.

Sroufe, A. (1997). Psychopathology as an outcome of development. *Development and Psychopathology, 9*(2), 251–266.

Stanley, L., & Wise, S. (1993). *Breaking out again: Feminist ontology and epistemology new edition.* Routledge.

Turner, S., & Masch, T. (2015). Feminist and empowerment theory and social work practice. *Journal of Social Work Practice, 29*(2), 151–162.

Warrington, C., & Brodie, I. (2019). Developing participatory practice and culture in CSE services. In H. Beckett & J. Pearce (Eds.), *Understanding and responding to child sexual exploitation.* Routledge.

Weatherall, A., & Priestley, A. (2001). A feminist discourse analysis of sex work. *Feminism and Psychology, 11*(3), 323–340.

Webb, J., & Holmes, D. (2015). *Working effectively to address CSE: An evidence scope. Research in Practice.* Dartington.

CHAPTER 17

Social Work Men as a Feminist Issue

Jason Schaub

Introduction

The primary aim of this chapter is to disturb the gender regimes currently occupying social work. In order to accomplish this aim, a critical examination of men's experiences may be helpful. Men who are social workers run counter to their gender norm by pursuing a social work profession, and social work student men experience a complex series of interlocking privileges and challenges (Schaub, 2017). Scrutinising their experiences can provide knowledge about the gender boundaries present in social work, and how social workers navigate these.

This work is situated within critical studies on men (Hearn et al., 2002) and mobilises feminist theories to illuminate the topic of men's position in social work. These theories require increased positionality descriptions from the researcher. Feminist methodologies are less commonly used in the study of men and masculinities, but can be useful to describe the intersecting issues experienced by men in education (Haywood et al., 2015).

J. Schaub (✉)
University of Birmingham, Birmingham, UK
e-mail: j.schaub@bham.ac.uk

C. Cocker, T. Hafford-Letchfield (eds.), *Rethinking Feminist Theories for Social Work Practice*,
https://doi.org/10.1007/978-3-030-94241-0_17

Because of the aim and definitions described above, it is important to present my position and outline my involvement in this work. I am a white, gay, cisgender middle-aged immigrant man. I have experience of practising social work in the United States of America, Republic of Ireland, and United Kingdom, primarily in children's social care, but with some experience of working with people with mental health difficulties. I have been an academic for a dozen years. I use research to develop knowledge to help understand the challenges that are as yet poorly understood. During the production of this knowledge, I believe it is essential to engage with people who use services, to ensure that the search for this knowledge is useful and has the least negative impact on participants. This chapter draws on reviews and updates research conducted in my doctoral study into the experiences and progression of social work students who were men (Schaub, 2017).

It is a complicated task to use the lens of feminist theories to research men and masculinity. Using a critical pro-feminist lens is helpful when working within this space of dissonance. A pro-feminist standpoint requires the researcher critique the privileged societal power of men and how men's power perpetuates gender inequality, but a commitment to destabilise inequality is an important element (Pease, 2001). Whitehead (2002) suggests that a man researching society should ensure that their work does not increase men's hegemony. Some suggest that men are better able to jeopardise their position, because they occupy the positions of power, arguing that men are well placed to critique their positions of power and are useful in problematising the continuation of gender inequality (Kristeva, 1981).

The construction of this chapter draws heavily on West and Zimmerman's theory of 'doing gender', where gender 'is the activity of managing situated conduct in light of normative conceptions of attitudes and activities appropriate for one's sex category' (West & Zimmerman, 1987: 127). It also draws on concepts of *hegemonic masculinity* (Carrigan et al., 1985), an omnipresent masculinity theory where society measures men against an ideal of masculinity; the further from this ideal a man is, the less societal worth and consequential power he has.

This chapter considers the presented gender and not the originating sex. Participants were recruited about their presented gender, not their sex. It is important to recognise that not all male bodies are men. Derrida (1976, 1982) draws on concepts of linguistic deconstruction where language is understood to be both a means and an end, attempting to represent, but also to structuralise, to find new ways to express complex, inadequately described ideas. As a result, the term 'men' is used throughout instead of both 'male' and 'man'. Lexically, this creates some cumbersome phrasing, since English practice usually prefers a switch to 'male' (e.g., male social workers instead of men social workers), but these terms are used deliberately to refer to the presented gender, rather than the possible underlying sex.[1]

This chapter will outline two issues: first, how and why the proportion and position of men in social work is a feminist issue; second, a more focussed and nuanced examination of men social work students' experience and what this can improve understanding about how 'social work processes involve the production of gender through practical means, which relate both to immediate, local, and wider, institutional contexts' (Hicks, 2015: 483).

Men's Position in Social Work as a Feminist Issue

When examining the position of men in social work, it is important to remember that while men are a *numerical* minority in the social work profession, they do not experience all of the challenges of a minority group, as a result of the benefit of the patriarchal dividend (Connell, 2009). The proportion of men in social work has been decreasing for the past 30 years. As an indication of this (although not an exact relationship), the ratio of men studying social work dropped from 35% in 1980 to 25% in 1991 (Lyons et al. 1995) and continues to drop, currently at about 12% (Skills for Care, 2019). There are challenges in accurately identifying the number and proportion of men in social work, primarily because workforce information is held and presented by a range of government departments. Men make up 18% of the registered social workers in England (HCPC, 2019). This proportion will continue to decrease as a higher proportion of qualified social work men are close to retirement age and there are fewer younger men joining the profession (GSCC, 2012). Policy makers and scholars have expressed concerns with the less number of men in social work in the United Kingdom (Ashcroft, 2014; Parker & Crabtree, 2014),

Although the proportion of men in social work is decreasing, they continue to hold disproportionately more (and greater) positions of power (Kullberg, 2013; McLean, 2003), and this unequal representation creates challenges about increasing the number of men in the profession (Hicks,

2015). Men predominate in management, academia and policy-leading positions (McPhail, 2004). For example, both the Chief Executive Officer and Chair of the newly formed Social Work England are men. This phenomenon, where men are numerically a minority, but experience a greater proportion of the power, is referred to as the 'glass escalator'. This concept is a play on the so-called *'glass ceiling'*, which is frequently used to describe women's struggle to reach the top of organisations. Williams (1992, 2013) found that men in women-majority occupations (including librarians, flight attendants and social workers) experience accelerated progression into management and other positions of power. There are several different elements that combine to create this effect and these elements are both in the workplace and in private lives. In the social work workplace, men are often encouraged into management (sometimes being told, 'you seem like you would be a good manager'), as well as sometimes seeking the distance from direct work with service users because of the increased risk for men of directly engaging with vulnerable people (Pringle, 2001). This acceleration is not uniformly experienced by all men, for example, men from minority ethnic backgrounds do not experience the same benefits (Wingfield, 2009).

At this point, it may be helpful to consider why the social work profession may want to increase the number of men. There are generally three arguments provided for creating a profession with more men: firstly, the positive effect of men as role models; secondly, improving the status and prestige of the social work profession; thirdly, creating a more diverse profession (Pease, 2011). The first of these arguments suggests that men are useful role models for some service users, usually identified as men and boys, and that having more men as social workers provides them with professionals to whom they can connect. This perception predominates in the societal discussions about the topic and is found in news media reports about the low percentage of men in teaching and social work (Ashcroft, 2014; Galley & Parrish, 2014). The support for this reason has been rebuffed with work showing that children and young people are able to draw examples from a range of people irrespective of their gender (Tarrant et al. 2015).

The second argument, increasing the number of men to improve the status and prestige of the profession, is problematic as it runs counter to gender equality. This argument relies on gender inequality to improve the status of the social work profession, by perpetuating the higher status of men compared to women. It suggests that without the assistance of more men, the profession will struggle to increase its status and become more valued by society. Given the social work profession's connection to anti-oppressive practice (Cocker & Hafford-Letchfield, 2014; IFSW, 2014), this argument does not align to the social work values of diversity and inclusion.

A call for a more diverse profession is the final argument, that diversity would mean that social work would better support service users. This argument is often applied to other identities such as people with disabilities and people from ethnic minority communities and is useful when considering whether social work should attempt to increase the number of men in the profession. Fiore and Facchini (2013) suggested that seeking a more diverse profession was 'something which many social workers hope for' (pg. 321). By following this, men social workers could be encouraged to consider their choices through a gender equality lens and determine if they are increasing or diminishing inequality by their choices (such as promotion to positions of management or power). If one follows this argument through, it provides a logical process. If we want to have more women in positions of power, such as chief executives and directors of social work services, then it is appropriate to have more gender diversity in these women-majority[2] areas. As noted previously, it is important that men, when coming into women-majority spaces, ensure that their actions do not further disadvantage women. In order to assist in reducing gender inequality men should be encouraged to take pro-active steps, such as foregrounding the work of women. These actions will help to address concerns that men entering these spaces (such as social work) are not usurping them.

The leadership of public services are gendered, even though the workforces of these contain a greater than average proportion of women staff (Kalaitzi et al., 2017). In particular, healthcare leadership has been criticised for the over-representation of men in positions of power with a contrasting concentration of women in the workforce, with the issue being noted across the EU (Fjeldsted, 2013). There is a knowledge gap for examining gender as a factor in leadership for social care organisations (Lawler, 2007), but there are indications that there are barriers for women to seek and gain these leadership positions (Kalaitzi et al., 2017).

Moving from thinking about men social workers generally to consider the more specific issue of men social work students, using men's experience of studying social work can help expose how gender inequality has effects for both women and men. There is some earlier work that considers the differing reasons men and women have for entering social work (Cree, 1996), with men suggesting they ‘fell into’ social work, rather than approaching the profession with a long-standing plan. This is contrasted with later work that shows that social care is a field with a higher-than expected proportion of men (Hussein & Christensen, 2017), with the supposition that social care provides a pipeline to social work qualification for some men.

Men Social Work Students: Engaging with the Boundaries of Feminism

There is a lack of knowledge about how men experience studying social work (Giesler & Beadlescomb, 2015; Schaub, 2015). There are a few notable examples, with increasing frequency over the past decade (Cree, 1996; Parker & Crabtree, 2014; Schaub, 2015, 2017). The literature is more developed about men’s progression, men do not progress as well as women on social work programmes (Hussein et al. 2009; Schaub, 2015). My own research has shown that, even when managing against other variables, such as ethnicity, age and disability, men had more progression issues during their social work programmes than women. This should not be shocking, as men and boys do not perform as well as women and girls in education (Hillman & Robinson, 2016); although this experience is not symmetrical across all education settings. Girls perform better at all levels of education, including GCSEs, A levels, and other examinations (OECD, 2015; Skelton, 2006). In higher education, in the United Kingdom, women comprise a higher percentage of students overall and gather a higher proportion of what are considered ‘good degrees’ (Firsts or 2.1). These findings are replicated in a number of studies, with some broad agreement among them that women have within the past decade begun achieving better results overall across a range of subjects in higher education (Woodfield, 2014). These studies show a gendered pattern to achievement. What is interesting to note, however, is that men on women-majority programmes (such as social work) are more likely to drop out of their programme than women in men-majority programmes (such as STEM programmes) (Severiens & ten Dam, 2012), suggesting that there is a gendered effect that is differently experienced.

Given their entry to the profession, the views of men that are social work students can potentially help reveal issues that other, more experienced, social work men no longer identify. It is likely that the issues have become so familiar for the experienced practitioner that they cannot see them in such detail. In particular, students can help identify when there are professionally specific approaches or situations that cause friction, as a result of joining the *community of practice* (Wenger-Trayner et al., 2014). This entry to the profession through training likely brings these issues into sharper relief for students than for those that have become acculturated to social work. The liminal space of studenthood allows the opportunity to gather the experiences of men that have recently started to engage with the profession and seek their perspective on where the difficulties lie.

Examining the reasons for these progression issues, it is useful to consider the ways that gender influences the choices that people make on a daily basis, and the far-reaching consequences of these choices. The gendered nature of a profession is likely part of the consideration when someone chooses a career (compare social work and policing, and the gender profile of these two professions in the United Kingdom). When men enter women-majority professions, they can experience retribution from friends and family as a result of crossing these gendered occupation boundaries (Weaver-Hightower, 2011).

Men in Social Work Education: A Study

I undertook an interview study of 21 undergraduate social work student men in England (Schaub, 2017), the first study considering how progression and experience interacted for social work student men. My research found that social work student men experienced a number of interlocking issues that increased their challenges with progression. The themes developed from the analysis were: feeling unwanted by the social work profession; concerns that men are not 'natural' social workers; feeling silenced; self-protection (including being cautious); and disengagement. These themes engaged with mitigating factors such as having a strong relationship with women in their family and being a father, as well as general student challenges such as time management and financial concerns. For the purposes of this chapter, to help illuminate how the findings can present issues of gender inequality, the following will explore the following themes: that men were not 'natural' social workers; concerns around physical contact; feeling the profession did not want them.

Men Are Not 'Natural' Social Workers

All of the participants in my study described a perception that men were not considered be 'natural' social workers. They talked about how social work was equated with caring, and that women were identified as caring. They each felt that they were doing something *against the grain* by choosing social work. Two different participants with different profiles (one a young A-level entry student, the other a mature student with children) described very similar concerns:

> People think [social work] is a caring profession and I don't want to sound oppressive to women or anything like that, but the fact that women seem to be more caring possibly than men. (Mike,[3] 41 years old)

> They are not going to look at social work and say, "Look, admire that man", maybe it's because it's a caring role. It may be more feminine than other jobs. (Dean, 20)

The participants were generally able to present a nuanced perspective of gender roles, often suggesting that they believed themselves to be 'non-traditional'. These participants were aware that they are transgressing a gender boundary by joining the profession. They believed that society associated social work with caring and femininity and, therefore, a career in social work was an unusual choice for a man. This was commented on by their friends and family, usually along the lines of 'there are not many men in social work', with these conversations serving to increase the sense of being out of place. Another participant suggested simply that 'I think [social work] was more difficult for me because of being a man' (Will).

Social work is not a 'traditional' occupation for a man (Christie, 1998), which is shown by the reducing proportion of men in the profession. Following West and Zimmerman (1987; 2009)conceptualising, our gender is presented through our actions, and according to a set of social rules, which are different for men and women (Goffman 1977). Our actions signal our gender to those around us. These actions allow for reactions, which create a dialogue that either supports or opposes the action (Chafetz, 1990). This means that when a man chooses social work, he is transgressing a gendered boundary. These boundaries are socially policed, with retribution if one transgresses too much or too often (Butler, 2004). Butler (1990) described the 'heterosexual matrix', a combination of sexuality and gender norms that combine to create a demanding set of rules for men and women to follow. This means that when men in social work hear, 'You're a social worker, that's surprising, isn't it mostly women that do that?', this is an experience of policing the gender boundary, reminding the man that he is presenting something that does not fit gender norms. Several scholars have explored this policing of men that undertake women-majority professions, with descriptions of challenges from friends, family and strangers (Simpson, 2009; Weaver-Hightower, 2011)

Anxieties About Physical Contact

In addition to this general observation, the participants felt strongly that there are some areas of social work where it is more problematic to be a man. Physical contact with a service user takes on different connotations when the social worker is a man; research into other helping professions describe men's touch as 'sexualised' (Harding et al., 2008). Most participants were concerned about this, and often mentioned that men are more likely to physically and sexually abuse others than women. In my research, for example, Owen said, 'If you hear about stuff going on, things like kids getting abused, it's more to do with men.' (Owen, 21). This is not without reason, since there are more men are dismissed from the professional regulator. An analysis of this statistic shows the regulator for the social work profession dismissed more men for inappropriate contact with service users (Furness, 2015; Melville-Wiseman, 2016). As illustrated in these empirical findings, these trends may well influence men student social workers concerns and given the wider societal issues regarding men's power relationships that lead to abusive situations this concern is justified, as described by Pringle (2001), with men's predominance as abusers, there are complications for them to be social workers.

Expanding on this theme, the participants were worried about physical contact with service users, particularly with children. One had a placement supervisor mention to him that a male employee had been sacked for having physical contact with a child (which was described as a 'side hug'), and Saban said, 'I learned straight away I have to be really cautious with how I present myself physically to the children' (Saban, 31). What was important about this exchange was that Saban had not been told whether the actions of the former employee were improper, only that he had been fired for having physical contact with a child. Generally, the participants felt that their concern for this was greater than that of their fellow students that were women. This concern about the impact of their being a man on their ability to engage with service users was a consistent and pressing worry for the participants. Whilst there are various stances about touch, often organisationally defined in procedures and culture, the knowledge base about the use of touch in social work is not very developed, even though the topic is frequently discussed (Green, 2017). This topic is more challenging for some men than others, with one participant noting that his identity as a gay man made him highly anxious about being around children, out of fear that colleagues or service users might accuse him of inappropriate touching. Jeyasingham (2014) suggests that this additional challenge is because 'queer identities… are positioned as overtly sexual and requiring management in public arenas' (pg. 219), requiring that gay men, in particular, have to undertake additional work to manage how their identity interacts with their professional work.

Feeling Social Work Did Not Want Them

Finally, the men in this study also believed that the social work profession did not 'want them'. They felt that they were sometimes related to as representatives for *all* men, often in situations or discussions where they did not feel their experience qualified them to represent men. These feelings were particularly prominent during discussions of abuse and domestic violence. More broadly, though, the participants often felt they could not disagree without being seen as being combative, meaning they were unsure of how to engage (and often were less engaged as a result). John said, 'So if there's a debate and we have a view from a different perspective, you have to be really careful what you're gonna say' (John, 34). He raised these concerns that were echoed by other participants, who suggested they sometimes felt the whole class (who were predominantly women, and usually led by a woman teacher) turned to them as if they could speak on behalf of the abuse perpetrated by all men. Dominelli argues that 'being identified as an oppressor can cause feelings of paralysis and guilt, especially where it is difficult for the individual concerned to individually extricate him or herself from a privileged status' (2002: 46). These discussions are important in social work education, as they are the focus of a significant amount of social work practice, and if these students are unable to engage fully, it is likely that their understanding of the topic is hindered. In addition, when thinking of some of the participants in my UK study, it would be unfair to expect a 19-year-old new university student to represent all men, particularly if one reason he chose social work was to address the abuse of other men.

One of the key questions emerging from my research was: Does social work actually want more men? Some of the experiences of the participants showed that there were some people in the profession who made them feel they were not welcome, which might lead one to think that there are some in the profession that do not wish there to be a greater proportion of men. Nick felt that his experience at university suggested that 'some of the older [lecturers] are a little bit institutionalised and think that male social workers aren't right … It's an observation that came to mind that's made my experience a bit more uncomfortable' (Nick, 26). What he highlighted from this situation was that some of his teachers felt that he (and other men) should not be social workers, but it is also possible that these teachers were reacting to the 'glass escalator' effect of men entering the profession.

Peter, a Black African father of two, provided an example from his first placement; he described this situation as the defining experience of his degree. He described feeling upset when his placement manager told him that the placement's staff and service users did not want a man on placement with them:

> The manager told me, 'I specifically asked for a male student.'… Every staff have rejected it, and all the service users have said they didn't want a male student. … The manager told me they don't welcome me, but she will support me. Anything happening, I should inform her. She had specifically asked for me, so she's going to see me through. It was very hard. I wasn't comfortable. I don't know who is my friend, who isn't my friend. The first few weeks were very, very challenging. (Peter, 38)

This situation was difficult to hear, and he felt it was difficult to experience. He described this experience as something that he had to 'get over' so he could engage with his first placement. While there may have been very good reasons for feeling uncomfortable with men, to expect a student to successfully engage in this environment would provide a challenge. Peter was aware of their discomfort and felt that this was a situation he needed to address to be successful on his placement. What was interesting was that he felt that this experience was likely to be similar to what he would experience across the rest of his career.

The men in this study clearly felt that they had made a choice that contrasted them with a hegemonic ideal of masculinity (Carrigan et al., 1985). Further, this study helped show how individuals can experience complicated privilege and marginalisation. Similarly, Coston and Kimmel (2012) found that some men can experience gendered privilege, but marginalisation because of another factor. They found that 'among members of one privileged class, other mechanisms of marginalisation may mute or reduce privilege based on another status' (Coston & Kimmel, 2012, p.110). In the social work context, it is important to remember that men are privileged generally but are expected to be able to easily avoid stigmatising situations (Crocker et al., 1998). This means that they are less likely to have developed the coping mechanisms that those with more marginalisation need to develop. Some of the quotes presented above suggest this, such as Nick's challenge in accepting that some lecturers may have conflicts about men's position and advancing in the social work profession.

It is important to recognise that gender inequality produces challenges for women and girls which are multi-layered and dynamic. Gender inequality places an unequal burden on women and girls. Behaviours are often codified into gender norms that support the division of labour for men and women into gendered forms that help perpetuate gender inequality (Chafetz, 1990). From my research, it is clear that we need to think more carefully about gendered expectations of 'care'. In particular, we need to consider how care relates to social work and gendered notions of 'suitable' professions. Given that men are challenged by gender stereotypes of not being 'caring' enough for social work, it seems reasonable that one mechanism to improve gender equality is by investigating these gender stereotypes to better understand the gendered notions of 'care'.

Conclusions

This chapter examined men and their engagement with social work as an issue that may be useful when attempting to understand gender inequality, a central tenet of feminist theory (Wendt & Moulding, 2016). The proportion of men in social work in the United Kingdom is dropping and will continue to do so. Men's position in social work is complicated, particularly as a result of their predominance and swifter advancement to holding positions of power. Similar issues are found in related occupations, such as nursing, and is called the *glass escalator*. This unequal representation creates challenges when considering increasing the number of men in social work. There are some arguments that can assist examinations about why and whether the social work profession wants to increase the number of men in social work.

Men as social work students present a more visible example of these issues, because they are joining the profession, and may not have developed the coping strategies of men with more experience. The men in the study presented in this chapter felt keenly that they were out of place, and many thought that their gender made them unwelcome to the profession. There were some sites of specific difficulty, such as physical contact and discussions about domestic abuse. Their experiences can help us identify when society in appropriately suggests that women are more naturally suited to caring roles and men are, therefore, not appropriate to undertake tasks related to caring. Social work is often described as a 'caring profession'. These gendered norms create challenges for male social work students, but the same norms also support the continued overburdening of women and girls with caring responsibilities. If we seek to address these gender norms in an attempt to improve women and girls' lives, then it is problematic and sexist to suggest that women are 'natural' carers and that men, therefore, are not suited to these roles. Examining how we engage and support men to undertake social work can help to address gender equality, as well as providing more nuanced understandings of how people (of all genders) can undertake caring roles.

Notes

1. I am grateful for the input of Dr. Elliot Evans in addressing the linguistic challenges created by presenting this study.
2. The phrase 'women-majority' is used decidedly here (instead of 'female-dominated' or 'traditionally female'), because while there are more women in these occupations, they do not predominate in positions of authority, power or financial reward (McPhail, 2004).
3. All the participant's names have been changed to pseudonyms and any identifying characteristics have been removed from their presented narrative, to protect their identities.

References

Ashcroft, J. (2014, March 4). Social care employers need to recruit more men [online]. *The Guardian*. Available from: http://www.theguardian.com/social-care-network/2014/mar/04/social-care-employers-recruit-more-men. Accessed 21 March 2014.

Butler, J. (2004). *Undoing gender*. Routledge.

Carrigan, T., Connell, B., & Lee, J. (1985). Toward a new sociology of masculinity. *Theory and society, 14*(5), 551–604.

Chafetz, J. S. (1990). *Gender equity: An integrated theory of stability and change*. Sage.

Christie, A. (1998). Is social work a 'non-traditional' occupation for men? *The British Journal of Social Work, 28*(4), 491–510.

Cocker, C., & Hafford-Letchfield, T. (Eds.). (2014). *Rethinking anti-discriminatory and anti-oppressive theories for social work practice*. Palgrave Macmillan.

Connell, R. (2009). *Gender: In world perspective* (2nd ed.). Polity.

Coston, B. M., & Kimmel, M. (2012). Seeing privilege where it isn't: Marginalized masculinities and the intersectionality of privilege. *Journal of Social Issues, 68*(1), 97–111.

Cree, V. E. (1996). Why do men care? In V. E. Cree & K. Cavanagh (Eds.), *Working with men. Feminism and social work* (pp. 65–86). Routledge.

Crocker, J., Major, B., & Steele, C. (1998). Social Stigma. In: D. T. Gilbert & S. T. Fiske, eds. The Handbook of Social Psychology. Boston: McGraw-Hill, 504–53.

Derrida, J. (1976). *Of grammatology*. Johns Hopkins.

Derrida, J. (1982). *Positions*. University of Chicago Press.

Dominelli, L. (2002). *Anti oppressive social work theory and practice*. Palgrave.

Fiore, B., & Facchini, C. (2013). Social work as gendered issue from a generational point of view. *International Review of Sociology, 23*(2), 310–325.

Fjeldsted, K. (2013). *Female leadership in health care: Leadership in health care organizations*. Standing Committee of European Doctors.

Furness, S. (2015). Conduct matters: The regulation of social work in England. *British Journal of Social Work, 45*(3), 861–879.

FSW, (2014). IFSW Definition of Social Work [Online]. Berne: International Federation of Social Workers. Available from: http://ifsw.org/policies/definition-of-social-work/ [Accessed 6 November 2015].

Galley, D., & Parrish, M. (2014, July 25). Why are there so few male social workers?. *The Guardian*.

Giesler, M. A. & Beadlescomb, T. (2015). Elevated or Erased? A Content Analysis of the Portrayal of Men in Introduction to Social Work Textbooks. *The Journal of Men's Studies, 23*(2), 147–160.

Goffman, E. (1977). The Arrangement between the Sexes. *Theory and Society, 4*(3), 301–331.

Green, L. (2017). The trouble with touch? New insights and observations on touch for social work and social care. *British Journal of Social Work, 47*(3), 773–792.

GSCC, (2012). *Regulating Social Workers, 2001–2012*. London: General Social Care Council.

Harding, T., North, N., & Perkins, R. (2008). Sexualizing men's touch: Male nurses and the use of intimate touch in clinical practice. *Research and Theory for Nursing Practice, 22*(2), 88–102.

Haywood, C., Ghaill, M. M. A., & Allan, J. A. (2015). Schools, masculinity and boyness in the war against boys. *Boyhood Studies, 8*(1), 15–21.

HCPC. (2019). *Gender breakdown of registered social workers in England-January 2019*. [online]. Available at: https://www.hcpc-uk.org/resources/freedom-of-information-requests/2019/gender-breakdown-of-registered-social-workers-in-england%2D%2Djanuary-2019/. Accessed 5 July 2020.

Hearn, J., Hearn, J., Pringle, K., Müller, U., Oleksy, E., Lattu, E., Chernova, J., Ferguson, H., Holter, Ø. G., Kolga, V., Novikova, I., & Ventimiglia, C. (2002). Critical studies on men in ten European countries. *Men and Masculinities, 5*(2), 192–217.

Hicks, S. (2015). Social work and gender: An argument for practical accounts. *Qualitative Social Work., 14*(4), 471–487.

Hillman, N., & Robinson, N. (2016). *Boys to men: The underachievement of young men in higher education – And how to start tackling it: HEPI report 84*. Higher Education Policy Institute.

Hussein, S., & Christensen, K. (2017). Migration, gender and low-paid work: On migrant men's entry dynamics into the feminised social care work in the UK. *Journal of Ethnic and Migration Studies, 43*(5), 749–765.

Hussein, S., Moriarty, J., & Manthorpe, J. (2009). Variations in Progression of Social Work Students in England: Using student data to help promote achievement: Undergraduate full-time students' progression on the social work degree. London: GSCC/Social Care Workforce Research Unit.

Jeyasingham, D. (2014). Deconstructing sexuality in anti-oppressive practice. In C. Cocker & T. Hafford-Letchfield (Eds.), *Rethinking anti-discriminatory and anti-oppressive theories for social work practice* (pp. 211–224). Palgrave Macmillan.

Kalaitzi, S., Czabanowska, K., Fowler-Davis, S., & Brand, H. (2017). Women leadership barriers in healthcare, academia and business. *Equality Diversity and Inclusion: An International Journal, 36*(5), 457–474.

Kristeva, J. (1981). Women's time. *Signs, 7*(1), 13–35.

Kullberg, K. (2013). From glass escalator to glass travelator: On the proportion of men in managerial positions in social work in Sweden. *British Journal of Social Work, 43*(8), 1492–1509.

Lawler, J. (2007). Leadership in social work: A case of caveat emptor? *British Journal of Social Work, 37*(1), 123–141.

Lyons, K., VALLE, I. L., & Grimwood, C. (1995). Career patterns of qualified social workers: Discussion of a recent survey. *The British Journal of Social Work, 25*(2), 173–190.

McLean, J. (2003). Men as minority: Men employed in statutory social care work. *Journal of Social Work, 3*(1), 45.

McPhail, B. A. (2004). Setting the record straight: Social work is not a female-dominated profession. *Social Work, 49*(2), 323–326.

Melville-Wiseman, J. (2016). The sexual abuse of vulnerable people by registered social workers in England: An analysis of the health and care professions council fitness to practise cases. *British Journal of Social Work, 46*(8), 2190–2207.

OECD. (2015). *Education at a glance 2015. Education at a glance*. OECD Publishing.

Parker, J., & Crabtree, S. A. (2014). Fish need bicycles: An exploration of the perceptions of male social work students on a qualifying course. *British Journal of Social Work, 44*(2), 310–327.

Pease, B. (2001). Developing pro feminist practice with men in social work. *Critical Social Work, 2*(1), 1–8.

Pease, B. (2011). Men in social work challenging or reproducing an unequal gender regime? *Affilia, 26*(4), 406–418.

Pringle, K. (2001). Men in social work: The double-edged sword. In A. Christie (Ed.), *Men and social work: Theories and practices* (pp. 35–48). Palgrave.

Schaub, J. (2015). Issues for men's progression on English social work honours and postgraduate degree courses. *Social Work Education, 34*(4), 315–327.

Schaub, J. (2017). *Making sense of men's experiences and progression through social work programmes*. PhD Thesis [unpublished]. University of Bath.

Severiens, S., & ten Dam, G. (2012). Leaving college: A gender comparison in male and female-dominated programs. *Research in Higher Education, 53*(4), 453–470.

Simpson, R. (2009). *Men in caring occupations: Doing gender differently*. Palgrave Macmillan.

Skelton, C. (2006). Boys and girls in the elementary school. In C. Skelton, B. Francis, & L. Smulyan (Eds.), *The SAGE handbook of gender and education* (pp. 139–165). Sage.

Skills for Care. (2019). *The state of the adult social care sector and workforce in England*. Skills for Care.

Weaver-Hightower, M. B. (2011). Male preservice teachers and discouragement from teaching. *The Journal of Men's Studies, 19*(2), 97–115.

Wendt, S., & Moulding, N. (Eds.). (2016). *Contemporary feminisms in social work practice*. Routledge.
Wenger-Trayner, E., Fenton-O'Creevy, M., Hutchinson, S., Kubiak, C., & Wenger-Trayner, B. (Eds.). (2014). *Learning in landscapes of practice: Boundaries, identity, and knowledgeability in practice-based learning*. Routledge.
West, C., & Zimmerman, D. H. (1987). Doing gender. *Gender & society, 1*(2), 125–151.
West, C., & Zimmerman, D. H. (2009). Accounting for doing gender. *Gender & Society, 23*(1), 112.
Whitehead, S. M. (2002). *Men and masculinities: Key themes and new directions*. Polity Press.
Williams, C. L. (1992). The glass escalator: Hidden advantages for men in the 'female' professions. *Social Problems, 39*(3), 253–307.
Williams, C. L. (2013). The glass escalator, revisited: Gender inequality in neoliberal times, SWS feminist lecturer. *Gender & Society, 27*(5), 609–629.
Wingfield, A. H. (2009). Racializing the glass escalator: Reconsidering men's experiences with women's work. *Gender & Society, 23*(1), 5–26.
Woodfield, R. (2014). *Undergraduate retention and attainment across the disciplines*. Higher Education Academy.

CHAPTER 18

A Relational Approach to Work with Couples Where Men Have Been Violent Towards Women: Feminist Dilemmas and Contributions to Social Work Practice

Rebecca Infanti-Milne, Richard Mc Kenny, and Lee Walton

Introduction

In this chapter we describe the relationship between feminist theory and family therapy practice. We show how this influence became less explicit from the turn of the century as practice concerns in relation to gender were folded into a larger project, a concern for 'social graces'. We describe aspects of our own systemic social work and therapeutic practice, working together with parents where at least one parent has used violence and abuse. We think of this work as a "concrete pointer" (Featherstone, 2010, p. 83) towards applying a feminist ethics of care in social work contexts.

R. Infanti-Milne (✉) • R. M. Kenny • L. Walton
Relational Third, London, UK
e-mail: info@relationalthird.com

C. Cocker, T. Hafford-Letchfield (eds.), *Rethinking Feminist Theories for Social Work Practice*,
https://doi.org/10.1007/978-3-030-94241-0_18

The work described here was undertaken as part of a larger project in an inner city area in east London from 2017 to 2020 (cf. Infanti-Milne & Walton, 2020). The context for this work, the particular community in east London, is one of the most diverse in the UK. Families live with the consequences of poverty, inadequate housing, ill-health, the influence of criminal gangs, racism, sexism, and Islamophobia. We are White British or Irish, heterosexual. To varying degrees we have had access to sources of privilege, and relative to many clients we live comfortable lives.

Feminisms

Family therapists Burck and Speed (1995) point out that talk of 'feminism', even as historical 'waves', overlooks the diversity and the "sometimes even contradictory ways in which feminists conceptualise and apply their thinking about gendered arrangements" (1995, p. 2). We locate our work within a feminist therapy tradition that acknowledges the complexity of clients' relationships, endeavours to respect women's decisions about who they live with, and acknowledges the harms caused by abuse and violence in intimate relationships. We also recognise that dangerous individuals, many of them men, use violence instrumentally in the context of family (historically referred to as 'batterers') and that the state must intervene to prevent their causing harm (Johnson, 2008; Wormith et al., 2020).

Systemic Social Work

We associate our practice with the 'reclaiming social work' (RSW) model introduced in the London borough of Hackney (a UK Local Authority which provides statutory services) in 2007 (Goodman & Trowler, 2012; Munro et al., 2010; Forrester et al., 2013; Bostock et al., 2017). Hackney is a highly diverse community with the fifth highest average Index of Multiple Deprivation score in England (Ofsted, 2006) and home, in the late 1990s and early 2000s, to "murder mile" (Mendick & Johnson, 2002). Goodman and Trowler joined the Council as senior managers in Children's Services in 2005 and by 2006 "outcomes for children and young people [were] improving from a historically low base" (Ofsted, 2006, p. 5). RSW replaced a single-discipline social work team with a multidisciplinary social care 'unit' led by a consultant social worker and including a family therapist (or clinical practitioner), social worker, children's practitioner, and unit coordinator. Cases are 'held' by the unit, necessitating shared understandings of family circumstance, aims, and goals for the work. Contemporary systemic theory and practice which welcomes collaboration with families (doing-with rather than doing-to) provides an overarching framework for the practices necessitated by unit organisation. The consultant social worker holds positional power and responsibility for the progress of work with the family.

This significant change in the delivery of child and family Social Services in the UK was subject to independent review, by Munro et al. (2010) and Forrester et al. (2013). Forrester et al. concluded that "if we were starting child protection from scratch ... there is no question that you would opt for the systemic unit model" (Forrester et al., 2013, p. 186). However, accounts of RSW rarely describe practices in relation to domestic violence and abuse. Contributors to Goodman and Trowler's edited volume mention domestic violence just twice; on neither occasion is a systemic intervention offered as a response (Goodman & Trowler, 2012) Munro et al.'s (2010) evaluation mentions domestic violence once. While Forrester et al. (2013) make 22 references to domestic violence, this is never to describe a systemic therapeutic intervention provided by an RSW unit. We agree with Hare that systemic therapeutic techniques in contexts of domestic violence and abuse should be given more prominence within systemic therapeutic skills trainings for social workers practicing within the RSW model (Hare, 2018). In the development of our project we worked with the Institute of Family Therapy (London) to design a domestic violence and abuse focused Postgraduate Diploma in Systemic Practice.

Early Feminist Critiques of Family Therapy

Systemic family therapy has engaged in a dialogue with feminist theory at particular points in its development. However, explicit discussion of feminism has all but disappeared in the last 20 years, suggesting that family therapy as a discipline understands itself to have incorporated feminist critique into contemporary practice. We will summarise key aspects of this dialogue and explain why feminism as a stand-alone issue is no longer a focus for explicit discussion within the professional literature.

Jones (1993, p. 149) notes that while family therapy thinks of itself as privileging *context* this initially meant attending to the intergenerational transmission of ideas, beliefs, and behaviours 'passed down' though generations and subject to variation and change (Taggart, 1985; James & McIntyre, 1983). Attending to other contexts was considered 'political' and so the antithesis of therapeutic activity. Drawing attention to the absence of gender as a relevant category, Goldner (1988) stated she "could not find a single reference to gender, let alone gender inequality, in any of the classic family texts", save for the works of Haley, the originator of strategic family therapy (1988, p. 22).

Goldner (1988, p. 20) describes how, within texts, Haley gradually collapses gender into generation—"mother" and "father" become "parents". Gendered differences in domestic power, potential and actual access to resources, are obscured: a process of mystification that amounts to justification of "socially structured inequalities in social relations" (MacKinnon & Miller, 1987, p. 140). This situation came about, Goldner argues, because the developing discipline required "the illusory division of the world into public and private domains" (1988, p. 22): the family as a haven from a heartless (capitalist) public sphere, precluding attention to gendered power differences and abuses within the family. Goldner thinks this ideological positioning of family supported its defence against processes of individualising and pathologising distress caused by social injustices.

There is less consensus today about some of Goldner's assertions, for example, that "gender … is not a secondary, mediating variable like race, class, or ethnicity, but, rather, a fundamental, organizing principle of all family systems" (1988, p. 17). We now think of categories such as gender and race as both unstable and inevitably both descriptive and prescriptive (e.g. Haslanger, 2012, p. 241). This shift illuminates the place of feminist theory in contemporary family therapy theory and practice, as we set out below.

Feminism and Neutrality

The Milan team, four psychiatrists and psychoanalysts turned family therapists, began working together in 1967, and in 1980 the team published a short paper setting out a trinity of guidelines for the in-room therapist working with the family: hypothesising-circularity-neutrality. The team sought to influence the family's way of understanding itself through an examination of the patterns of interaction family members find themselves in and part of. Neutrality, "the instrumental counterpart of the Systemic view" (Tomm, cited in MacKinnon & Miller, 1987), is defined as "a specific pragmatic effect that [the in-room therapist's] other total behaviour during the session exerts on the family (and not his intrapsychic disposition)" (Palazzoli Selvini et al., 1980, p. 11). The therapist is allied with everyone and noone (Jones, 1993, p. 16), seeking to be evenhanded (Jones, 1993, p. 102) and, by remaining 'meta' to the family, is able to be more helpful to them.

Neutrality requires the therapist to monitor their thoughts and feelings for signs of non-neutrality. MacKinnon and Miller argue that therapists cannot transcend their own, inevitably gendered, worldview and cannot be 'meta' to themselves, and so neutrality essentially collapses. Cecchin further developed the concept of neutrality in a 1987 paper attending to the feminist critique in his opening remarks: "Numerous discussions over the years have convincingly pointed out that it is impossible to be neutral with regard to language. All behaviour, including language, is politically laden. Any particular action helps to organize and constrain the possible patterns of social interaction" (1987, p. 405). The importance of this statement becomes clear in the light of MacKinnon and Miller's comments about Cecchin being "challenged at a conference [in April 1984] because he did not explore the concerns of a woman who described herself as an oppressed wife-mother in a role-play family. He maintained [at that time] that 'therapy is not the place for politics'" (MacKinnon & Miller, 1987. p. 143).

Feminist Family Therapy in the UK

The 1990s heralded a serious engagement between family therapy and feminism with landmark publications in 1990 (Miller & Perelberg, 2019) and 1995 (Burck & Speed, 1995). These texts mark a highpoint and the turn of this century saw a reduction in explicit discussion of, and writing about, feminism by family therapists. During this period the profession underwent a dual self-examination in relation to both sexism and racism. The racist murder of Black teenager Stephen Lawrence in 1993, and subsequent public inquiry in 1998, highlighted the issue of institutional racism in all public services. Some of those who had catalysed discussion of feminist family therapy turned their attention to the problem of institutionalised racism. For example, Miller and colleagues, including Krause at the Marlborough Family Service, took steps to establish a more racially and ethnically representative clinical service (Miller, 1990; Malik, 2006). Krause later moved to lead similar developments at the Tavistock Clinic.

The decline in explicit discussion of feminist family therapy happened because consideration of gender became part of a larger process of self-examination that led to new understandings of what it means to practice as a family therapist. The concepts of *reflexivity* and *social graces* emerged at this time as key articulations of this change.

Reflexivity and the Social Graces

Tomm developed the Milan team's ideas in three papers published in 1987 and 1988 (Tomm, 1987a, b, 1988), introducing the idea of reflexivity as "an inherent feature of the relationships among meanings within the belief systems that guide communicative actions" (Tomm, 1987b, p. 168). The Milan approach facilitates changes in meanings at the level of a family's belief system so family therapy becomes a catalysing of reflexive processes. At some point in the early 1990s Tomm's use of reflexivity, and feminist and anti-racist critiques, came together so that it became more important for the therapist to ask (self-) reflexive questions about socially constructed differences in relation to their practice with families.

Together with Alison Roper Hall, Burnham is responsible for the 'open source' and evolving mnemonic, *social graces*, or *Social GGRRAAAACCEESSSs*, standing for gender, geography, race, religion, age, ability, accent, appearance, class, culture, ethnicity, education, sexuality, sexual orientation, and spirituality (Burnham et al., 2008). This mnemonic is intended to encourage the therapist's interest in a broad range of socially constructed differences. Whilst we will have unique and personal relationships to different *graces*, gender and race, as visible and organising, have particular import. This mnemonic has been subject to critique, as too-easily affording a separating out of differences, creating the illusion that each is examinable on its own and in turn, and so as failing to acknowledge the intersection of different forms of discrimination and prejudice in lived experience (Butler, 2017; McDowell & Hernández, 2010).

Family Therapy and Couple Violence

While the twenty-first-century literature has not featured much explicit discussion of feminism, models and techniques for working with couples in contexts of violence and abuse have developed. Our own model of practice, developed in a local authority children's social care context, borrows heavily from the work of Cooper and Vetere (2001, 2005; Scerri et al., 2017), and Jenkins (1990, 2009). We will illustrate how these different accounts of practice can be read as embodying key messages from feminist psychology.

Cooper and Vetere attend carefully to "the complexity of the situations, dilemmas and binds within relationships, past and present, where violence, coercion and abuses of power intersect with attachments and dependence" (2001, p. 386). We understand Vetere and Cooper are seeking to move towards "a reading of family relationships at two levels of description: one elucidating the paradoxes of circularity, the other confronting the realities of domination" (Goldner, 1988, p. 24). We see strong connections between this line of thinking and Gilligan's work on moral development. Gilligan challenged Kohlberg's claim that a universal 'ethics of justice' underpins mature moral reasoning (e.g. Reed, 1997). She developed the idea that girls and women tend to develop an 'ethics of care', as a basis for making ethical judgements (Gilligan, 1993).

While Cooper and Vetere's citations of Gilligan's work are limited in scope, recent accounts of their therapeutic work (Scerri et al., 2017, p. *77ff*) rely on contemporary understandings of attachment theory, emphasising the importance of emotional commitment, relational responsibility, and felt safety (i.e. care). In work with couples in contexts of violence and abuse this focus supports the naming of "troubling emotional experiences" and developing "a repertoire for regulating unhelpful emotion and for interactive regulation. … Couple therapy requires co-presence and inter-subjectivity by showing our connections to each other with emotional mirroring" (Scerri et al., 2017, p. 84).

Gilligan's research with colleagues at the Harvard Project on Women's Psychology and the Development of Girls (e.g. Gilligan, 1993) explores the theme of *voice*. Listening to diverse groups of adolescent girls speak, Gilligan and colleagues "heard a … change in voice … girls at the edge of adolescence [described] impossible situations—psychological dilemmas in which they felt that if they said what they were feeling and thinking no one would want to be with them, and if they didn't … no one would know what was happening to them" (1993, p. *xx*). For these adolescent girls the result is a kind of dissociation between experience and "what is generally taken to be reality" (1993, p. *xxi*). This dissociation of adolescent female experience from gendered anticipations provides a mechanism for understanding adult-gendered responses to intimate partner violence and abuse. Dissociation helps explain women's emerging responses and strategies for living that social workers think of as 'minimising', 'lacking in insight', 'failing to place children's needs first', and so on. While Gilligan's work must be located in its cultural context, we find these ideas useful in understanding the responses of women we have met whose origins and life experiences are very different from those women and girls Gilligan got to know through her studies.

We trace a line back from Goldner through Gilligan's attention to *voice* to the activity of New England feminist Margaret Fuller (e.g. Fuller, 1998, Marshall, 2013). In addition to writing the founding text of American feminism Fuller organised weekly 'Boston Conversations', starting in 1839, bringing together "a 'circle' of women 'desirous to answer the great questions. What were we born to do? How shall we do it?'" (Marshall, 2013, p. 132). These conversations were collaborative inquiries (2013, p. 164), an impetus to action, (2013, p. 136), leading ultimately to women's liberation (2013, p. 188). We discern a common theme running through Fuller's conversations, Gilligan's research, and Cooper and Vetere's therapy—a careful excavation of women's voices and experiences in relationship that we seek to develop in our clinical practice. In the context of feminist family therapy this excavation encompasses the voices of men as partners and fathers.

We also discern a commitment to excavating unheard voices in the work of Jenkins (2009). Jenkins works with men within a restorative framework, adding an ethics of protest (a partial reconstruction of Kohlberg's ethic of justice), remorse, and generous love to an ethics of care. Jenkins thinks of violence as a possibility of all relationships, and that male violence, as gendered violence, "is commonly enacted as a form of 'legitimate' self-defence against perceived disrespect or threat from others" (2009, p. 4). Control and coercion are not necessarily harmful and can be ethical when enacted to safeguard others. Harmful coercive and controlling actions constitute abuse when enacted with the intent to cause harm, in relational contexts of power and advantage, where an aggressor experiences a sense of entitlement and abdicates responsibility for the well-being of the other person. As Jenkins poignantly notes, these "criteria for abusive behaviour … constitute prescriptions or recipes for competence, adequacy and success in dominant [hegemonic] masculine culture" (Jenkins, 2009, p. 6). Jenkins argues that all forms of violence can prompt "apparently noble and honourable justifications", and labelling these accounts as evidence of 'misguided blueprints', he seeks to separate out ethical intent from action (Jenkins, 2009, p. 4). It is in this separation that we locate the uncovering of dissociations at the heart of the lived experience of men who use violence and abuse. (We are excluding 'instrumental' or 'cold' forms of violence from this account.)

Jenkins engages men in conversations about their past, present, and future intentions for living ethically in relationship with others, against a background understanding of abuse as a strategy of "over-conforming to dominant cultural interests and practices" (2009, p. 7). Attending to abuse in this way foregrounds responsibility and accountability: "when a man ceases abusive behaviour and embraces more respectful ways of relating, it will be on the basis of consideration of his own ethics, not by the imposition of someone else's ethics or some external moral code" (2009, p. 12).

Combining elements of these models we seek to draw out aspects of men and women's lived experiences, which have been subject to dissociative processes, to support women to voice their experience and men to acknowledge harms done. Forgiveness is not an entitlement, and relational repair and restoration are ongoing tasks for which men take responsibility. We think this is more likely to be achieved when men reconnect their ethical intent with actions that "[enable] justice, education, love, security and other honourable aims" (Jenkins, 2009, p. 4).

Our Practice

This description of core principles of our practice downplays how we make use of contemporary psychological theory. While models for working with couples after violence and abuse remain rare, they tend to contain core common ingredients including: (1) elements of attachment theory and pro-mentalising practices, (2) recognition of how past trauma shows itself in the present (via dissociation, hyperarousal, hyper-vigilance, etc.), and (3) systemic understandings of patterns of interaction in relationships (e.g. Asen & Fonagy, 2017; Scerri et al., 2017; Tavistock Relationships, 2020).

Domestic violence and abuse is identified as a concern in over 50% of children's social care assessments (DfE, 2018). While there is strong evidence that women and men use violence in intimate relationships (Johnson, 2008; Bates, 2019), there are marked differences in prevalence rates and harms done. About 8.4% of women (compared to 4.2% of men) aged 16–59 years experienced domestic abuse in the year to March 2019, and 74% of female victims, as compared to 26% of male victims, sought medical attention during the previous 12 months (ONS, 2019) For 38% of female homicide victims aged 16 years or over, the suspect was their partner or ex-partner, compared to 4% of male homicide victims (ONS, 2020). Our discussion is not intended to minimise other forms of intimate violence but reflects the generality of therapeutic work in our particular practice setting. The impact of domestic violence and abuse on children in multiple domains of development is significant and lasting (Laing & Humphreys, 2013). Similarly, the harms caused to adults through intimate partner violence, particularly women, are considerable (Feder & Howarth, 2014). A common feature of referrals to children's social care is a lack of contact with the person who was violent (Philip et al., 2018).

Men are often absent from social work assessments (Maxwell et al., 2012; Scourfield, 2006). Often mothers are expected to speak for themselves and on behalf of the person who has committed the violence and abuse (Stanley et al., 2011), responsible for explaining the violence and abuse they have experienced and responsible for safeguarding themselves and their children in the future. Social workers' interventions tend to become organised by a need to reduce risk and one common measure of risk reducing is the social worker's perception of the mother's 'insight' concerning her partner's behaviour. This leads social workers into a preoccupation with explaining to themselves why particular clients stay in violent and abusive relationships rather than asking themselves how women's experiences, circumstances, and commitments can prevent them exiting these relationships. Social workers' answers to these questions often obscure the complexity of women's lives.

Social workers often focus on a mother's behavioural change with expectations set out in a 'working agreement' or 'written agreement' that outlines how the mother should safeguard her children (Ofsted, 2017). In our experience agreements are often confused with safety plans and rarely encourage women to share their own expertise, build on strengths, or engage their networks sufficiently. Social workers can lose sight of who is responsible for the violence, often a man, and then not engage with men to understand, explain, and challenge violent actions and reactions.

Since all our actions are also communications, we believe that unless social workers clearly and unambiguously hold users of violence and abuse responsible for their actions, they will communicate a willingness to minimise risks to parents and their children and inadvertently collude with men's minimising and other-blaming violence-justifying accounts of their actions (e.g. O'Sullivan, 2013; Humphreys & Campo, 2017).

We recognise that a range of "philosophical, sociopolitical positions have created huge controversies in the domestic violence field" (Scerri et al., 2017, p. 4). We recognise both the usefulness and limitations of typologies of violence (e.g. Johnson, 2008). We carefully risk assess parents before we decide if it is safe to propose couple work, reminding ourselves that our judgements are fallible, and in particular that we might be 'taken in' by more 'cold', instrumental users of abuse. We assume that women want to live in safety and have agency to act to achieve this, to exit abusive relationships, with support. This assumption helps us to focus attention on the particular push/pull factors that particular women take into account when weighing up decisions. We carefully attend to the way women may have adapted to traumatic experiences, which may mean their lived experience of violence and abuse is not as we anticipate. We think about how women might talk about their experiences through this 'lens' of trauma (Scerri et al., 2017) and, by listening for the 'not-said', the aspects of self that Gilligan suggests can become unvoiced in the course of girls' development (Gilligan et al., 1990). We take responsibility for both trying to make sense of some of the apparently contradictory responses of families, women, and children who may have experienced multiple traumas (in childhood and as adults) and for finding ways of voicing experiences when relevant (it is our intention that this should be; in ways which are not experienced as re-traumatising (Infanti-Milne & Walton, 2020).

We are open to the possible emergence of isomorphic patterns, or 'parallel processes', to apparently similar processes emerging in professional networks, professional-client interactions, and in couples and families (Koltz et al., 2012; White & Russell, 1997). For example, we notice when men hold women responsible (blame women) for their (male) violence and social workers appear to do the same thing. We think that the social work task can overwhelm the emotional resources of any individual and that social workers can use dissociative strategies to help cope with their duties and responsibilities (Somer et al., 2004). One way we see this happening is when social workers' accounts of their clients lives, and their explanations for their clients' actions, no longer fit with their clients' lived experience.

Case Study

We work as teams of two, following the model developed by Cooper and Vetere (see Vetere & Cooper, 2001, p. 385). In this case example Rebecca worked with a Bengali female co-practitioner in-room consultant, Jasmin Choudhury. Jasmin provided Rebecca with informal 'cultural consultation', supporting her to develop a therapeutic alliance with both parents.

Rebecca and Jasmin met and worked with Aiysha and Mohammed (not his real name), Muslim parents of Bangladeshi heritage. Aiysha, not her real name, has read this account and agreed to its publication. Aiysha, in her thirties, was married to Mohammed, in his mid-forties. Aiysha has some family in the UK. They have two children; the youngest has both physical and learning disabilities. When they married Aiysha could not give consent as she was a child at the time. The couple lived in southern Europe and while Mohammed had acquired EU citizenship he did not seek to regularise Aiysha's status so that later, when they relocated to the UK, Aiysha had 'no recourse to public funds'.

The sessions were conducted in Bengali-Sylheti and so Rebecca required an interpreter.[1] Children's Social Services were concerned about reports of domestic violence and abuse and that Aiysha had been forced into marriage. The older child was struggling with school. When Rebecca and Jasmin met the family the children were subject to a child protection plan. Aiysha had previously fled with her children but struggled to manage two vulnerable children while negotiating complex systems when excluded from universal entitlements and so had she returned to Mohammed.

Aged 13, Aiysha had been 'sent' by her family to marry Mohammed. Mohammed and his family kept her 'inside' because the police had told Mohammed she was too young to be married and should return to Bangladesh. She remained effectively imprisoned for years, repeatedly raped, physically assaulted, and psychologically abused, all of which continued through her first pregnancy. Aiysha's only meaningful contact outside the family was a downstairs neighbour without whose emotional support she is not sure she could have coped. Some other neighbours expressed concern but were reassured by Mohammed.

Aiysha's telling of her story took many hours and would not have been possible without the support Jasmin provided—not just cultural consultation to Rebecca but a safer context for Aiysha to speak and be heard. On one occasion both Rebecca and Jasmin had to hold Aiysha off the floor when she collapsed, unable to breathe through her distress. On occasion Aiysha would dissociate in the room and have to be helped back into the present with grounding techniques. Aiysha wanted to tell her story despite the impact on her, emphasising the importance of being listened to and believed. Aiysha's observation that in her second home, in southern Europe, the community and authorities "never had a problem with the domestic violence" was heard by Rebecca and Jasmin as an extreme form of the dissociation between experience and "what is generally taken to be reality" that Gilligan identifies in the accounts of young women in the Harvard study (Gilligan, 1993, p. xxi).

In Rebecca and Jasmin's conversations with Mohammed he would not, or could not, talk about the harms he caused Aiysha beyond vague minimising allusions to some of his actions, perhaps reflecting worries that he might be held criminally accountable, and the impact criminal sanctions might have on his children. It was difficult to gauge the extent of any shame he felt and perhaps he could not contemplate or imagine renegotiating his relationship with Aiysha. Often tearful and distressed in sessions, Mohammed seemed to want to make amends for the past but was unwilling or unable to take the risks required to move towards a more just and equitable way of being.

At the end of the assessment phase the wider team together concluded that they could not offer a couple intervention to Aiysha and Mohammed. Aiysha's feelings of contempt for Mohammed were so strong she could not imagine, and did not want, a future with him. At the same time, she was not yet ready to leave. In the context of a 'no violence contract' (Vetere & Cooper, 2005) the team worked with Aiysha on agreed tasks, to access practical support and improve her financial independence. Aiysha and Mohammed were offered housing in a neighbouring borough, closer to her sisters, and at their last session together she brought Rebecca and Jasmin gifts, small gold hearts, whose meaning and significance is immediately apparent. Aiysha said that telling her story and being believed had changed her life. A year later, when Jasmin encountered Aiysha by chance in the street she had left Mohammed (taking her children).

Conclusion

Feminist family therapy, a distinct position in the debates of the 1980s and 1990s, became subsumed into a wider discourse concerning 'social graces' in the 2000s, a broad attention to difference and alterity in the therapy room, underpinned by an emphasis on the therapist's self-reflexivity. Systemic child and family social work has not, until recently, paid close attention to domestic violence and abuse (Hare, 2018). This is surprising given the prevalence of the issue in child and family social work and the theoretical insights and practice techniques a systemic model makes available. Our own practice seeks to import into a social care context a feminist family therapy tradition developed in the US north east, by Goldner and colleagues, and in Reading (UK), by Vetere and Cooper. Central preoccupations of practice include creating safety for women in couple work contexts, in order to pay close attention to *voice,* to the unvoiced experiences-in-context, of violence, control, threat, and fear, that women, mothers of children with social workers, dissociate from in order to survive and cope day-to-day.

Engaging men in social work, particularly after reports of domestic violence and abuse, can give rise to considerable professional anxiety. The risk of injury and death for women and children increases at, and after, separation (e.g. Almond et al., 2017). Keeping men at a distance makes sense given this evidence and as a way of managing difficult feelings that such work can evoke, including about professionals' own experiences of violence. However, when social workers do not engage with men they inevitably transfer responsibility for safety to women, and their assessments of risk are incomplete. We seek to work with men to help them voice and live out an ethics of protest, remorse, and generous love (Jenkins, 2009).

We believe that social workers are still not well enough trained and supported in the task of assessing the risk that men present in contexts of violence and abuse, and rarely feel able to talk about the feelings that arise for them in the work.

Note

1. To the extent that we cannot avoid the issue, we prefer to locate any language skills deficit with the person who speaks the least number of languages.

References

Almond, L., McManus, M., Brian, D., & Merrington, D. P. (2017). Exploration of the risk factors contained within the UK's existing domestic abuse risk assessment tool (DASH): Do these risk factors have individual predictive validity regarding recidivism? *Journal of Aggression, Conflict and Peace Research, 9*(1), 58–68.

Asen, E., & Fonagy, P. (2017). Mentalizing family violence part 2: Techniques and interventions. *Family Process, 56*(1), 22–44.

Bates, E. (2019). Why change current practice? In E. Bates & J. Taylor (Eds.), *Intimate partner violence: New perspectives in research and practice* (pp. 1–7). Routledge.

Bostock, L., Forrester, D., Patrizo, L., Godfrey, T., Zounouzi, M., Antonopoulou, V., Bird, H., & Tinarwo, M. (2017). Scaling and deepening the Reclaiming Social Work model: Children's Social Care Innovation Programme Evaluation Report 45. Department for Education.

Burck, C., & Speed, B. (1995). *Gender, power and relationships.* Routledge.

Burnham, J., Alvis Palma, D., & Whitehouse, L. (2008). Learning as a context for differences and differences as a context for learning. *Journal of Family Therapy, 30*(4), 529–542.

Butler, C. (2017). Intersectionality and systemic therapy. *Context, 151*, 16–18.

Cecchin, G. (1987). Hypothesizing, circularity, and neutrality revisited: An invitation to curiosity. *Family Process, 26*(4), 405–413.

Department for Education (2018). Characteristics of Children in Need: 2017–2019. *Crown Copyright*. Retrieved on 19 February 2022: https://assets.publishing.service.gov.uk/government/uploads/system/uploads/attachment_data/file/762527/Characteristics_of_children_in_need_2017-18.pdf

Featherstone, B. (2010). Ethics of care. In M. Gray & S. Webb (Eds.), *Ethics and value perspectives in social work* (pp. 73–84). Palgrave Macmillan.

Feder, G., & Howarth, E. (2014). The epidemiology of gender-based violence. In S. Bewley & J. Welch (Eds.), *ABC of domestic and sexual violence* (pp. 1–4). Wiley.

Forrester, D., Westlake, D., McCann, M., Thurnham, A., Shefer, G., Glynn, G., & Killian, M. (2013). *Reclaiming social work?: An evaluation of systemic units as an approach to delivering children's services*. UOB.

Fuller, M. (1998). *Woman in the nineteenth century*. W. W. Norton and Company.

Gilligan, C. (1993). *In a different voice*. Harvard University Press.

Gilligan, C., Lyons, N., & Hanmer, T. (Eds.). (1990). *Making connections: The relational worlds of adolescent girls at Emma Willard School*. Harvard University Press.

Goldner, V. (1988). Generation and gender: Normative and covert hierarchies. *Family Process, 27*, 17–31.

Goodman, S., & Trowler, I. (2012). *Social work reclaimed: Innovative frameworks for child and family social work practice*. Jessica Kingsley.

Hare, R. (2018). *An evaluation of the implementation of systemic ideas into child safeguarding social work practice with domestic violence and abuse using the principles of realistic evaluation* (unpublished doctoral dissertation). University of Sussex.

Haslanger, S. (2012). *Resisting reality: Social construction and social critique*. Oxford University Press.

Humphreys, C., & Campo, M. (2017). Fathers who use violence options for safe practice where there is ongoing contact with children. *CFCA, Paper, 43*, 2017.

Infanti-Milne, R., & Walton, L. (2020). NewDAy, an innovation project: Bringing systemic ideas to working with domestic violence in a child protection context. *Context, 168*, 20–24.

James, K., & McIntyre, D. (1983). The reproduction of families; the social role of family therapy? *Journal of Marital and Family Therapy, 9*, 119–129.

Jenkins, A. (1990). *Invitations to responsibility: The therapeutic engagement of men who are violent and abusive*. Dulwich Centre Publications.

Jenkins, A. (2009). *Becoming ethical: A parallel, political journey with men who have abused*. Russell House Publishing.

Johnson, M. P. (2008). *Typology of domestic violence: Intimate terrorism, violent resistance, and situational couple violence*. Northeastern University Press.

Jones, E. (1993). *Family systems therapy: Developments in the Milan-systemic therapies*. Wiley.

Koltz, R. L., Odegard, M. A., Feit, S. S., Provost, K., & Smith, T. (2012). Parallel process and isomorphism: A model for decision making in the supervisory triad. *The Family Journal, 20*(3), 233–238.

Laing, L., & Humphreys, C. (2013). Social work and domestic violence: Developing critical and reflective practice. Sage.

MacKinnon, L. K., & Miller, D. (1987). The new epistemology and the Milan approach: Feminist and sociopolitical considerations. *Journal of Marital and Family Therapy, 13*, 139–155.

Malik, R. (2006). British or Muslim: Creating a context for dialogue. *Youth & Policy, 92*, 91–105.

Marshall, M. (2013). *Margaret Fuller: A new American life*. Houghton, Mifflin, Harcourt.

Maxwell, N., Scourfield, J., & Featherstone, B. (2012). Engaging fathers in child welfare services: A narrative review of recent research evidence. *Child & Family Social Work, 17*, 160–169.

McDowell, T., & Hernández, P. (2010). Decolonizing academia: Intersectionality, participation, and accountability in family therapy and counselling. *Journal of Feminist Family Therapy, 22*(2), 93–111.

Mendick, R., & Johnson, A. (2002, January 6). Eight men shot dead in two years. Welcome to Britain's murder mile. *Independent Newspaper*. Retrieved on 18 November 2020: https://www.independent.co.uk/news/uk/this-britain/eight-men-shot-dead-in-two-years-welcome-to-britains-murder-mile-662314.html

Miller, A. C. (1990). Introduction II. In A. Miller & R. Perelberg. (Eds.) (2019). *Gender and power in families*. Karnac.

Miller, A. C., & Perelberg, R. J. (2019). *Gender and power in families*. Karnac.

Munro, E., Cross, S., & Hubbard, A. (2010). *Reclaiming social work*. LSE.

O'Sullivan, L. (2013). Engaging with male perpetrators of domestic violence: An exploration of the experiences and perspectives of child protection social workers. *Critical Social Thinking: Policy and Practice, 5*, 111–131.

Ofsted. (2006). *Joint area review of Hackney children's services authority area*. HM Government.

Ofsted. (2017). *The multi-agency response to children living with domestic abuse*. HM Government.

ONS. (2019). Domestic abuse in England and Wales overview: November 2019. HM Government.

ONS. (2020). *Homicide in England and Wales: Year ending March 2019*. HM Government.

Palazzoli Selvini, M., Boscolo, L., Cecchin, G., & Prata, G. (1980) Hypothesizing-circularity-neutrality: Three guidelines for the conductor of the session. *Family Process, 19*, 3–12.

Philip, G., Clifton, J., & Brandon, M. (2018). The trouble with fathers: The impact of time and gendered-thinking on working relationships between fathers and social workers in child protection practice in England. *Journal of Family Issues, 40*(16), 2288–2309.

Reed, D. (1997). *Following Kohlberg: Liberalism and the practice of democratic community*. University of Notre Dame Press.

Tavistock Relationships. (2020). *Parents in dispute, putting children first: Summary report of a programme delivered by Tavistock Relationships in partnership with Cafcass, and funded by the DWP*. Tavistock Relationships.

Scerri, C. S., Vetere, A., Abela, A., & Cooper, J. (2017). *Intervening after violence: Therapy for couples and families*. Springer.

Scourfield, J. (2006). The challenge of engaging fathers in the child protection process. *Critical Social Policy, 26*, 440–449.

Somer, E., Buchbinder, E., Peled-Avram, M., & Ben-Yizhack, Y. (2004). The stress and coping of Israeli emergency room social workers following terrorist attacks. *Qualitative Health Research, 14*(8), 1077–1093.

Stanley, N., Miller, P., Richardson Foster, H., & Thomson, G. (2011). A stop–start response: Social services' interventions with children and families notified following domestic violence incidents. *British Journal of Social Work, 41*, 296–313.

Taggart, M. (1985). The feminist critique in epistemological perspective: Questions of context. *Journal of Marital and Family Therapy, 11*, 113–126.

Tomm, K. (1987a). Interventive interviewing: Part I. strategizing as a fourth guideline for the therapist. *Family Process, 26*(1), 3–13.

Tomm, K. (1987b). Interventive interviewing: Part II. Reflexive questioning as a means to enable self-healing. *Family Process, 26*(2), 167–183.

Tomm, K. (1988). Interventive interviewing: Part III. Intending to ask lineal, circular, strategic, or reflexive questions? *Family Process, 27*(1), 1–15.

Vetere, A., & Cooper, J. (2001). Working systemically with family violence: Risk, responsibility and collaboration. *Journal of Family Therapy, 23*(4), 378–396.

Vetere, A., & Cooper, J. (2005). *Domestic violence and family safety*. Whurr Publishing Limited.

White, M. B., & Russell, C. S. (1997). Examining the multifaceted notion of isomorphism in marriage and family therapy supervision: A quest for conceptual clarity. *Journal of Marital and Family Therapy, 23*(3), 315–333.

Wormith, J. S., Craig, L. A., & Hogue, T. E. (2020). An overview of violent behaviour from aggression to homicide: Theory, research, practice (pp. 3–32). In J. S. Wormith, L. Craig, & T. Hogue (Eds.), *The Wiley handbook of what works in violence risk management* (pp. 99–118). Wiley.

CHAPTER 19

Feminist Perspectives on Social Work Leadership

Adi Cooper and Lyn Romeo

Introduction

In this chapter, we discuss feminist leadership in social work. We suggest that there is a strong alignment between the principles and values of feminism and the principles and values of social work. Wendt argues that 'feminism is core to social work knowledge and hence cannot be separated from the values and ideals of social work', as 'both come from a similar foundation of shared responsibility for social change and development' (2016, p. 13). However, this does not mean that all social workers and social work leaders are feminists. Swigonski and Raheim (2011) argue that although there is considerable overlap between feminism and social work in terms of theories, knowledge, skills, values and actions, they are also unique and distinct.

A. Cooper (✉)
Independent Consultant, London, UK

L. Romeo
Department of Health and Social Care, London, UK

C. Cocker, T. Hafford-Letchfield (eds.), *Rethinking Feminist Theories for Social Work Practice*,
https://doi.org/10.1007/978-3-030-94241-0_19

Feminist social work aims to achieve social justice and inclusion for those that social work serves. Its practice is informed by, and has its roots in, the women's movement of the 1960s, 1970s and 1980s and is based on principles arising from feminist political and social analyses (Hyde, 2013). The different ideological strands that developed were, to varying degrees, integrated into social work theory and practice (Dominelli, 2002). Concern for the social, political and economic situation of individuals, families and communities remains as valid today as it was in the last century. Despite the progress made, women continue to experience injustice and violation of their human rights (U.N. Progress of the World's Women, 2020). Feminist social work aims to achieve an egalitarian world where differences are understood and respected through addressing the economic, social and cultural inequalities experienced, recognising and addressing the intersectional issues that perpetuate the marginalisation of individuals and groups arising from class, race, skin colour, ethnic origin, disability, LGBTQ+, experience of state care, poverty and gender (Valesco, 2017).

There is a long history of social work aiming to address the impact of structural inequalities on peoples' lives (Cocker & Hafford-Letchfield, 2014). Diversity and equality are key social work principles that underpin activity at every level. Social work at its core is concerned with 'understanding and responding to hardship and oppression' (Wendt & Moulding, 2016 p. 1), so feminist insight regarding gender as well as other forms of oppression is highly relevant. Feminists argue that structural inequalities and social exclusion prevents people from participating fully in the life of the society in which they live. The inherent tension for social workers is that they are part of the system but charged with the responsibility of challenging the system to support the people with whom they work, who are disadvantaged by that system. This provides a unique opportunity for feminist social workers to position and assert themselves as agents of change.

Despite the large proportion of women in social work, men remain over-represented in senior leadership and management positions in the sector: 33% of senior managers are male although overall comprise only 17% of the adult social care workforce (Skills for Care, 2019, p. 66). With this in mind, we look at concepts such as gender equality, gender mainstreaming and the practical barriers for women moving into senior roles. In considering how feminists lead, we reflect on the criticality of the core feminist concept that 'the personal is political' in relation to the 'use of self' in leadership. Further, for feminists moving into management and leadership roles (especially in large hierarchical organisations), fundamental challenges about the nature and place of 'feminist leadership' arise. This includes how to 'lead' as an activist and how to enable collective decision making as opposed to hierarchical decision making across organisations. Challenging rather than reinforcing unequal power structures and relationships is also an issue, particularly for those operating in hierarchical organisations that maintain the status quo. Together these challenges can be conceptualised as 'double or triple jeopardy' issues (Lazzari et al., 2009, p. 349). We do not have all the answers to these dilemmas; however, researchers have offered some insight into how feminist leadership can be demonstrated and these challenges engaged with positively.

What is the role of feminist leadership and how are core feminist and social work principles expressed in feminist social work leadership? Leadership is a process of influencing others to achieve change, so feminist leadership can effect positive change. This can be achieved through relationships and relational practice, which requires engagement and communication. Feminist leadership is demonstrated through inclusive and participatory styles, which privileges collaboration, shared decision making and continuous reflexivity to consider why and what is being actioned and whether the goals of inclusion and social justice remain at the heart of practice (Gray & Schubert, 2016).

Although this chapter is primarily concerned with leadership, management is relevant. Leaders articulate vision and strategy, provide meaning, inspire and influence others, leading the changes necessary to deliver the strategy and achieve that vision (Obholzer & Miller, 2004). Managers have operational responsibilities; they manage work and people. Managers may demonstrate leadership, for example, when they motivate others to achieve the organisation's aims and objectives, manage change and develop new initiatives (Rusaw, 2005). However, leadership can be demonstrated in other roles, including as advanced or consultant practitioners, or Principal Social Workers. (Each Local Authority Adult Social Services department in England has a Principal Social Worker responsible for social work [Department of Health and Social Care, 2019c], and Case Study below.) These roles provide practice leadership, advice and support without necessarily carrying management responsibilities.

Our experiences as women leaders in social work, as feminists and as people who are committed to promoting diversity, inclusion and equality, have informed our view that feminist distributive and transformative leadership styles are essential. We suggest that explicit acknowledgement and further development of these approaches to leadership can deliver significant improvements to the experiences of those who are served by social work and those who work in social work in England. The case study illustrates professional leadership developments from a feminist perspective: This describes the leadership roles of the Chief Social Worker for Adults who is the senior civil servant responsible for professional leadership of adult social work in England and Principal Social Workers. Their impact in strengthening social work professional leadership across the sector has been informed by feminist social work approaches.

Feminist Core Principles

The core principles of feminist social work include the following features: firstly, a gendered lens, which informs an understanding, analysis and solution of problems and also acknowledges power dynamics; secondly, 'the personal is political', which emphasises the social reproduction of power relationships in lived experiences; democratised structures and processes, which forces attention onto processes of decision making as well as outcomes; thirdly, inclusivity and diversity, which recognises the multiplicity and complexities of the range of oppressions and subordinations, privileges and power that impact on peoples' lives; and fourthly, transformational, which means that feminist practice challenges all oppressions and aims to contribute to structural and cultural change (Hyde, 2013). Appreciating and valuing the experiences of others, particularly those who are marginalised, can be conceptualised as 'standpoint theory' (Rusaw, 2005, p. 388) and further informs feminist practice. Anti-oppressive feminist practice underpins feminist leadership in seeking to genuinely understand the lived experience of people together with the social, cultural and historical context within which their lives are lived and to act to overcome the structural obstacles which prevent them from achieving the best possible lives (Valesco, 2017).

Feminism: Definitions and Multiple Types

There are many forms of feminism, categorised in a variety of ways, including 'post-structuralist feminism, radical feminism, liberal feminism, intersectional feminism, feminist appropriations of critical race theory, structural feminism and material feminism' (Wendt & Moulding, 2016, p. 3) and 'psychosocial, Marxist, socialist, postmodernist' (Rusaw, 2005), also 'womanist, cultural' (Hyde, 2013). They all share an understanding of the world in terms of the relative power and privilege of men, which negatively impacts on women, but also on men and children. Common themes that cross over all categories include women being denied power and privilege because of gender norms, roles, responsibilities, theories or positioning of gender as a structural inequality. Feminists share commitment to challenge the subordinate status of women in society, in their communities and in their lives (Hyde, 2013), and this informs their leadership in social work.

Across the range of feminisms there is a shared understanding that gender inequality is structural, and as such, criticising dominant organisational power structures and challenging discrimination is core to feminism (Rusaw, 2005). All feminist approaches share the proposition that the position of women will improve if gender inequality is addressed and there is 'transformation of these social relations of advantage and disadvantage' (Wendt & Moulding, 2016, p. 3). How this is to be achieved varies according to the different feminist perspective adopted.

Ethics of Care and Feminist Leadership

Debates on the ethics of care are relevant to a discussion of feminism and leadership. Managers and leaders make moral judgements based on their ethical approach. A 'care orientation' or ethic of care was identified from an analysis of women's decision making in response to moral dilemmas, as a 'different voice'. This approach was demonstrated through an authentic relationship, whereby decision makers aimed to understand the subjective experiences and needs of others and then try to be responsive to these experiences and needs (Simola et al., 2010). Such a 'care orientation' is fundamental to social care, social work values, as well as feminist leadership. The concept of care can extend from compassion to social responsibility for others, to the environment as well as social justice (Rusaw, 2005). Further, research has found that 'leader propensity towards using an ethic of care was significantly positively related to follower perceptions of transformational (but not transactional) leadership' (Simola et al., 2010, p. 179), so there is a link between care ethic and leadership style.

Feminists have long argued that 'ethics of care have been excluded from organisational discourse' (Miller et al., 2019, p. 218) and emphasise the importance of a feminist contribution to social work leadership using an ethics of care approach. Feminist leaders 'sincerely care, encourage mutuality and create environments that inspire worker growth professionally and personally'; they support others as well as promote relational approaches to practice and reciprocal interactions between women and the environment (Malinger et al., 2017, p. 85). This can be considered as resulting from the social construction of gender, expressed through behaviour (Rusaw, 2005). Both gender and organisational environments shape management decision making regarding ethical dilemmas in work (Miller et al., 2019). However, there is a danger of binary notions reinforcing gender stereotyping, confusing 'feminine' with 'feminist'.

Pullen Sansfacon (2016) argues for an ethics of care that is a form of 'virtue ethics', which means a focus on developing traits that help others but can also address power struggles and inequalities (2016, pp. 45–6). She argues that 'virtue ethics combined with a feminist standpoint' provide the basis for 'ethical feminist social work practice' that achieves the goals of international social work to promote 'social change and development, social cohesion, and the empowerment and liberation of people' (2016, p. 51). Rather than perceiving ethics of care and ethics of justice (typified as male/masculine, see Simola et al., 2010) as mutually exclusive, we would argue that both have a place. A feminist approach to leadership in social work would encompass and promote an ethic of care within relational practice, with a self-conscious critique, alert to negative gender stereotyping.

Feminist Leadership: Transformational, Shared, Relational and Intersectional

Lazzari et al. have reviewed models and approaches to leadership and argue that feminist leadership is a form of transformational leadership in which leaders are 'change agents' who transcend their own needs to garner support and action on behalf of the needs of the greater collective (2009, p. 350), in response to the changing landscape of society and organisations (Newman, 2012). This approach to leadership and management is compared to more 'masculine' transactional styles where the manager/leader is motivated by personal success and focused on performance management, which privileges outcomes and key performance indicators over relationships, and compliance over professional autonomy and critical thinking (Gray & Schubert, 2016; Lazzari et al., 2009). Transformational leaders work to empower excluded or marginal groups, introducing new person-centred management and participative leadership styles, and develop 'new spaces of possibility for others' (Newman, 2012, p. 105). We would argue that feminist leadership is, by definition, essentially transformational and can be demonstrated in different social work roles.

A feature of effective transformational leadership involves shared or distributed power through relationships that pay attention to the processes involved. This approach promotes a move away from hierarchical leadership structures to flatter structures and valuing collaborative and inclusive leadership styles (Lazzari et al., 2009). Distributed leadership has been analysed in terms of 'team working'. Konradt (2013) describes the mechanics and impact of dispersed leadership, and how the influence of the single leader decreases as leadership functions are distributed among different people. We would agree that dispersed, distributive or shared leadership is consistent with feminist principles and attitudes towards power.

These concepts of leadership assist feminists in addressing the fundamental tensions implicit in the leader-follower power relationship. If power comes from consensus, then collaboration and relational working are key to achieving goals (Rusaw, 2005). Post-structuralist feminist approaches are helpful in supporting 'de-centred notions of authority … they can allow for and enable a diversity of perspectives … to effectively embrace a multiplicity of voices, subjectivities and ways of knowing and doing' (Carey & Dickenson, 2015, p. 513). These authors argue further that feminist theories can enable and support cross-boundary working, through insight into its pluralities and complexities: cross-boundary or multi-disciplinary and multi-agency working are essential features of current social work leadership both in terms of front-line practice and in strategic or senior management.

In the public sector, feminist approaches to leadership bring capabilities in working with multiple and interconnected stakeholders to achieve common purpose, using relational skills and 'catalytic conceptions of power' as well as recognition of structural inequalities (Rusaw, 2005, p. 386). Bringing an emphasis on the relational dimensions of leadership, feminist leaders can build networks, partnerships and support participation, which have become fundamental requirements of leadership in the public sector (Newman, 2012). Women can 'transgress' 'dominant modes of being and acting for individual leaders' both in terms of moving beyond simple dynamics of relationships between leaders and followers towards dispersed leadership, as well as seeing leadership as an individual's capacity to work relationally with others (Page, 2011, p. 320).

Fundamental to social work values and practice is the recognition of multiple forms of structural oppression; anti-discriminatory and anti-oppressive practices have been core elements of social work. However, Cocker and Hafford-Letchfield (2014) challenge us to develop beyond these conceptual frameworks and incorporate different ways of thinking and understanding inequalities and diversity. Crenshaw (1991) argues that it is problematic to treat race and gender as mutually exclusive experiences and developed feminist theory and antiracist politics by embracing this and providing a concept of intersectionality, which can be applied to a range of structural inequalities and the complexities of multiple interfaces. Intersectionality is therefore a 'methodological lens' that helps to understand the complexities of multiple identities, of power and privilege and reproduction of inequalities (Bryant, 2016, p. 85). The notion of 'white privilege' is gaining recognition in social care with the imperative to address unequal power over resources, decision making, relationships and information (Farqueson, 2020).

As well as women generally being under-represented in leadership and senior management positions in England, the number of Black, Asian and ethnic minority men and women is also limited, despite the emphasis on equalities duties and managing diversity in the public sector. People with Black, Asian and minority ethnic backgrounds comprise 25% of the adult social care workforce yet are only 15% of registered managers and 17% of senior managers (Skills for Care, 2019). They face barriers of process discrimination and challenges regarding complex expectations (Spillett, 2017). It is not clear how many disabled people are in senior positions; however, they are slightly over-represented in the workforce at 21% compared to 18% in the population (Skills for Care, 2019).

As organisations recognised the value of diversity, through opening up new markets, audiences and customer groups, promoting diversity has been seen as an asset (Newman, 2012; Conley & Page, 2017; Obholzer & Miller, 2004). We would argue that there are significant benefits of an intersectional approach to social work leadership and management. Promoting leaders and managers who are representative of the variety of different communities they serve is essential. We would define communities to include communities of place, interest and identity. They provide diverse contributions to the leadership, development and understanding of social work practice. These leaders can offer multiple and different lenses through which people see the world, understand problems and find solutions. However, the blocks and barriers to achieving this ambition are structural and further accentuated for women who are Black, Asian or from Minority Ethnic Communities (Spillett, 2017). Recognition of these blocks and barriers, and measures to reduce them, are required to achieve representative social work leadership.

Gender Equality/Mainstreaming and Barriers to Leadership in the Public Sector

Statutory social work in England operates predominantly in the public sector and reflections on feminist leadership in this area indicate how feminist leaders can be effective. The modernisation or new public management agenda in 1990s local government in England brought an accompanying new discourse on leadership (Ford, 2006) as women were encouraged into leadership and management. Women brought 'feminism, anti-racism and redistributive concepts of social justice ... notions of participation, redistributive power and a strong ethical and value base'; they represented a 'marker of modernity' *and* were seen as 'assets in the management of change' (Newman, 2012, p. 98, p. 94, p. 105). Consequently, some women were able to achieve senior leadership positions, enabling them to influence policy and service delivery (Newman, 2012). Their contribution positively impacted on management culture and practice, bringing different management styles and approaches to the public sector. Gender equality in management continued to be desirable to improve management, performance and productivity (Haile et al., 2016).

Liberal feminists have advocated for gender equality in the workplace as a social justice issue (Malinger et al., 2017). Evans et al. (2015) argue that gender equality in public service leadership is both a 'moral and instrumental imperative', being 'a key component of how we understand "Good Society"' (p. 502); it is critical, otherwise 'it is unlikely that the interests of women can be fully taken into account' (Carey & Dickenson, 2015, p. 509). Further, they point out that diversity, including gender, is a key policy instrument for achieving social as well as economic wellbeing (Evans et al., 2015, p. 502). Karroff et al. (2019) argue for more women in social work leadership roles to provide powerful role models for other women.

'Gender mainstreaming' continues to be presented as combining social justice and commercial interests, but it is 'an uneasy alliance between the dual agendas of business-driven efficiency and the moral case for women's equality' (Page, 2011, p. 318; Conley & Page, 2017). It has been criticised as bureaucratising and depoliticising feminism (Newman, 2012). Through implementing statutory public sector equality duties, 'equality' has become 'locked into the bureaucracy of performance management', in a paradigm of 'managing diversity'. The impact on austerity from 2007 has further limited and undermined the delivery of these duties as well as disproportionately impacting on women and marginalised groups (Conley & Page, 2017).

Despite the modernisation agenda influencing a shift from heroic leadership discourses to discourses which are relational and distributed, macho-management discourses have continued through the 2000s in local government (Ford, 2006). Legal-rational conceptions of organisation in the public sector have favoured 'masculine' leadership styles due to hierarchical structures and regulatory functions (Rusaw, 2005). This legacy is perpetuated in an environment where, at the time of writing, austerity and the Covid-19 pandemic mean that heroic behaviours are valued, and meeting savings and performance targets are paramount. However, leadership such as that demonstrated by New Zealand Prime Minister Jacinda Arden during the pandemic have offered an alternative, distinguished by authenticity, transparency, empathy, valuing people and inclusivity. Additionally, social workers' response to Covid-19 has been recognised as innovative and inspiring, through ways they have supported people in need, safeguarded those at risk and worked together with people in their communities (Brindle, 2020).

'The Personal is political': Barriers and Enablers

The gender mainstreaming debate centres on what the barriers and obstacles are for women achieving positional power in management hierarchies. These are categorised as organisational culture, gender stereotyping and male chauvinism (Haile et al., 2016). Cultural biases continue, such as assuming women must make choices between family and career. Negative male perceptions remain about women's abilities to lead. Further, these unconscious biases benefit men. Workplace cultures and practices still undermine women's self-confidence and self-belief (Evans et al., 2015). In the context of the gender profile in social work leadership and management, these barriers limit women applying for senior leadership and management roles. For Black, Asian and minority ethnic people, there are additional blocks and barriers, preventing recruitment, retention and promotion into senior role (Spillett, 2017).

None of this is new knowledge; women (feminists and feminist leaders) have 'worked the spaces of power generated through contradictions in the ruling relations of their time, mobilizing new spaces of agency, prefiguring alternative rationalities and opening spaces for those that followed' (Newman, 2012, p. 3). Sometimes women have succeeded in leadership roles. What makes the difference between women as leaders and feminist as leaders, we would argue is the use of self, or rather how we apply 'the personal is political'. This core slogan of the women's liberation movement linked personal experiences to broader social structures and trends, enabled an articulation of how roles are socially reproduced and must be deconstructed (Hyde, 2013). Mullaly (2007) describes how this recognition that social problems are connected with larger embedded structures in society continues to marginalise particular groups. In this respect, social work and leadership practice must recognise the intersecting nature of oppressions and act to address these. Applying a post-structuralist lens, every individual has multiple identities, or 'contradictory selves' (Ford, 2006, p. 80). In this context, 'consciousness raising' regarding the blocks and barriers preventing personal progress as a leader or manager, understanding how internalised oppression works, as well as how structural inequalities operate on a personal level, can be helpful. This approach applies to all forms of structural oppression and their intersections.

Specific initiatives in the UK, to address these barriers, include the Association of Directors of Adult Social Services (ADASS) Accelerate Programme, which supports senior leaders and is explicitly committed to attracting women and unrepresented groups (ADASS, 2019). The Skills for Care 'Moving Up' programme is aimed at promoting diversity of leadership in social care for people from Black, Asian and minority ethnic backgrounds (Skills for Care, 2020). These have supported more women leaders from different backgrounds into leadership roles. Following the Local Government Association annual adult social work health check survey, in addressing diversity and inclusion, there has been widespread take up of coaching and mentoring programmes in social care departments, including reverse mentoring programmes (Local Government Association, 2019).

Leadership is both socially constructed and a performative process (Ford, 2006). The 'use of self' in leadership is relevant to feminist leaders as congruence between what a leader says and how they behave is essential for building trust, inspiring and influencing their behaviours to achieve the shared goals. In this sense, the use of self in leadership mirrors the use of self in social work practice (Trevithick, 2018). Transformational leadership relies on relationships and investing in relationships, demonstrating a care ethic, and this provides the essential housework of feminist leadership.

Making sense of personal biography interconnects with roles inside and outside the workplace (Ford, 2006). We are 'experts' in our own lives, and analysing personal challenges with others is essential tactic for achieving change (Hyde, 2013). As feminists we aim to integrate our understanding of the world, perceptions of structural oppressions and commitment to challenge them throughout our professional careers and in our various social work leadership roles.

Despite, or because of, the barriers that women face (and any person facing structural oppression), we would suggest that feminist social work leaders seek to achieve change and improvement, rally others to achieve shared goals, make visions a reality, while always putting people first. This requires the ability to engage, connect, and be empathetic and motivate others. Specifically, feminist social work leaders understand that 'the personal is political' and this is what separates them from others and provides a core element of their leadership practice. We would further argue that leadership can be demonstrated by different people at different positions within organisations and does not solely rely on positional power or status. The next section explores feminist leadership approaches through a case study.

Case Study: Chief Social Worker and Principal Social Worker Leadership Roles

The establishment of the Chief Social Worker for England role as a senior civil service official within the Department of Health and Social Care (DHSC) and the establishment of Principal Social Workers in each local authority with responsibility for social work were the result of Professor Eileen Munro's report (2011). Munro called on local authorities to designate a Principal Social Worker to report the views and experiences of frontline practitioners to all levels of management, with the aim of enabling 'social workers to exercise more professional judgement' and to 'improve their expertise' (2011, p. 12). Additionally, she recommended the establishment of a Chief Social Worker within government to advise on social work practice and effectiveness, to ensure that social work would have 'greater visibility and voice within Government', provide professional leadership, raise the status of social work and 'send out a strong message that this work is valued and important' (Munro, 2011, pp. 12, 121). Subsequently, two Chief Social Workers were appointed, one for children and families, and one for adults, in September 2013. Working together on joint aims, they provide leadership and work with the sector to improve practice, with the network of Principal Social Workers (DHSC, 2017a, 2018, 2019a, 2020b).

Transformative Leadership

The Chief Social Worker for Adults, based in the Department of Health and Social Care, provided leadership through a participative, collaborative and partnership approach, predicated on her feminist and social work values and principles. This was evidenced by explicit and relentless valuing of the experiences of social workers, social work leaders and people with whom they worked, in agreeing priorities and solutions in achieving the shared vision of improved social inclusion and social justice. This resulted in recognition of social work's role in the statutory guidance accompanying the Care Act 2014, putting the role of the Principal Social Worker on a statutory footing to provide practice leadership in all 152 local authority adult social care departments in England (DHSC, 2020a, paragraphs 1.27 & 1.28). This marked a shift in focus from process, procedural, target and performance management approaches to a focus on valuing professional relational practice in enabling people to achieve what matters to them, and utilising distributed leadership to achieve change.

Transparency and Distributed/Shared Leadership

The Chief Social Worker for Adults prioritised engaging with social workers across England and visited social workers and people using services in every local authority, to facilitate their involvement in developing priorities and undertaking actions to improve practice (Department of Health and Social Care, 2018, p. 21). The explicit focus on personalised, relational, strengths-based practice and working together with people was consolidated through work undertaken by the Chief Social Worker for Adults, Principal Social Workers, social workers and sector organisations to develop and publish a report on strengths-based social work practice

(Department of Health and Social Care, 2017b). Subsequently, the strengths-based practice framework and practice handbook was published to support social work practice and leadership in working with adults, their families and communities (Department of Health and Social Care, 2019b). This has led to a genuine focus on people and their communities: to prioritise what matters to them, to listen deeply and to support real choice and control for people to achieve their wellbeing outcomes. The Chief Social Worker's annual reports demonstrate participative and dispersed leadership with video contributions and case studies developed by social workers and people using services (Department of Health, 2014–16; Department of Health and Social Care, 2017a, 2018, 2019a).

Addressing Structural Inequalities

The Principal Social Workers Network has provided opportunities to support professional leadership, through the co-chairs and regional chairs, supported by the Chief Social Workers for Adults. The network has initiated activities including action days on mental capacity and human rights practice, publications (James et al., 2020) and the Ethical Framework for Adult Social Care in response to Covid-19 (Department of Health and Social Care, 2020c). These demonstrate the professional leadership role in advocating human rights and strengths-based practice approaches in different adult social work settings. Calls to action from Principal Social Workers and the Chief Social Worker in relation to gender and diversity have also helped to raise awareness of diversity issues, including a report on gender and diversity in social work leadership (APSW network, 2017) and the Chief Social Workers' blog regarding Black Lives Matter (Chief Social Worker June, 2020). They led on the Black and Minority Ethnic sub-group for the Covid-19 Taskforce to address health and social inequalities (Department of Health and Social Care, 2020d).

Leadership by the Chief Social Worker and the Principal Social Workers has resulted in a renaissance for social work in adult social care in England. The influence of feminist-informed social work practice and leadership approaches is evidenced in both personal style and articulated ambitions, such as the prioritisation of citizenship outcomes for those served by social care. From October 2019, the Interim Chief Social Workers for Adults continued to build on the transformative approach established by the first Chief Social Worker for Adults and to enhance the development of

Principal Social Workers and their distributed professional leadership network (Department of Health and Social Care, 2020b).

Conclusion

Feminism and social work share common values and principles. Feminist social work leaders demonstrate transformational, shared, relational and intersectional leadership styles and practice. This challenges hierarchical organisations and traditional cultures which favour transactional styles of leadership and are not conducive to feminist leadership, feminist and social work values and approaches. This presents choices for feminist social workers—to work outside conventional statutory services or work inside and become managers and senior leaders or stay at the front line, advocate, and achieve change in peoples' lives. We suggest that feminist leadership is valid in all these roles. Over the past few years, the development of the Principal Social Worker role is an example of shared leadership driven by a Chief Social Worker for Adults demonstrating a feminist leadership style, as illustrated by the Case Study providing professional social work leadership roles in adults services. Whichever pathway is chosen, the feminist social work leader needs courage to criticise and challenge, to build support and allies through networking, and to demonstrate through their behaviour and decision making that we are feminists who care and are ambitious to achieve change.

References

Adults Principal Social Worker Network. (2017, October 10). *Who is leading the way?* https://adultpswnetwork.files.wordpress.com/2017/10/fringe-report-4.pdf. Accessed 31 Aug 2020.

Association of Directors of Adult Social Services (ADASS). (2019, May 29). *Accelerate: Empowering social care leaders learning from the first cohort.* https://www.adass.org.uk/home/accelerate

Brindle, D. (2020, November 25). They don't see themselves as heroes: England's social workers hailed for Covid response. *The Guardian.*

Bryant, L. (2016). Repositioning social work research in epistemology, research and praxis. In S. Wendt & N. Moulding (Eds.), *Contemporary feminisms in social work practice.* Routledge.

Carey, G., & Dickenson, H. (2015). Gender in public administration: Looking back and moving forward. *Australian Journal of Public Administration, 74*(4), 509–515.

Chief Social Worker for Adults. (2020, June 5). *Social work with adults blog: Black lives matter*, GOV.UK. https://socialworkwithadults.blog.gov.uk/2020/06/05/black-lives-matter/. Accessed 31 Aug 2020.

Cocker, C., & Hafford-Letchfield, T. (2014). *Rethinking anti-discriminatory and anti-oppressive theories for social work practice.* Palgrave.

Conley, H., & Page, M. (2017). Revisiting Jewson and Mason; The politics of gender equality in UK local government in a cold climate. *Gender, Work and Organization, 24*(1), 7–19.

Crenshaw, K. (1991). Mapping the margins: Intersectionality, identity politics, and violence against Women of color. *Stanford Law Review, 43*(6), 1241–1299.

Department of Health (2014-16) *Annual Report by the Chief Social Worker for Adults, One Year on*, Department of Health, Crown copyright.

Department of Health and Social Care. (2017a). *Annual report by the chief social worker for adults 2016–2017, being the bridge.* Department of Health and Social Care, Crown copyright.

Department of Health and Social Care. (2017b). *Strengths-based social work practice with adults.* Department of Health and Social Care, Crown copyright.

Department of Health and Social Care. (2018). *Chief social worker for adults annual report, 2017–2018 from strength to strength, strengths based practice and achieving better lives.* Department of Health and Social Care, Crown Copyright.

Department of Health and Social Care. (2019a). *Chief social worker for adults annual report: 2018–2019 – Social work leadership in changing times.* Department of Health and Social Care, Crown copyright.

Department of Health and Social Care. (2019b). *Strengths-based social work: Practice framework and handbook.* Department of Health and Social Care, Crown copyright.

Department of Health and Social Care. (2019c). *Roles and responsibilities: Adult principal social workers (PSW).* https://www.gov.uk/government/publications/principal-social-workers-in-adult-services-roles-and-responsibilities

Department of Health and Social Care. (2020a). *Care and support statutory guidance, updated 24 June 2020.* DHSC.

Department of Health and Social Care. (2020b). *Chief social workers' annual report: 2019 to 2020.* Department of Health and Social Care, Crown copyright.

Department of Health and Social Care. (2020c). *Responding to Covid 19: The ethical framework for adult social care.* Department of Health and Social Care, Crown copyright.

Department of Health and Social Care. (2020d). *Social care sector Covid-19 support taskforce: Report on first phase of Covid-19 pandemic.* BAME Advisory Group report.

Dominelli, L. (2002). *Feminist social work theory and practice.* Palgrave.

Evans, M., Edwards, M., Burmiester, B., & May, D. (2015). Not yet 50/50 – Barriers to the progress of senior women in the Australian public service. *Australian Journal of Public Administration, 73*(4), 501–510.

Farqueson, C. (2020, October 5). *Intersectionality: A smokescreen for exclusion?* Skills for Care.

Ford, J. (2006). Discourses of leadership: Gender, identity and contradiction in a UK public sector organisation. *Leadership, 2*(1), 77–99.

Gray, M., & Schubert, L. (2016). Do something, change something: Feminist leadership in social work. In S. Wendt & N. Moulding (Eds.), *Contemporary feminisms in social work practice*. Routledge.

Haile, S., Emmanuel, T., & Dzanthor, A. (2016). Barriers and challenges confronting women for leadership and management positions: Review and analysis. *International Journal of Business and Public Administration, 13*(1), 36–51.

Hyde, C. (2013, June). *Clinical and direct practice: Gender and sexuality macro practice race, ethnicity, and culture. encyclopedia of social work*. National Association of Social Workers Press and Oxford University Press Online publication data Jun 2013. https://oxfordre.com/socialwork/view/10.1093/acrefore/9780199975839.001.0001/acrefore-9780199975839-e-151. Accessed 3 July 2020.

James, E., Mitchell, R., & Morgan, H. (2020). *Social work, cats and rocket science, stories of making a difference in social work with adults*. Jessica Kingsley Publications.

Karroff, M., Norton, C., Tucker, A., Gass, M., & Foerster, E. (2019). A qualitative gender analysis of Women's field guides' experiences in outdoor behaviour healthcare: A feminist social work perspective. *Affilia: Journal of Women and Social Work, 34*(1), 48–64.

Konradt, U. (2013). Towards a theory of dispersed leadership in teams: Model, findings, and directions for future research. *Leadership, 10*(3), 289–307. Sage.

Lazzari, M., Colarossi, L., & Collins, K. (2009). Feminists in social work: Where have all the leaders gone? *Affilia: Journal of Women and Social Work, 24*(4), 348–359.

Local Government Association. (2019). *Adult social work health check survey 2018 – Report of findings*. Local Government Association.

Malinger, G., Stareks, S., & Tarter, K. (2017). Women social workers: A road map to gender equality. *Affilia: Journal of Women and Social Work, 23*(1), 81–91. Sage.

Miller, Y., Kark, R., & Zohar, N. (2019). Her/his ethics? Management ethics in moral decision making from a contextual, gendered and relational perspective. *Sex Roles: A Journal of Research, 80*(3–4), 218–233. Springer.

Mullaly, B. (2007). *The new structural social work*. Oxford University Press

Munro, E. (2011). *The Munro review of child protection: Final report, a child centered system*. The Stationery Office.

Newman, J. (2012). *Working in the spaces of power: Activism, neoliberalism and gendered labour*. Bloomsbury Academic.
Obholzer, A., & Miller, S. (2004). Leadership, followership and facilitating the creative workplace. In C. Huffington et al. (Eds.), *Working below the surface: The emotional life of contemporary organisations*. Karnac.
Page, M. (2011). Gender mainstreaming – Hidden leadership? *Gender, Wok and Organisation, 18*(3), 318–336.
Pullen Sansfacon, A. (2016). Ethics and feminist social work. In S. Wendt & N. Moulding (Eds.), *Contemporary feminisms in social work practice*. Routledge.
Rusaw, C. (2005). A proposed model of feminist public sector leadership. *Administrative Theory and Practice, 27*(2), 385–393.
Simola, S. K., Barling, J., & Turner, N. (2010). Transformational leadership and leadership moral orientation: Contrasting ethic of justice and ethic of care. *Leadership Quarterly, 21*(1), 179–188.
Skills for Care. (2019). *The state of the adult social care workforce*. Skills for Care.
Skills for Care. (2020). *Moving up programme, 2019/20*. Skills for care. https://www.skillsforcare.org.uk/Leadership-management/developing-leaders-and-managers/Moving-Up.aspx
Spillett, M. (2017). *Leadership imbalance: Black and Asian leaders missing in action A think piece*. https://www.ppma.org.uk/wp-content/uploads/2017/12/Leadership-imbalance.pdf. Accessed 31 Aug 20.
Swigonski, M., & Raheim, S. (2011). Feminist contributions to understanding women's lives in the social environment. *Affilia: The Journal of Women and Social Work, 26*(1), 10–20.
Trevithick, P. (2018). The 'self' and 'use of self' in social work: A contribution to the development of a coherent theoretical framework. *The British Journal of Social Work, 48*(7), 1836–1854.
UN Women. (2020). *Progress of the world's women 2019–2020 families in a changing world*. UN Women.
Valesco, J. R. (Ed.). (2017). *Feminism: Past present and future*. Nova Science Publishers.
Wendt, S. (2016). Conversations about theory: Feminism and social work. In S. Wendt & N. Moulding (Eds.), *Contemporary feminisms in social work practice*. Routledge.
Wendt, S., & Moulding, N. (Eds.). (2016). *Contemporary feminisms in social work practice*. Routledge.

Index[1]

A
Abuse
  domestic, 216, 218–221, 315, 329
Adoption, 28, 102, 184, 271, 275, 276, 278–281
  lesbian and gay, 276, 278
Advocacy, 7, 9, 47, 53, 54, 77, 88, 98–103, 105–111, 148, 177, 203, 207, 218, 221, 223, 243, 249
African
  colonialism, 7, 127–129, 260
  feminism, 7, 8, 124, 129, 130, 133
  women, 123–135, 137
Afrocentric, 7, 123–137
Ageing, 9, 207, 215, 217, 222
Agency, 10, 11, 22, 26, 32, 48–50, 53, 62, 88, 124–126, 132, 134, 135, 137, 144, 147, 161, 192, 193, 195, 196, 198, 199, 201, 202, 264, 276, 278, 287, 288, 292–294, 296, 297, 299, 300, 331, 349
Anti-racist, vii, 8, 24, 141, 178, 326, 346
Asylum-seeker, 171
Australia, 7, 80, 126, 247
Autoethnography, 98, 100–101, 103, 109, 250

B
Bisexual, 27, 197, 217, 238
Black
  women, 5, 19, 23–25, 41
Blame, 11, 287, 292, 293, 296, 331
Bodies, 7, 9, 28, 46, 64, 89, 91, 110, 125, 130, 148, 149, 154n1, 167, 177–182, 184, 190–192, 198–201, 203, 219–221, 236, 239, 241, 242, 293, 304

[1] Note: Page numbers followed by 'n' refer to notes.

C. Cocker, T. Hafford-Letchfield (eds.), *Rethinking Feminist Theories for Social Work Practice*,
https://doi.org/10.1007/978-3-030-94241-0

Bodily performance, 9, 190
Boys, 2, 165, 262, 274, 306, 308
Buddhist, 8, 141–143, 147, 149
Butler, Judith, 26, 209, 310

C

Caring, 44, 83, 89, 91, 125, 132, 151, 213, 215–217, 220, 222, 261, 310, 314, 315
  women, 213, 310
Case studies
  leadership, 341, 342, 350–353
Children, vii, 6, 7, 10, 11, 42, 44, 45, 52, 53, 66, 97–100, 102, 106–111, 133, 134, 142–148, 150–154, 165, 167, 170, 184, 213, 216, 217, 256, 258, 260, 261, 264, 266, 272–279, 281, 282, 287–289, 292, 294–296, 298, 300, 304, 306, 310–312, 322, 326, 327, 329–334, 343, 351
Choice, 10, 27, 28, 31, 33, 45, 62, 69, 70, 82, 88, 127, 164, 183, 201, 213, 236, 237, 240, 273, 274, 276, 277, 282, 287, 288, 292, 294–300, 307, 309, 310, 314, 348, 352, 353
  feminism, 5, 19, 26–28
Class, vii, 2, 4, 21, 23, 24, 28, 41, 42, 64, 66, 79, 81–83, 85, 89, 100, 102, 105, 111, 124, 126, 135, 136, 154, 216, 217, 239, 275, 278, 296, 312, 314, 324, 326, 340
Cocker, Christine, 5, 10, 274, 276–281, 307, 340, 346
Co-construct, 7, 77
Coercion, 3, 161, 295, 299, 327, 328
Collaborative, 6, 7, 10, 50, 60, 65, 77, 89, 90, 98, 102, 110, 193, 194, 256, 264, 328, 345, 351
  autoethnography, 7, 97–112
Colonialism, 2, 7, 64, 123, 124, 127–129, 136, 177, 182, 185
Community building, 7, 101, 102, 243
Context
  institutional, 6, 305
Crenshaw, K., 2, 4, 5, 8, 25, 40, 42, 170, 239, 346

D

Disability, vii, 2, 4, 6, 8, 9, 38, 46, 52, 82, 110, 141–154, 183, 184, 189–203, 210, 213, 214, 217, 223, 278, 307, 308, 332, 340
Discourse, vii, viii, 3–6, 10, 11, 26, 27, 37, 60, 64, 65, 80, 82, 86, 89, 90, 125, 129, 130, 146, 149, 184, 193, 203, 208, 209, 239, 248, 257, 262, 271, 272, 274, 277, 279, 287, 288, 290, 299, 300, 333, 347, 348
Domestic violence, 22, 51, 60, 65, 216, 218, 248, 260, 312, 323, 329, 330, 332–334
Dominelli, Lena, 3, 63, 84, 86, 88, 89, 92, 289, 295, 312, 340

E

Employment, 1, 4, 20, 54, 66, 150, 160, 212, 214, 265
Empowerment, 6, 9, 37–54, 82, 89, 111, 133, 135, 143, 160, 171, 192, 201, 203, 207, 211, 213, 255, 266, 300

Engagement, 22, 49, 51, 60, 70, 72, 87, 99, 103, 106, 108, 110, 130, 145, 152, 222, 236, 246, 249, 263, 314, 325, 341
Equalities, 3–5, 12, 22, 29, 168, 259, 272, 278–280, 282, 340, 341, 346, 348
Ethics, 7, 8, 31, 64, 65, 70, 78, 141, 147, 154, 195, 242, 244, 245, 248, 249, 321, 328, 329, 334, 343–344, 350
Ethnography, 7
Explanatory, 5, 7, 78

F
Family
lesbian and gay, 274, 277
LGBT, 276, 282
therapy, 11, 321, 323–329, 333
Feminism
African, 7, 8, 124, 129, 130, 133
challenge, 19–33, 82
choice, 5, 19, 26–28
contested, 7, 123, 124
controversy, 19–33
critical feminism, 6, 67
fourth wave, 5, 19, 26–28
Islamic, 6, 19, 28–30
leadership, viii, 339–353
pedagogy, vii, 7, 77–93
practice, 1–13, 37–54, 59–72, 123–137, 189–203, 235–250, 255–267, 321–334
research, 59–72, 97–112, 189–203, 235–250
second wave, 5, 19–24, 26–28
third wave, 5, 19, 26–28
transfeminist, viii, 175–185
Fostering, 49, 136, 162, 170, 183, 271
adoption, 275, 278, 280, 281
Foucault, M., 4, 86
Freedom, 4, 28, 41, 62, 80, 82, 127, 201, 242
women, 3, 82

G
Gay, 6, 23, 191, 200, 217, 238, 271, 274–278, 280, 312
Gayby boom, 10, 275
Gender
disparaties, 169
education, 2, 3, 46, 52, 80, 214, 308, 326
equality, 2, 4, 29, 73, 92, 168, 171, 211, 257, 259, 306, 307, 314, 315, 340, 347–348
inequalities, 2, 4, 10, 32, 41, 82, 161, 167, 169–171, 208, 209, 212, 214–216, 257, 304, 306, 307, 309, 314, 324, 343
Genderism, 4
Geography, 2, 102, 111, 130, 131, 326
Gerontology, 2, 102, 111, 130, 131, 326
critical, 208–210
Girls, 2, 3, 8, 72, 100 142, 150, 160–162, 165, 168–171, 215, 259, 298, 300, 308, 314, 315, 327, 328, 331
Glass escalator, 11, 313, 314
Globalisation, 130, 210

H
Harassment
sexual, 3, 165
Heterosexism, 4, 185, 217, 272
Heterosexuality, 3, 184, 192, 271, 273
Hicks, Stephen, 229–234, 275, 277, 278, 280
Hierarchy, 3, 29, 142, 144–145, 147, 150, 154, 184, 348

I

Identities
- bisexual, 27, 191, 217
- categorisation, 4
- gay, 191, 312
- lesbian, 25, 191, 277
- sexual, 5, 9, 25–27, 38, 190–192, 194, 195, 198, 202, 203, 239, 240, 245, 248, 277, 280
- trans, 27, 176, 180, 181, 191, 237

Ideology, vii, ix, 7, 20, 61, 62, 82, 88, 123, 124, 127, 129, 210, 289
- western, 7, 123

International women's day, 3
Intersectionality, viii, 4, 5, 7, 8, 12, 25, 31, 92, 111, 170, 178, 216, 222, 256, 257, 260, 275, 346
Islamic feminism, 6, 19, 28–30
Islamophobia, 11, 322

K

Knowledge, vii, viii, 4, 5, 7, 8, 11, 12, 22, 24–26, 59, 60, 62–65, 67, 69–72, 79, 84–89, 92, 101, 109–111, 123–129, 132–137, 152, 154, 162, 183, 190, 192, 193, 198, 202, 203, 218, 220, 222, 235, 240, 241, 244, 246–250, 278, 288, 290, 291, 300, 303, 304, 307, 308, 312, 339, 349

L

Lajja-baya, 8, 141, 142, 148, 149
Later life, 9, 10, 207–223
Leadership, vii, viii, 12, 50, 307, 339–353
Lesbian, vii, 3, 5, 6, 10, 23, 41, 217, 238, 271–282
Lived experiences, viii, 2, 6, 9, 10, 12, 67, 78, 80, 82, 84, 86–87, 90, 92, 131, 143, 160, 162, 163, 167–169, 171, 180, 183, 190, 192, 193, 196, 202, 235–250, 281, 292, 326, 328, 329, 331, 342

M

Managerialism, 32, 61, 63, 80, 182, 185, 299
Managers, 1, 12, 306, 313, 322, 340, 341, 343, 344, 346, 349, 353
Marginalisation, 6, 17, 23, 82, 89, 209, 221, 222, 240, 314, 340
Metaphor braid, 6, 60, 72
Mothers, 8, 44, 66, 99, 100, 102, 108, 126, 133, 135, 141–154, 165, 256–266, 272–279, 281, 294, 324, 330, 334

N

Narrative, 9, 11, 26, 27, 44, 87, 90, 98, 99, 101, 109, 110, 125, 126, 144, 160, 163, 181, 189–203, 239, 243, 248, 262, 287, 288, 294, 315n3
Neoliberalism, 6, 59–64, 69, 70, 80, 82, 182, 185
Non-binary, 102, 191, 198, 237, 271, 275, 277–279

O

Oppression, viii, 6, 10, 20–25, 31–33, 40–42, 60, 72, 88, 91, 100, 108, 112, 127, 129, 130, 168, 170, 177–182, 184, 185, 190, 192–194, 196, 201–203, 211, 257, 282, 340, 342, 346, 349, 350

Organisations
hierarchical, 341, 353
social care, 3, 307
social work, 4, 306, 323, 341, 351

P

Parents
LGBT, 282
of trans children, 110, 111
Pedagogy, vii, 7, 77–93, 129, 184
Performativity, 10
Pluralism, 256
Positionality, 11, 202, 303
Power, 4–7, 9, 11, 20, 21, 23–26, 28, 31, 37, 38, 40–47, 49, 53, 59–61, 64, 67–72, 78, 79, 81, 85, 86, 88–90, 92, 98, 105, 110, 123, 124, 128, 129, 132, 133, 144, 146, 149, 150, 160, 161, 168–170, 177, 181, 183, 184, 190, 203, 207, 209, 211, 219, 243, 244, 247, 249, 250, 257, 258, 289, 291, 295, 304–307, 311, 314, 315n2, 323, 324, 327, 328, 341–350
colonial, 129
Practice, vii–ix, 3–12, 20, 22, 28–32, 37–54, 59–72, 77–86, 88, 90–93, 98, 99, 102, 106–108, 124, 125, 127, 132, 135, 141, 145–149, 152, 154, 168, 176–179, 181–185, 189–203, 208, 210, 211, 214, 221, 222, 255–263, 265, 267, 273, 278, 280, 289, 290, 292, 299, 300, 305, 307, 309, 321–324, 326–331, 333, 340–342, 344–353
social work, vii, viii, 1–13, 30, 38, 43, 52, 59–64, 67, 68, 72, 78–80, 89, 111–112, 123–137, 175–185, 190, 193, 207, 210, 221, 222, 255–267, 288, 289, 293, 295, 312, 321–334, 344, 346, 350–352
Principles, vii, 6, 31, 38, 40, 48–53, 59, 60, 63, 67–69, 72, 77, 78, 80, 84, 86, 88, 89, 92, 93, 127, 131, 132, 162, 178, 184, 289, 290, 298, 324, 329, 339–342, 345, 351, 353
Professional
identities, 6, 53, 235–250
social work, 124, 353

Q

Qualitative
methodology, 7, 101
research, 44, 101, 160, 162
Queer
identities, 9, 190, 193, 195, 199, 312
theory, 26

R

Race, vii, 6, 38, 42, 46, 52, 79, 81, 82, 85, 89, 100, 105, 111, 130, 136, 170, 171, 209, 217, 239, 245, 277, 298, 324, 326, 340, 342, 346
Racism, 11, 24, 25, 60, 185, 216, 243, 257, 296, 322, 325
Reflexivity, 4, 70, 106, 108, 194, 326, 341
Refugees, 8, 159–171
Relationality, 7
Religion, 2, 21, 29, 79, 82, 83, 111, 136, 160, 184, 326
Reproduction, 88, 143, 183, 211, 342, 346
rights, 22, 273
Research
autoethnographic, 98, 100–101, 236

S

Second wave, 5, 19–24, 26–28, 41
Sex, 10, 11, 68, 71, 72, 142, 154, 165, 169, 176, 184, 198, 199, 200, 201, 203, 235–250, 276, 277, 278, 279, 280, 288, 289, 290, 291, 293–300, 304, 305
  sex work, 10, 11, 71, 235–245, 247–249, 288–291, 293, 294, 296, 297, 299, 300
Sexism, 4, 11, 24, 25, 83, 124, 170, 184, 185, 194, 199, 203, 209, 217, 221, 322, 325
Sexual orientation, 2, 46, 52, 100, 136, 191, 202, 236, 276, 281, 326
Shame, 8, 44, 45, 91, 141–154, 333
Social relations, 4, 26, 30, 50, 82, 148, 149, 247, 324, 343
Social work
  education, 5, 62, 79–81, 179, 181, 183–185, 222, 309–314
  feminist, vii, 7, 41, 42, 52, 53, 67, 77–93, 123, 137, 194, 211, 255–267, 295, 340–342, 350, 353
  men, 305, 307–309, 340
  pedagogy, 7, 77, 79, 84–89, 91–93, 184
  skill, 339
  systemic, 11, 321–323
Sri Lanka, 142, 143, 146–148, 150, 154
Syria, 163, 165–168

T

Theories
  critical, 5
  explanatory, 7, 78
Transgender, 68, 97, 99, 100, 102, 176, 191

U

United Nations (UN), 2, 3, 38, 48, 161, 162, 168, 211
United States (US), 6, 21, 23, 24, 60, 85, 178, 179, 273, 304, 333

V

Values, vii, 6, 7, 11, 24, 28, 31, 32, 38, 48, 53, 61–63, 65, 67, 70, 78, 82, 84, 89, 98, 109, 124, 125, 128, 130, 131, 133–135, 137, 144, 145, 148, 149, 160, 177, 182–185, 193, 211, 221, 236, 238–240, 242, 246, 259, 267, 289, 291, 300, 307, 339, 343, 346, 347, 351, 353
Victim, 50, 53, 54, 131, 137, 161, 168, 218, 220, 221, 287–300, 329
  domestic abuse, 329
Violence
  against women, 3, 20, 30, 65, 160, 161, 170
  sexual, 60, 78, 161, 163, 218–221, 241

W

Welfare
  systems, 66, 146
Western ideology, 7, 123
Wollstonecraft, Mary, 2
Women
  bodies, 191, 192
  coercion, 3, 161
  freedom, 3, 82
  liberation, 177, 328
  refugees, 8, 159–171
  rights, 2, 4, 20, 22, 41, 82, 161, 169, 179, 256, 260, 273
Work-life balance, 4

The manufacturer's authorised representative in the EU is Springer Nature Customer Service Centre GmbH, Europaplatz 3, 69115 Heidelberg, Germany. If you have any concerns regarding our products, please contact ProductSafety@springernature.com

Printed and bound by CPI Group (UK) Ltd, Croydon, CR0 4YY

30/09/2026

02225492-0010